Behavioural Ecology of Teleost Fishes

Behavioural Ecology of Teleost Fishes

Edited by

Jean-Guy J. Godin

Department of Biology, Mount Allison University

Sackville, New Brunswick, Canada

OXFORD
UNIVERSITY PRESS

This book has been printed digitally in order to ensure its continuing availability

OXFORD
UNIVERSITY PRESS

Great Clarendon Street, Oxford OX2 6DP

Oxford University Press is a department of the University of Oxford.
It furthers the University's objective of excellence in research, scholarship,
and education by publishing worldwide in

Oxford New York

Auckland Bangkok Buenos Aires Cape Town Chennai
Dar es Salaam Delhi Hong Kong Istanbul Karachi Kolkata
Kuala Lumpur Madrid Melbourne Mexico City Mumbai Nairobi
São Paulo Shanghai Singapore Taipei Tokyo Toronto
with an associated company in Berlin

Oxford is a registered trade mark of Oxford University Press
in the UK and in certain other countries

Published in the United States
by Oxford University Press Inc., New York

A catalogue record for this book is available from the British Library

Library of Congress Cataloging in Publication Data
Behavioural ecology of teleost fishes / edited by Jean-Guy J. Godin.
Includes indexes.
1. Osteichthyes-Behaviour. 2. Osteichthyes-Ecology. I. Godin, Jean-Guy J.
QL639.3.B46 1997 597.5'045—dc20 96-43788

ISBN 0-19-854784-6 (Hbk)

ISBN 0-19-850503-5 (Pbk)

To the memory of Gerard J. FitzGerald and R. Jan F. Smith
and
To Miles H.A. Keenleyside on the occasion of his official retirement

Preface

This book is about the behavioural ecology of teleost (bony) fishes, the most abundant and diverse group of vertebrates, with over 24 000 recognised living species. They are found in almost all conceivable types of aquatic environment and are extremely diverse morphologically, physiologically, ecologically and behaviourally. Their evolutionary success as a taxonomic group is partly owing to their striking behavioural plasticity. Considerable progress has been made over the past decade in understanding how behaviour, ecology and genetics interact to determine individual survival and reproductive success in fishes, and thus the evolution of their behaviour. Moreover, fishes have historically been useful model organisms to test mechanistic and functional models of behaviour. However, published studies on fish behaviour are surprisingly underrepresented in general texts on behavioural ecology. Therefore, I felt that a volume entirely dedicated to the behavioural ecology of teleost fishes was both needed and timely. The idea for this book originated from a conversation I had with Gerard FitzGerald in the summer of 1991. We had initially planned to coedit the volume, but unfortunately Gerry decided later to withdraw because of a serious illness which eventually led to his death in early 1994. I am grateful for our friendship and for his support during the early gestation of this book. Since the publication of the hardback edition of this book, one of our contributors, Jan Smith, has also died. I therefore wish to honour his memory by including him in the dedication to this paperback edition.

This multiauthor book aims to provide insightful reviews and syntheses of recent theoretical concepts and empirical knowledge on behavioural adaptations for survival and reproduction in teleost fishes, and to identify promising future directions and approaches for research. The book thus focuses on a coherent theme and, as such, is not a comprehensive treatise of all aspects of the behavioural ecology of fishes. The intended audience includes advanced undergraduates, graduate students and researchers interested in the behaviour and ecology of fishes; the book should also appeal to anyone generally interested in fish biology, including fisheries biologists, aquaculturists and conservationists.

The volume is organised into five sections, each with a varying number of chapters written by active researchers in the field. The first four sections concern habitat selection and space use, foraging, antipredation and reproduction, respectively. These topics were chosen because of their importance in determining the lifetime reproductive success of individuals and their prominence in the discipline of behavioural ecology. The fifth section considers the implications of behavioural flexibility in individual fish for ecological processes at the population and community levels, a relatively neglected aspect of the behavioural ecology of animals. The general approach followed throughout the book is a functional (evolutionary) one, with constraints on and proximate mechanisms of behaviour addressed only if they help explain the existence of the diverse behavioural strategies and tactics observed in fishes. I have attempted to impose a relatively uniform chapter format, style and terminology throughout the volume, and to enhance its cohesiveness and integration by cross-referencing the component chapters.

Each contributed chapter was reviewed by at least two referees of my choice, revised and edited. I am grateful for the expert advice provided by these external reviewers, who are Mark Abrahams, George Barlow, Mary Bremigan, Joe Brown, Nick Donovan, Douglas Fraser, Gary Grossman, Gene Helfman, Anne Houde, Felicity Huntingford, Miles Keenleyside, Jens Krause, Rob McLaughlin, Anne Magurran, Gary Mittlebach, Tom Quinn, John Reynolds, Derek Roff, Andy Sih, Roy Stein, William Tonn, Eric van den Berghe, Rena Vandenbos, Robert Warner and Brian Wisenden. In addition, all authors had colleagues review their manuscripts prior to submission, and they are acknowledged at the end of the respective chapters. Most of my editorial work was carried out whilst I was on sabbatical leave; I thank Mount Allison University for granting me this leave and the Marjorie Young Bell Faculty Fund at this institution for partially supporting this work. Finally, on the home front, I am grateful to Heather, Danielle and Jennifer for their support and understanding.

Sackville   Jean-Guy J. Godin
January 1996

Contents

Contributors

Anders Berglund Department of Zoology, University of Uppsala, Villav. 9, S-752 36 Uppsala, Sweden

Lauren J. Chapman Department of Zoology, Bartram Hall, University of Florida, Gainesville, Florida 32611, USA

Bent Christensen Department of Animal Ecology, University of Umeå, S-901 87 Umeå, Sweden

Sebastian Diehl Department of Animal Ecology, University of Umeå, S-901 87 Umeå, Sweden (Present address: Zoologisches Institut, Ludwig Maximillians Universität, Postfach 202136, D-80021 Munich, Germany)

Julian J. Dodson Département de biologie, Université Laval, Ste-Foy, Québec G1K 7P4, Canada

Lee A. Dugatkin Department of Biology, University of Louisville, Louisville, Kentucky 40292, USA

Peter Eklöv Department of Animal Ecology, University of Umeå, S-901 87 Umeå, Sweden

Gerard J. FitzGerald Département de biologie, Université Laval, Ste-Foy, Québec G1K 7P4, Canada (deceased)

Jean-Guy J. Godin Department of Biology, Mount Allison University, Sackville, New Brunswick E0A 3C0, Canada

James W.A. Grant Department of Biology, Concordia University, 1455 blvd de Maisonneuve Ouest, Montréal, Québec H3G 1M8, Canada

Paul J.B. Hart Department of Zoology, University of Leicester, University Road, Leicester LE1 7RH, UK

Roger N. Hughes School of Biological Sciences, University of Wales, Bangor, Gwynedd LL57 2UW, UK

Donald L. Kramer Department of Biology, McGill University, 1205 avenue Docteur Penfield, Montréal, Québec H3A 1B1, Canada

Lennart Persson Department of Animal Ecology, University of Umeå, S-901 87 Umeå, Sweden

Robert W. Rangeley Department of Biology, McGill University, 1205 avenue Docteur Penfield, Montréal, Québec H3A 1B1, Canada (Present address: Biologial Station, Fisheries and Oceans Canada, St. Andrews, New Brunswick E0G 2X0, Canada)

Robert Craig Sargent Center for Ecology, Evolution and Behavior, T.H. Morgan School of Biological Sciences, University of Kentucky, Lexington, Kentucky 40506-0225, USA

R. Jan F. Smith Department of Biology, 112 Science Pl., University of Saskatchewan, Saskatoon, Saskatchewan S7N 5E2, Canada (deceased)

1 *Behavioural ecology of fishes: adaptations for survival and reproduction*

Jean-Guy J. Godin

1.1 Introduction

Behavioural ecology, as a scientific discipline, has been concerned primarily with understanding the evolution of animal behaviour and thus how behaviour, ecology and genetics interact to determine individual inclusive fitness. Perhaps to a lesser extent, it is also concerned with the consequences of the behaviour of individuals on ecological processes at the population and community levels (Krebs and Davies 1991, 1993; Gross 1994). Although still a relatively young discipline, behavioural ecology has none the less contributed significantly to our current understanding of behavioural adaptations (*sensu* Reeve and Sherman 1993) and their importance in ecology, and has experienced a rapid growth in new ideas, concepts, and methodological and analytical techniques over the past decade (Krebs and Davies 1991, 1993; Gross 1994). During this period, the subject areas of 'reproductive strategies' and 'survival strategies' have received considerable attention from behavioural ecologists (Gross 1994). Reflecting the current importance of these topics to the discipline, this book focuses on behavioural adaptations for survival and reproduction in teleost fishes. More specifically, the book aims to provide insights into the evolution of the diverse strategies and tactics (*sensu* Gross 1984, 1996) for survival (i.e. those concerning habitat selection, territoriality, foraging and antipredation) and reproduction (i.e. concerning mating systems, sex allocation, competition for mates and mate choice, and parental care) exhibited by fishes.

Studies using teleost fishes have played an important role in the development of behavioural ecology, including early demonstrations of mate choice, territoriality, strategic fighting, diet selection, antipredator adaptations and parental care, among others, and early tests of optimality and game-theoretical models of behaviour (reviewed in Barlow 1993; Huntingford 1993; Pitcher 1993). Despite their important contributions to the discipline and the fact that teleost fishes represent the most abundant and diverse group of vertebrates with over 24 000 recognised living species (Nelson 1994), published studies on fish behaviour are surprisingly underrepresented, compared with birds and mammals in particular, in general texts on behavioural ecology.

Fishes have much to offer as model animals for the study of behavioural ecology, as they have for other biological disciplines (Powers 1989). They are extremely diverse taxonomically, morphologically, physiologically, ecologically and behaviourally, and are found in almost all conceivable types of aquatic habitat (e.g. Keenleyside 1979; Wootton 1990; Sale 1991; Pitcher 1993; Nelson 1994). They thus offer excellent opportunities for both intra- and interspecific comparative studies and for understanding the origins and evolution of behavioural adaptations and life histories (e.g. Barlow 1993; Bell and Foster 1994; Endler 1995; Magurran *et al.* 1995). Because of their indeterminant growth (which is unusual among vertebrates), teleost fish typically experience marked ontogenetic changes in their structure, behaviour and physiology. They are thus particularly suitable subjects for investigating behavioural trade-offs and the role of constraints on the evolution of behaviour, for example (e.g. Dill 1987; Milinski 1993). Teleosts are also highly suitable for experimental laboratory work for several reasons (e.g. most species are small, readily available, and relatively easy and cheap to maintain) and are readily observed in a number of aquatic habitats, despite limitations imposed by water depth and turbidity and by difficulties in identifying individuals for prolonged periods (Keenleyside 1979; Sale 1991; Barlow 1993). Although there exist books specifically on the behaviour of fishes (e.g. Keenleyside 1979; Pitcher 1993) and others containing sections on fish behaviour (e.g. Wootton 1990; Keenleyside 1991; Sale 1991; Bell and Foster 1994), only one (Huntingford and Torricelli 1993) seems to have been solely dedicated to their behavioural ecology to date; this latter volume constitutes the proceedings of a conference held in 1991.

1.2 Structure and contents of the book

The lifetime reproductive success (i.e. fitness) of individual animals is determined mainly by

(1) where they live and reproduce,

(2) their foraging abilities,

(3) predator avoidance and evasion, and

(4) their reproductive activities.

This volume is divided into five sections, the first four of which generally reflect each of these categories. The fifth section considers how the behaviour of individual fish affects ecological processes at the population and community levels.

1.2.1 Part I: Use of space

Teleost fishes not only occupy a great diversity of habitats, they exhibit considerable variation in their distributional patterns both within and between species.

Because the physiology, structure and behaviour of fish change throughout ontogeny, the particular habitat most suitable for (maximising) the survival and reproductive success of an individual is therefore expected to change over its lifetime. Moreover, because all habitats are characterised by particular suites of abiotic and biotic factors that change over different time scales, they will vary in their suitability for individual survival and reproduction. Therefore, which particular habitat individual fish occupy and how they use space should have fitness consequences and, in turn, determine their distribution and abundance. For these reasons, the study of habitat selection and space use provides an important link between behavioural ecology and population and community ecology. In this first section, patterns of distribution and behavioural mechanisms of habitat (space) choice and use in fishes are considered. Its three chapters are sequentially ordered in a quasi-hierarchical manner, beginning with seasonal migratory movements between geographically distant habitats, to selection of habitats over smaller spatial scales and shorter time periods, and ending with territoriality within a given habitat.

Dodson (Chapter 2) addresses, within the framework of life-history theory, the evolution of patterns and strategies of large-scale movement resulting in changes in habitat occupancy. He attempts to relate the seasonally-timed migratory behaviour of primarily diadromous fishes (which move between freshwater and marine habitats) to their reproductive success, thereby asking how such movements influence the fitness of migrants compared with non-migrants. His treatise essentially concerns selection of reproductive habitat. Dodson argues that many migratory patterns in fishes have evolved as the result of selection favouring adults that migrate into reproductive habitats in which the early survival and growth of their offspring are maximised.

Much of the past work on habitat selection in fishes has consisted of little more than documenting where they occur. Kramer *et al.* (Chapter 3) distinguish between active habitat selection (non-random use of space resulting from active movement) from passive distribution by considering intraspecific patterns of distribution among habitats used for growth and survival in fishes. With the goal of understanding the adaptive significance of habitat-selection 'decisions' and the constraints that limit these decisions, they effectively summarise the theoretical approaches used to explain and predict distributional patterns arising from behaviour, and indicate where specific habitat-selection models have been tested on fishes. Kramer *et al.* also provide an overview of the principal types of habitat occupied by fishes over a broad range of temporal and spatial scales, and briefly review the empirical evidence for environmental factors influencing habitat selection in fishes.

Once an individual fish has chosen to occupy a particular habitat, it will be in competition with others for use of space and resources in that habitat. Fishes often use aggression to compete for (i.e. defend) space/resources. Grant (Chapter 4) reviews territoriality in fishes, with particular emphasis on defence behaviour, in an attempt to gain new and broader insights into this phenomenon. He first compares the frequency of occurrence of territoriality in families of three distinct fish faunas. Although selective and incomplete, this comparison reveals some interesting contrasts and similarities in the incidence of defence behaviour between these faunas, which presumably reflect differences in their habitats and in the economic defendability of various resources (e.g. food, spawning sites, mates). Within the framework of resource-defence theory, Grant then provides an informative overview of the ecological conditions that favour territorial defence and how certain ecological variables (e.g. competitor density, resource density and resource dispersion), which are assumed to influence the benefits and costs of territorial behaviour, should affect territory size. Empirical studies using fishes that provide tests of theoretical predictions about, and evidence for fitness benefits and costs of, territoriality in fishes are reviewed.

1.2.2   Part II: Foraging

Teleost fishes exhibit perhaps the greatest diversity in foraging behaviour and diet composition of all vertebrates (e.g. Keenleyside 1979; Gerking 1994). This reflects in large part their behavioural flexibility and diverse ecology, mode of locomotion, body form, sensory physiology and oral morphology (e.g. Dill 1987; Wootton 1990; Gerking 1994). This section of the book addresses theoretical and empirical knowledge on two important aspects of foraging behaviour in teleosts, namely, (1) foraging tactics (modes) and (2) prey selection. How an individual fish searches for, captures and handles prey and which of available prey types it selects both potentially affect its foraging success and presumably its fitness. The relationship between foraging behaviour and fitness is particularly interesting in fishes because of their indeterminant growth. Therefore, the study of these aspects of foraging behaviour is important to understanding the evolution of diverse foraging behaviours observed in fishes.

Most optimality models of foraging behaviour (Stephens and Krebs 1986) say little about the tactical behaviours that a forager should use to realise a specified optimal foraging strategy. Hart's essay (Chapter 5) is thus about the nature of behavioural tactics for effectively searching for, capturing and handling prey in fishes. When environmental conditions vary unpredictably and prey exhibit a wide range of antipredator defences, evolution of flexible and diverse foraging tactics in animals is expected. Hart reviews evidence showing that fish facultatively alter their prey searching, capturing and handling tactics in response to a wide variety of ecological conditions in ways that appear adaptive, and that the evolution of such tactics is constrained in particular by their body form. He concludes that flexible foraging tactics, as exemplified by fishes, violate some of the assumptions underlying basic foraging theory, and that any new foraging theory should take into account more explicitly such behavioural flexibility in individual foragers.

The diet of most fishes typically constitutes a non-random subset of prey types available in the environment. The phenomenon of diet (prey) selectivity is addressed by Hughes (Chapter 6) within the framework of optimal foraging theory (OFT; Stephens and Krebs 1986). He briefly reviews the major predictions of the basic prey model, one of the models of OFT, and provides examples of studies on fishes that are in accordance with some of them. However, contrary to one important prediction, fishes commonly show partial preferences, that is, acceptance of some proportion of prey that fall outside the optimal (energetically profitable) set. Hughes demonstrates that such partial preferences can result from changes in relative prey abundance and in the handling time (and thus energetic profitability) of different prey types, owing to concomitant changes in the forager's hunger level, learning and memory, over the course of a foraging bout, for example. Constraints (e.g. mechanical, physiological, genetical) on diet selection and major ecological factors (e.g. predation risk, competition) affecting prey choice in fishes are also reviewed. Throughout this chapter, emphasis is placed on the advantages of combining behavioural, ecological and physiological approaches to studying diet selection and on the importance of recognising the different time scales involved in each case.

1.2.3 Part III: Antipredation

Most animals face predators during their lifetime. Failure to avoid or evade a predator results in death and consequently in greatly reduced future fitness (if the prey is not already in the post-reproductive phase). Therefore, the study of adaptations that increase individual survival in the face of predation hazard is central to behavioural ecology (Lima and Dill 1990; Krebs and Davies 1993; Gross 1994).

Teleost fishes inhabit a wide spectrum of habitats in which they encounter different types and densities of predators. Moreover, an individual fish's vulnerability to predation may change dramatically with growth throughout its lifetime. It is therefore not surprising that teleosts have evolved diverse behavioural, morphological and chemical antipredator defences. The two chapters in this section review defences that occur prior to and after a predator attack, respectively, in fishes. Antipredator defences may be categorised either as pre-attack defences, which reduce the likelihood of predators detecting, correctly identifying and attacking prey, or post-attack defences, which reduce the probability of a prey being captured and eaten once an attack has taken place. Such a dichotomy is not purely arbitrary as pre-attack defences are used more frequently and appear to be relatively less costly per event (at least in terms of time, energy expenditure and risk) than post-attack defences in general.

Smith (Chapter 7) examines defences that serve primarily in prey avoiding predators and in deterring predator attack. The exchange of information (e.g. visual, chemical signals) between participants in predator–prey interactions is an important theme in the chapter. Alarm signalling among conspecifics, predator inhibition signalling and aposematic signalling from prey to predator, and information-gathering behaviours (e.g. predator inspection) are all potentially capable of determining the outcome of predator–prey encounters. However, it is emphasised that too often the fitness benefits and costs associated with particular defences are simply assumed without experimental verification. In the companion Chapter 8, I review behavioural, morphological and chemical traits possessed by teleost fishes that serve as defences mitigating predator capture and ingestion once a predator approach/attack has been initiated. Much evidence is presented showing that individual fish are able to assess the fitness-related benefits and costs associated with alternative defence options in making decisions about when and how to respond to an approaching predator. Particular emphasis is placed on the value of using functional models and of quantifying the fitness benefits and costs of alternative (facultative) antipredator tactics, and how these tactics might conflict with other behavioural activities, to ultimately understand their evolution and current existence in fish populations. This echoes Smith's (Chapter 7) assessment for pre-attack defences.

1.2.4   Part IV: Reproduction

Longevity, fecundity, mating success and offspring survival are major components of reproductive success. Because selection operates differently on the sexes, males and females have evolved different strategies to maximise their respective reproductive success (e.g. Potts and Wootton 1984; Clutton-Brock 1991; Andersson 1994). Such sexual differences in reproductive strategy are reflected in the diverse mating and parental care patterns exhibited by teleost fishes which, with sexual selection, are the subjects of interest in this section comprising three closely inter-related chapters. Collectively, reproductive strategies currently represent the 'hottest' topic in behavioural ecology (Gross 1994).

Chapter 9 by Berglund is about the strategies and tactics used by fishes to reproduce. He provides an overview of the types of mating system exhibited by teleost fishes in relation to mating partners (e.g. monogamy, polygyny, polyandry, promiscuity) and to time (e.g. breeding cycles, lifetime frequency of reproductive events). Ecological conditions favouring the evolution of certain mating systems over others are comprehensively reviewed, and the relationship between mating systems and sexual selection explored. Some mating systems allow for alternative reproductive strategies within fish populations; these include various behavioural strategies, sex change and hermaphrodism. Berglund identifies the conditions that should favour the evolution and maintenance of such alternative strategies within populations and reviews the empirical evidence for them.

Empirical advances in knowledge of both intrasexual and intersexual selection in fishes are reviewed by Dugatkin and FitzGerald (Chapter 10), with emphasis placed on the evolution of traits specifically selected to enhance individual reproductive success in males and females. They review contemporary evolutionary models of sexual selection and studies on fishes that provide tests of model predictions. The proximate mechanisms underlying both intra- and intersexual selection, as well as the fitness-related benefits and costs accrued to individuals of both sexes in such interactions, are also addressed. These are viewed as fundamentally important to understanding the observed variability in mating success among individuals in populations and thus the process of sexual selection.

In those species in which females oviposit eggs that are fertilised externally, one or both parents may care for the eggs until hatching or for some time thereafter. According to parental investment theory (Clutton-Brock 1991), allocation of parental resources to current offspring potentially reduces the survival and future reproductive success of the parents. Therefore, individual fish face decisions regarding whether to give care to a current brood and, if so, how much. Sargent (Chapter 11) reviews both the theory and empirical evidence for parental care patterns observed in teleost fishes within the context of life-history evolution. He provides a thoughtful overview of parental investment theory in terms of optimal resource allocation by parents that is constrained by trade-offs among fitness components (e.g. between offspring number and offspring survival, between current and future reproduction) and summarises recent empirical evidence for these trade-offs and for fitness-related benefits and costs of care to parents in fishes. Sargent further shows how life-history theory provides new insights into diverse parental-investment phenomena, such as filial cannibalism and sexual conflict over parental care. In addressing the behavioural ecology of parental care in fishes within the broader context of life-history evolution, this chapter bridges in a general way Chapters 9 and 10.

1.2.5 Part V: Ecological implications of individual behaviour

The previous chapters reveal considerable intrapopulation variation in the behaviour of individual fish with regards to habitat choice, space use, resource defence, foraging, predator avoidance/evasion, mating and parental care. This single-chapter section explores the consequences of such variation in the behaviour of individual fish for population dynamics and community structure. Persson *et al.* (Chapter 12) review current theoretical and empirical knowledge of the consequences of flexible behaviour in individual fish for interactions between competitors and between predator and prey at the levels of the population and community. The importance of size-structure in fish populations (i.e. differences in body size among individuals), and of morphological constraints that limit the performance of individual fish, for such ecological interactions is emphasised throughout. By considering the implications of individual flexibility in habitat choice, space use, foraging and antipredator behaviours in particular, this chapter as such integrates many of the behavioural phenomena addressed in the other chapters of this volume. In doing so, it illustrates the important role behavioural ecology can play in extending our understanding of the dynamics and stability of fish populations and communities.

Acknowledgements

I thank Lee Dugatkin, Cathy Kennedy, Jens Krause and Dan McClary for their comments on previous versions of this chapter.

References

Andersson, M. (1994). *Sexual selection*. Princeton University Press, Princeton.

Barlow, G.W. (1993). Fish behavioral ecology: pros, cons and opportunities. In *Behavioural ecology of fishes* (eds. F.A. Huntingford and P. Torricelli), pp. 1–5. Harwood Academic Publ., Reading

Bell, M.A. and Foster, S.A. (eds.) (1994). *The evolutionary biology of the threespine stickleback*. Oxford University Press, Oxford.

Clutton-Brock, T.H. (1991). *The evolution of parental care*. Princeton University Press, Princeton.

Dill, L.M. (1987). Animal decision making and its ecological consequences: the future of aquatic ecology and behaviour. *Can. J. Zool.*, **65**, 803–811.

Endler, J.A. (1995). Multiple-trait coevolution and environmental gradients in guppies. *Trends Ecol. Evol.*, **10**, 22–29.

Gerking, S.D. (1994). *Feeding ecology of fish*. Academic Press, New York.

Gross, M.R. (1984). Sunfish, salmon, and the evolution of alternative reproductive strategies and tactics in fishes. In *Fish reproduction: strategies and tactics* (eds. G.W. Potts and R.J. Wootton), pp. 55–75. Academic Press, London.

Gross, M.R. (1994). The evolution of behavioural ecology. *Trends Ecol. Evol.*, **9**, 358–360.

Gross, M.R. (1996). Alternative reproductive strategies and tactics: diversity within sexes. *Trends Ecol. Evol.*, **11**, 92–98.

Huntingford, F.A. (1993). Behaviour, ecology and teleost fishes. In *Behavioural ecology of fishes* (eds. F.A. Huntingford and P. Torricelli), pp. 1–5. Harwood Academic Publ., Reading.

Huntingford, F.A. and Torricelli, P. (eds.) (1993). *Behavioural ecology of fishes*. Harwood Academic Publ., Reading.

Keenleyside, M.H.A. (1979). *Diversity and adaptation in fish behaviour*. Springer-Verlag, Berlin.

Keenleyside, M.H.A. (ed.) (1991). *Cichlid fishes—behavior, ecology and evolution*. Chapman and Hall, New York.

Krebs, J.R. and Davies, N.B. (eds.) (1991). *Behavioural ecology: an evolutionary approach*, 3rd edn. Blackwell Scientific Publ., Oxford.

Krebs, J.R. and Davies, N.B. (1993). *An introduction to behavioural ecology*, 3rd edn. Blackwell Scientific Publ., Oxford.

Lima, S.L. and Dill, L.M. (1990). Behavioral decisions made under the risk of predation: a review and prospectus. *Can. J. Zool.*, **68**, 619–640.

Magurran, A.E., Seghers, B.H., Shaw, P.W. and Carvalho, G.R. (1995). The behavioral diversity and evolution of guppy, *Poecilia reticulata*, populations in Trinidad. *Adv. Study Behav.*, **24**, 155–202.

Milinski, M. (1993). Predation risk and feeding behaviour. In *Behaviour of teleost fishes*, 2nd edn. (ed. T.J. Pitcher), pp. 285–305. Chapman and Hall, London.

Nelson, J.S. (1994). *Fishes of the world*, 3rd edn. John Wiley and Sons, New York.

Pitcher, T.J. (ed.) (1993). *Behaviour of teleost fishes*, 2nd edn. Chapman and Hall, London.

Potts, G.W. and Wootton, R.J. (eds.) (1984). *Fish reproduction: strategies and tactics*. Academic Press, London.

Powers, D.A. (1989). Fish as model systems. *Science*, **246**, 352–358.

Reeve, H.K. and Sherman, P.W. (1993). Adaptation and the goals of evolutionary research. *Q. Rev. Biol.*, **68**, 1–32.

Sale, P.F. (ed.) (1991). *The ecology of fishes on coral reefs*. Academic Press, New York.

Stephens, D.W. and Krebs, J.R. (1986). *Foraging theory*. Princeton University Press, Princeton.

Wootton, R.J. (1990). *Ecology of teleost fishes*. Chapman and Hall, London.

2 Fish migration: an evolutionary perspective

Julian J. Dodson

2.1 Introduction

The study of fish migration has been dominated by the description of migratory trajectories and of the proximate control of orientation behaviour (e.g. Harden Jones 1968; McCleave *et al*. 1984; McKeown 1984; Smith 1985). Although fascinating in themselves, these themes are generally not concerned with the most central assumption of behavioural ecology, that individuals behave so as to maximise their reproductive success or, rather, their inclusive fitness (Krebs and Davies 1993). The purpose of this chapter is to relate migratory behaviour to reproductive success in fishes. The coverage of the subject is more selective than exhaustive. Although I focus principally (but not exclusively) on 'diadromous' migrations (between freshwater and marine habitats), I have not dealt with the ultimate causes of variability in the rates of homing to and straying from natal habitats. The reader is referred to Quinn (1985) and Kaitala (1990) for a treatment of the latter. The population genetic consequences of varying levels of gene flow among locally-adapted populations is beyond the scope of this chapter.

Migration means to move from one place to another. Landsborough Thompson (1942) classified three types of movement patterns: local and seasonal movements, dispersals, and true migrations. Harden Jones (1968) qualified local and seasonal movements as 'merely changes of ground at a particular time of the year' that may involve short or long distances. This category involves foraging behaviour, defined in the context of spatial displacement as 'reiterative locomotor activity that is readily interrupted by an encounter with a resource item of one particular kind' (Kennedy 1985). Such movements may occur at time periods ranging from seasonal to semi-diurnal and in both the horizontal and vertical planes. Dispersal involves wandering away from the breeding habitat over a wide area, but with the majority of individuals remaining near the breeding habitat. This is in contrast to a 'true' migration, which involves movement between widely-separated and well-defined areas, 'a shift in what may be called the centre of gravity of the population' (Landsborough Thompson 1942).

Given the great diversity of animal movements, attempts to place specific examples into discrete categories are inevitably unsatisfactory (McKeown 1984). I will not attempt such a categorisation here. Rather, I will limit my discussion of migration to the regular coming and going of fish on a seasonal basis that involves a shift in the distribution of all or part of a population between habitats (Harden Jones 1984). Such movements include a return to habitats previously occupied. From an evolutionary perspective, migration involves specialised behaviours that have arisen through natural selection (Dingle 1980). An appropriate starting point for behavioural ecology is to ask how migration influences the fitness of individuals that express such behaviour. Migration confers many advantages, including increased feeding, avoidance of adverse environmental conditions, and improved current reproductive success (Northcote 1978). Migration also has associated costs, including the actual energetic cost of displacement, risks of increased predation, the energetic and developmental costs of any special migratory adaptations, and potential reproductive costs due to decreased lifetime reproductive effort (Rankin and Burchsted 1992). Obviously, benefits must exceed costs on average for migratory behaviour to evolve. Fitness may be defined as the lifetime summation of an individual's probability of surviving to reproduce at any age x (l_x) multiplied by its fecundity or male fertility and breeding success at that age (m_x). Therefore, a migratory life style will be favoured by natural selection if the contribution to $l_x m_x$ of each habitat occupied by a migrant, minus the cost of moving between habitats in terms of $l_x m_x$, exceeds the fitness of a non-migrant who completes the entire life cycle within one habitat (Gross 1987). This model, developed by Gross (1987) to explain the evolution of diadromous migrations, is based on the premise that a migratory strategy would be favoured by selection only if the adult feeding habitat provides a fitness benefit that exceeds the fitness advantage of staying in the juvenile rearing habitat. This assumes that the evolutionary origin of the migratory lifestyle is to be found in an advantage of leaving the habitat used for reproduction and early rearing.

The evolution of many migratory patterns may also be the result of reproductive adults placing their eggs in habitats in which the early survival and growth of their offspring are maximised. As density-dependent mortality occurs early in the life of fishes and declines with age (Cushing 1975), the greatest contributions to reproductive success in many migratory species may thus be achieved by maximising early survival. The location of spawning sites and the timing of reproduction in fishes is often population-specific and is thought to be linked, through natural selection, to ecological conditions (e.g. predation risk, habitat productivity) that favour early growth and survival (Leggett 1985, and references therein). Any evaluation of the evolution of migration should therefore begin with a consideration of the reproductive environment. This is not a new idea; Fulton (1889, cited in Sinclair 1988) originally proposed that offshore spawning areas were chosen as a function of the safety of floating eggs and larvae and their transport to places suitable for the welfare of the juveniles. This point of view is most clearly illustrated by examples of reproductive habitats in which species could never have completed a non-migratory lifestyle. Capelin, *Mallotus villosus*, and grunions (e.g. *Leuresthes tenuis*) are marine fishes that spawn on beaches and their eggs develop in the gravel (Taylor 1990). Other high-intertidal spawners typically place eggs in protected sites near the high-tide line, including the mummichog, *Fundulus heteroclitus*, whose eggs develop in air (Taylor 1990). These species could not exist as non-migratory populations in such habitat. Ephemeral aquatic habitats such as flood plains and tide pools that are subject to periodic dessication, but used by many fishes as reproductive habitat, could never support non-migratory populations. Goulding (1980) identified 11 genera of large characins that descend nutrient-poor tributaries of the Rio Madera basin and spawn either adjacent to or in the nursery habitats of the expanded Rio Madera flood plain. On both Atlantic and Pacific coasts, sticklebacks (Family Gasterosteidae) spawn in tidal salt-marsh pools from which the young must migrate before the pools dry up or are choked by the growth of macroalgae (FitzGerald 1983). I am not implying that all reproductive and juvenile rearing habitats are unsuitable for adult fishes. Rather, the above examples serve to illustrate that the evolution of fish migration may best be understood by considering the fitness benefits that accrue to individuals who migrate to and exploit specific reproductive habitats, as well as the fitness benefits that accrue to fish following their emigration from the reproductive habitat.

2.2 Patterns of migration and the evolutionary significance of the reproductive habitat

Migration patterns have traditionally been described with reference to freshwater and marine habitats: oceanodromous migrations occurring entirely in the oceans, potamodromous migrations occurring entirely in fresh water, and diadromous migrations that involve crossing the saline–freshwater boundary (Harden Jones 1968). These simple categories hide a broad range of migration patterns and life-history adaptations and are not sufficient for an evolutionary consideration of migration. Since the salinity of aquatic habitats is only one of a large number of variables that may influence the reproductive fitness of fishes, the continued categorisation of migrations according to salinity serves no useful purpose beyond providing the most superficial of descriptive categories.

A large body of literature supports the contention that the associations of larvae and juveniles with specific environmental conditions are adaptations to increase fitness. Although such studies have mainly been the purview of population biologists, much of this work illustrates that spawning is coordinated in time and space with conditions that favour the survival of offspring during early life-history stages (see below). Stochastic fluctuations in these conditions may generate variations in reproductive success among individuals, as well as causing variations in population size. Migration of adults to particular spawning sites and the subsequent passive or active migration of larvae to nursery sites may be considered adaptations to reduce the impact of environmental fluctuations on the survival of larvae and, consequently, on lifetime reproductive success (Leggett 1985).

2.2.1 Survival of the early life-history stages

The major determinants of survival during early life-history stages are considered to be starvation and predation (Wooster and Bailey 1989). Cushing's (1975) match–mismatch hypothesis emphasises the importance of starvation, and states that larval mortality is determined by the availability of food when fish larvae switch to exogenous feeding after yolk-sac absorption. Yolk-sac absorption should therefore coincide with the abundance peak of their food resources. Since planktonic crustaceans (mainly copepods) are the most important single source of food for most fish larvae (Bone and Marshall 1982), the magnitude of primary production and the fraction of it consumed by copepods may well determine survival during early life-history stages. Ephemeral and localised mixing events in the water column allow rapid bursts in production of large phytoplanktonic cells, which may account for most production that is grazed by copepods (Legendre and Le Fèvre 1989). Such events, including the pycnocline on the vertical axis, eddies, fronts and upwelling areas on the horizontal axis, and temporal transitions in vertical stability with periodicities ranging from annual to tidal along the time axis, may be tracked by the reproductive cycles and migratory patterns of fishes if their occurrence is predictable in both time and space.

A number of field studies have demonstrated that the spawning of marine fish is linked to tidal and coastal shelf fronts, and the accumulation of larval fish in these areas has been interpreted as a process likely to enhance survival (reviewed by Heath 1992). The relationship between fish spawning patterns and the timing of annual plankton production cycles is not clearly established and remains controversial. For example, peak spawning for several important marine species in continental shelf waters of northeastern United States is synchronised temporally with increasing abundance levels of their copepod prey, whereas other species continuously produce larvae over more prolonged periods, possibly allowing these populations to respond rapidly to favourable conditions (Sherman *et al.* 1984). The timing of cod, *Gadus morhua*, spawning is coupled to the timing of plankton production in the Northwest Atlantic ocean, but only a general correspondence between larval fish production and planktonic production exists rather than a precise one as required by the original match/mismatch hypothesis (Myers *et al.* 1993).

In temperate coastal areas, patterns of migration and spawning may be adaptive responses to historical patterns of predation (Frank and Leggett 1985). The inverse correlations observed between the abundance of macroinvertebrate predators and ichthyoplankton in coastal Newfoundland (Canada) result from their occupation of discrete water masses, whose presence in the coastal area oscillates temporally in response to changes in wind direction. Frank and Leggett suggested that such adaptations involve active behavioural responses of fish to reliable physical and/or biological signals indicative of the existence of ecological 'safe sites'. For example, capelin spawn on gravel beaches at the head of bays along the eastern coast of Newfoundland, where their eggs and young are exposed to different types of coastal water masses. Prevailing westerly winds cause warmer water to be blown seaward and to be replaced by cold, subthermocline waters that upwell along the coast. Easterly winds re-establish the warmer coastal water masses, but are infrequent, unpredictable and of short duration. The biomass of the principal food of larval capelin is 2–3 times greater in the warm surface water mass, whereas the densities of their major predators are 3–20 times lower. Upon hatching, capelin larvae accumulate in the top 20 cm of the beach gravel and then emerge synchronously during onshore, wind-induced coastal water-mass exchanges. As a result, larvae are associated with water masses that are rich in food and poor in predators; these conditions should enhance larval survivorship (reviewed by Leggett 1985).

Refuge from predation may also influence patterns of migration and spawning in tropical waters. In the coastal tropics, many marine species move offshore to spawn in deep waters and larvae are subsequently transported onshore by prevailing currents (Johannes 1978). Whereas offshore tropical surface waters are relatively plankton-poor and provide less food for pelagic larvae, the threat of predation there is greatly reduced because these waters contain fewer planktonic and pelagic predators than inshore waters and are devoid of demersal and benthic predators. Johannes (1978) thus views offshore spawning as an adaptation to remove offspring from the intense predation pressure occurring in nearshore regions.

The transport of eggs and larvae away from favourable nursery areas may result in increased predation or starvation and has been demonstrated to reduce survival (Wooster and Bailey 1989). Parrish *et al.* (1981) hypothesised that spawning by fishes in the upwelling zone of the California current is adapted to counteract the adverse effects of offshore transport of eggs and larvae. They provided examples of species that spawn at particular times and places to minimise offshore transport. In contrast, Sinclair (1988) proposed that starvation and predation (identified as food chain processes) are not the principal determinants of spawning strategies, presenting examples of populations whose spawning times were not related to the annual production cycle. Sinclair (1988) viewed the aggregation and retention of planktonic larvae within specific areas as principally an adaptation to insure population persistence. Individuals that are 'lost' (vagrancy) from their population at any point in the life cycle and do not find a mate that is a member of the same population do not contribute to population persistence. Loss of individuals from the rearing area may thus influence individual fitness in two ways: mortality due to unfavourable conditions, and the possibility that lost individuals are unable to reproduce with members of their own population. Implicit in the latter proposal is that individuals who spawn with members of other populations suffer a reduction in fitness. However, if vagrants successfully reproduce in new areas, great selective value may accrue to such individuals if their offspring are associated with areas that are favourable for survival. It seems evident that vagrancy (*sensu* Sinclair 1988), or straying, is necessary to avoid extinction. As climatic fluctuations alter the availability of suitable habitat, straying promotes the colonisation of new habitats (Northcote 1978; Bowen *et al.* 1992).

2.2.2 The evolution of diadromy

Few topics in fish migration have generated more debate than the evolutionary origins of diadromy. Anadromous fish reproduce in fresh water and migrate to sea where they grow to maturity, whereas catadromous fish reproduce at sea and grow to maturity in fresh water. Amphidromous fish migrate between the sea and fresh water, but not for the purpose of breeding. The migrations occur regularly at some other stage of the life cycle. Two forms of amphidromy have been recognised: marine amphidromy, in which spawning is marine and the larvae or juveniles reside temporarily in fresh water before returning to sea to grow to maturity, and freshwater amphidromy, in which spawning is in fresh water and the larvae or juveniles are temporarily marine before returning to fresh water to grow to maturity (see McDowall (1988) for definitions and a review of diadromy). Baker (1978) noted a latitudinal trend in the occurrence of diadromy, with anadromy occurring more frequently at higher latitudes and catadromy being more frequent in the tropics. Northcote (1978) proposed that this trend may be a result of oceanic waters usually being less productive in the tropics than in temperate or polar areas. Expanding on this hypothesis, Gross (1987) and Gross *et al.* (1988) postulated that diadromy evolved as a response to the productivities of the marine and freshwater environments. Assuming that access to food resources for the growth of adults is the most important fitness benefit accrued from diadromous migrations, these authors demonstrated that oceans are more productive than fresh waters in temperate latitudes where anadromous species predominate. In contrast, catadromous species generally occur in tropical latitudes where freshwater productivity exceeds that of the ocean.

Although appealing in its apparent explanatory power, the simplicity of this food-availability hypothesis hides the complexity of the diadromous life style and the possibility that diadromy may have evolved for different reasons. An important problem concerns the classification of fishes as anadromous or catadromous, a categorisation that is overly simplistic and at times misleading. The hypothesis was based on the global geographic distribution of diadromous species as classified by McDowall (1988). A reading of the detailed analyses of anadromous and catadromous species presented by McDowall (1988) reveals that only 34% of anadromous and 34% of catadromous species actually match the generally accepted definitions of the two groups (Fig. 2.1). The classification of 24 and 42% of

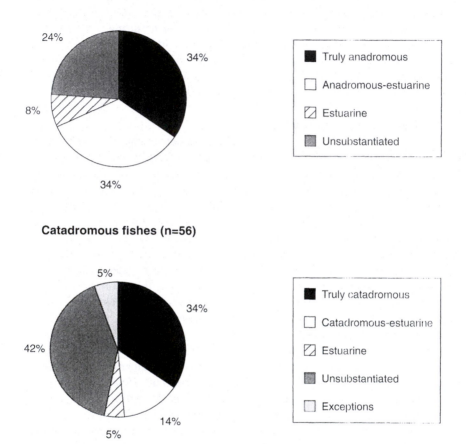

Fig. 2.1   The reclassification of 110 anadromous species and 56 catadromous species classified by McDowall (1988). See text for definition of categories.

species as anadromous and catadromous, respectively, was not substantiated in the detailed analyses presented by McDowall (1988). The life histories of these latter species are so poorly documented that insufficient knowledge precludes any reliable classification beyond euryhaline or 'marginally' diadromous. More detailed information on the life cycle of these species and the habitats that they exploit are needed to improve our understanding of diadromy.

A significant proportion of fishes classified as anadromous or catadromous exhibit life cycles that occur largely in estuaries or coastal areas influenced by freshwater outflow. The life history descriptions of 8 and 5% of species classified as anadromous and catadromous, respectively, suggest that these fishes were estuarine residents, spawning and growing to maturity in estuaries. Another 34% of species classified as anadromous spawn in fresh water but grow to maturity largely in estuaries (anadromous–estuarine, Fig. 2.1). Similarly, 14% of species classified as catadromous spawn at sea, but grow to maturity in estuaries (catadromous–estuarine, Fig. 2.1). Finally, 5% of species classified as catadromous exhibit life histories that are so exceptional that they do not fit any of the above categories. *Galaxias maculatus* experiences half its growth in fresh water and half at sea. *Lates calcarifer*, the barramundi, grows to maturity in fresh water, but post-spawners never return to fresh water. They continue to feed and spawn in tidal waters, where some males change to females following the first spawning. *Rhombosolea retiaria* grows to maturity in fresh water, spawns at sea and stays there to complete its life cycle (McDowall 1988).

A second potential shortcoming of the food-availability hypothesis is that primary productivity data used to quantify the productivity of 'freshwater' and 'ocean' waters may not reflect the production actually exported to higher trophic levels or represent the habitat actually occupied by many diadromous species. In open oceanic waters, two production systems exist within the euphotic system (Goldman 1988). The first (microbial food loop) consists of tiny phytoplankton cells (< 5 μm, ultraplankton), heterotrophic bacteria and protozoa. The primary production of this system is the one measured by standard incubations at sea. An unknown proportion of this production is exported, but much of the energy stored by the ultraplankton is dissipated within the loop (recycled production). The second system is based on larger cells (i.e. diatoms) growing in response to ephemeral and localised mixing events. These events allow rapid bursts in production that remain mostly undetected by bottle incubations. These large cells may account for most production that is exported from the euphotic zone by sedimentation or grazed by copepods (Legendre and Le Fèvre 1989). In fresh water, the young of many anadromous salmonids exploit a rich insect fauna that is largely independent of primary aquatic productivity (Groot and Margolis 1991). For example, over 50% of the energy obtained by young coho salmon, *Oncorhynchus kisutch*, in wooded streams in Oregon (USA) was estimated to be derived from primary production that did not occur in the water (Chapman 1965, cited in Hynes 1970). Because the production calculations used to support the food-availability hypothesis were based on published annual primary productivity estimates of freshwater and ocean habitats at different latitudes (Gross *et al.* 1988), they may not be accurate estimates of production that is available to higher trophic levels. Furthermore, the predominance of estuarine habitat occupied by many diadromous fishes (Fig. 2.1) is not reflected in the calculations of production. Estuaries are among the most productive systems in the world (Lalli and Parsons 1993). Since estuaries and coastal areas account for 10% of the world's oceans, latitude-averaged annual estimates of primary productivity greatly underestimate the importance of these areas to diadromous species.

Diadromy may have evolved to improve fitness, relative to a non-migratory strategy, by improving the survival of the early life-history stages rather than by improving the growth of older fish. Greater adult body size may be an important determinant of reproductive success in species that compete strongly for mates (Andersson 1994; see also Berglund, Chapter 9 and Dugatkin and FitzGerald, Chapter 10). These species are generally characterised by conspicuous secondary sex traits and sexual dimorphism. Among diadromous fishes, the seven species of Pacific salmon are all sexually dimorphic. Males are territorial and aggressive towards other males on the spawning grounds, with larger males gaining access to females (Groot and Margolis 1991). However, the reproductive success of individuals of species that do not build nests and compete over mates, but rather broadcast their gametes, may be far less influenced by sexual selection. In these species, improving the survival of the vulnerable egg and larval stages may represent a significant contribution to lifetime reproductive success. For example, anadromous American shad, *Alosa sapidissima,* and and alewife, *A. pseudoharengus,* release and scatter eggs in open waters (Balon 1975). Although sperm competition may be important, differences in reproductive success due to competition over mates may be far less important than differences in reproductive success due to variations in the mortality of the larvae. As the Clupeidae are primarily a marine group (Nelson 1994), anadromy in shad, alewife and other anadromous clupeids may be considered a derived evolutionary state to exploit fresh waters for reproduction. Similarly, anadromy among species belonging to predominantly marine families such as the Gadidae, Percichthydae and Gasterosteidae (McDowall 1993) may be considered a derived evolutionary state to exploit fresh waters for reproduction. This is in contrast to the salmonines, in which freshwater residency is the ancestral state (Stearley 1992) and migration to sea may be considered a derived state to exploit marine feeding grounds (Gross 1987).

The exploitation of estuaries as nursery areas by a large number of diadromous fishes (Fig. 2.1) may best be understood as adaptations to increase survival of the vulnerable larval stages. Estuaries have long been recognised as important nursery areas, providing high concentrations of food and possible refuge from predators because of their greater turbidity (Miller *et al.* 1985). Estuarine nursery areas are exploited by species spawning in rivers flowing into estuaries, offshore marine spawners and estuarine resident species (Miller *et al.* 1985). Two different larval migratory patterns have been observed: (1) larvae of freshwater or brackish spawners drifting downstream, and (2) larvae of some marine spawners 'drifting' inshore. For example, over 85% of the annual commercial catch of fish off North Carolina (USA) is composed of five species that spawn offshore in the winter. Larvae migrate about 100 km to inlets in the barrier islands and another 100 km to estuarine nursery areas, a migration that may be largely achieved by drifting passively in warm onshore currents at intermediate depths. These currents arise to compensate for offshore surface currents associated with prevailing offshore winter winds (Miller *et al.* 1984; Govoni and Pietrafesa 1994).

The exploitation of the estuary of the St. Lawrence River (Canada) by four migratory fish species illustrates the variety of migratory patterns that may evolve in relation to estuarine frontal zones. The salinity front formed by the upstream limit of salt-water intrusion in this river represents an important early-summer nursery area for larvae and juveniles of rainbow smelt, *Osmerus mordax*, and Atlantic tomcod, *Microgadus tomcod*, (Laprise and Dodson 1989, and references therein). Smelt spawn in the main river and tributaries adjacent to the estuary in the spring, and are concentrated at the frontal zone by a combination of residual circulation and active vertical migration. Tomcod spawn 150 km upstream in winter and, following spring hatching, larvae drift passively to the frontal zone where they occupy deep waters and are retained by residual circulation. Larvae and juveniles of both species feed on mesozooplankton that occur at peak abundances in the frontal zone as temperatures increase in early summer. Moving 150 km downstream from the salinity front, a tidally-induced stratification front associated with an abrupt change in bottom topography separates shallow, well-mixed inshore waters from deep, stratified offshore waters. This area is the spawning site of Atlantic herring, *Clupea harengus*, that migrate upstream from the Gulf of St. Lawrence in the spring and fall. The stratification front appears to limit the seaward drift of yolk-sac larvae, and the subsequent colonisation of inshore mixed waters located upstream of the stratification front is largely due to upstream residual circulation. Suitable prey for first-feeding larvae is initially produced uniquely at the front and is subsequently transported upstream toward shallower areas of the estuary in concert with the upstream movement of postlarvae (Fortier and Gagné 1990, and references therein). Located even further downstream along the Gaspé peninsula is a quasi-permanent 150 km-long estuarine front. In the spring, capelin migrate upstream from the Gulf of St. Lawrence and spawn on beaches of the estuary located between the upstream limit of salt-water intrusion in the river and the tidally-induced stratification front where herring spawn. Upon hatching, larvae occupy surface waters and are rapidly transported downstream where, as first-feeding larvae, they encounter high concentrations of copepod eggs and nauplii on which they feed. The high concentration of copepods is triggered by the production and accumulation of large diatoms related to freshwater runoff and mixing along the Gaspé frontal zone (Fortier *et al.* 1992, and references therein). In all of these cases, the spawning season and the migratory trajectories of adults and larvae in the St. Lawrence Estuary appear to have evolved to exploit estuarine frontal systems that offer predictable and dense mesozooplankton food resources favourable for larval survival.

In conclusion, fitness may be improved by migrating to habitats that provide food resources for greater growth and survival, or that provide better protection from unfavourable conditions for offspring. For pelagic or demersal spawners that broadcast gametes, differences in reproductive success among individuals may be caused mainly by early life-history mortality. Thus, the greatest contribution to reproductive success may be achieved by improving survival during the vulnerable egg and larval stages. In species that protect their young by building nests and/or by providing parental care, differences in reproductive success may be caused mainly by competition over mates. In such cases, a significant contribution to fitness may be achieved by improving growth during the adult stages. In the seasonally fluctuating environments of temperate and polar regions, variation in adult and juvenile mortalities due to periodic unsuitable conditions in the feeding habitat may be sufficient to favour the evolution of migration (Northcote 1978). The formulation of general evolutionary models of migration patterns must take into account the relative contributions to individual fitness of all habitats occupied during the migratory cycles of the species under study. Knowledge of the phylogenetic relationships between the species under study provides insights about their ancestral habitats and the part of the migratory cycle that may best be defined as the derived state.

2.3 Migration and life-history evolution

The evolution of migratory behaviour cannot be understood without considering the evolution of life history. Migratory behaviour and other life-history traits should coevolve to form adaptive strategies that convey selective advantage in specific environmental settings (Hutchings and Morris 1985). The complexity of the interaction between migration and other life-history traits is well illustrated by the American eel, *Anguilla rostrata*. Juvenile American eel (elvers), upon metamorphosis at sea, enter freshwater systems from latitudes 9°N to 60°N, with body size and the proportion of females increasing to the north (Helfman *et al.* 1987). These authors suggest that males and females have evolved different life histories, and hence different migratory strategies, in response to different selection pressures. Male eels mature rapidly at small sizes and are most common in estuaries of southeastern United States that are relatively close to their spawning grounds in the Sargasso Sea. In contrast, females maximise fecundity through increased size and are generally found further upstream in rivers and farther north than males within the species' range (Helfman *et al.* 1987). Females grow to maturity in fresh water further upstream and/or at higher latitudes (with greatest sizes being attained farthest to the north), where estimates of freshwater production are below that of temperate marine waters and tropical freshwaters (Gross *et al.* 1988). Greater size is associated with longer periods of residence in fresh water and late maturity (Helfman *et al.* 1987). In contrast, males occupy highly productive estuaries close to their oceanic spawning grounds, grow rapidly and mature at small sizes. Thus, for males, highly productive waters are occupied to grow rapidly and mature early, not to attain larger body sizes (Helfman *et al.* 1987). This time-constrained strategy contrasts with that of females, which increase their fecundity through increased body size and thus are selected to reproduce at the largest size possible. Large body size has little relative effect on male fitness, assuming that size-related social interactions do not affect spawning (Helfman *et al.* 1987).

The evolution of migratory and other associated life-history strategies may depend on (1) the energetic cost of migration, (2) the relationship between fecundity and body size, and (3) environmental uncertainty. These factors are addressed below.

2.3.1 The energetic cost of migration

The energetic cost of returning to spawning grounds can be great enough to reduce the somatic reserves of reproductive adults and contribute significantly to the cost of reproduction. Current reproduction may reduce lifetime reproductive success because of its adverse effect on future adult survival and breeding success (Rankin and Burchsted 1992). In the case of anadromous migrations, energy expenditures can be high and contribute significantly to the cost of reproduction. During the freshwater migration of iteroparous anadromous coregonines of James Bay (Canada), the energetic cost of migration was 1.6 times higher than the total energy invested in gonads by female cisco, *Coregonus artedii*, and equal to the total energy invested in gonads by female lake whitefish, *C. clupeaformis* (Lambert and Dodson 1990a, b). This cost of migration represented 20 to 29% of the somatic energy content of reproductive fish of both species. Iteroparous York River (USA) shad expended about 30% and iteroparous Connecticut River shad 40–60% of their energy reserves to migrate, spawn and return to sea (Glebe and Leggett 1981a, b). Evidence from several anadromous species show that expenditures in excess of approximately 70% of stored energy reserves results in complete spawning mortality (Glebe and Leggett 1981a, b).

Given the consequences of migration for current and future reproductive success, selection should favour the evolution of behavioural traits that minimise energetic expenditures (assuming that such traits are 'simple' life-history traits which evolve to maximise reproductive success in response to environmental conditions (*sensu* Gross 1987)). Because any reduction in current energy expenditure will permit an accumulation of stored energy that may enhance future reproductive success, selection may be expected to favour traits that minimise energetic cost per unit distance travelled.

2.3.2 The energetic efficiency of migration

To empirically test the hypothesis that fish migrants minimise energetic cost per unit distance travelled, Bernatchez and Dodson (1987) documented the migration energetics of 15 anadromous fish populations involving nine species. Anadromous fishes do not generally feed in fresh water, but derive energy required for migration from somatic reserves stored in muscles and viscera while at sea. Thus, studies of the energetic cost of anadromous migrations, based on observations of lipid and protein metabolism during the freshwater phase of the migration in combination with observations of migration length and duration, provide reliable estimates of energy cost per unit distance for a variety of species and hydrological conditions.

Table 2.1 Characteristics of the freshwater reproductive migrations of 15 anadromous fish populations. Populations are presented in ascending order of body weight. Adapted from Bernatchez and Dodson (1987).

Species	(No.)	Population-river	Mean weight (g)	Mean length (cm)	Migration distance (km)	Migration distance (Body lengths)	Vertical gradient (m km⁻¹)	Total migration cost (kJ kg⁻¹)	Cost per unit distance (kJ kg⁻¹ km⁻¹)	Mean swimming speed (bl s⁻¹)	Mean displacement rate (km d⁻¹)
River lamprey (*Lampetra fluviatilis*)	(1)	Severn	55	30	100	3.30×10^5	0.10	5109	51.09	0.33	0.6
Cisco (*Coregonus artedii*)	(2)	Eastmain	310	29	33	1.10×10^5	0.45	2114	64.06	0.86	0.6
Alewife (*Alosa pseudoharengus*)	(3)	La Have	327	32	32	1.00×10^5	-	884	27.62	-	5.4
Sea lamprey (*Petromyzon marinus*)	(4)	St. John (N.B.)	896	77	140	1.80×10^5	0.02	2280	16.29	0.13	1.8
American shad (*A. sapidissima*)	(5)	St. Johns (Florida)	1000	40	370	5.90×10^5	0.00	6463	17.47	1.50	8.8
American shad	(6)	York	1300	45	80	1.86×10^5	-	1520	19.00	1.00	4.6
American shad	(7)	Connecticut	1800	55	137	2.49×10^5	0.22	2863	20.89	1.09	6.0
Sockeye salmon (*Oncorhynchus nerka*)	(8)	Skeena	2200	57	380	6.30×10^5	1.84	3778	9.94	2.77	18.0
Sockeye salmon	(9)	Fraser (Stuart Lake run)	2545	58	1152	1.74×10^6	0.66	4555	3.95	2.71	42.6
Sockeye salmon	(10)	Fraser (Chilko Lake run)	2545	58	596	1.00×10^6	1.75	5070	8.51	2.25	33.0
Atlantic salmon (*Salmo salar*)	(11)	Northwest Miramichi	2900	74	113	1.50×10^5	1.77	2032	17.98	0.10	0.9
Chum salmon (*O. keta*)	(12)	Amur	3900	64	1193	1.86×10^6	0.25	6140	5.15	2.43	45.0
Atlantic salmon	(13)	Spey	4100	77	105	1.40×10^5	-	1262	12.02	-	-
Chinook salmon (*O. tshawytscha*)	(14)	Sacramento (Mc Cloud R.)	10000	86	1000	1.15×10^6	1.00	3856	3.86	0.67	7.1
Chinook salmon	(15)	Columbia (Snake R.)	15000	104	1134	1.10×10^6	0.53	2852	2.52	1.40	26.0

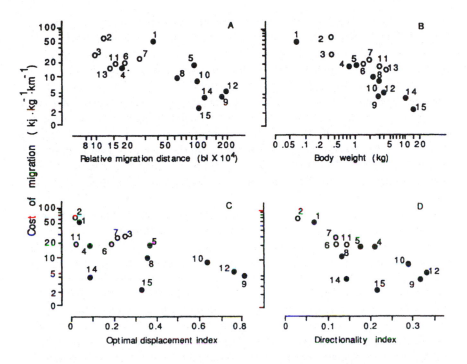

Fig. 2.2 Relationship between the energetic cost of migration (logarithmic scale) and (A) relative migration distance (log scale), (B) body weight (log scale), (C) optimal displacement index (arithmetic scale), and (D) directionality index[1] (arithmetic scale). Numbers refer to anadromous populations identified in Table 2.1. Open circles identify iteroparous species and closed circles, semelparous ones. Adapted from Bernatchez and Dodson (1987).

The anadromous freshwater migrations summarised in Table 2.1 are very different in terms of the body size of migrating fish, type of river, distance travelled and cost of migration. To compare the energetic efficiency of observed displacement rates (km per day), an index of optimal displacement was calculated for each population. The index was calculated by dividing observed displacement rate by a calculated optimal displacement rate, the empirically-determined ground speed at which energy cost per per unit distance is minimised (see Bernatchez and Dodson 1987 for calculations). An optimal displacement index of 1 corresponds to the

[1]An index of directionality was calculated by dividing the mean rate of displacement (km per day) by the mean swimming speed relaive to the water (calculated from regressions relating energy expenditure per unit time with swimtming speed that are empirically determined using respirometry techniques). The maximum value of the index is 1 and corresponds to the ideal situation where all swimming activity is translated into upstream displacement. Values less than 1 imply meandering (low degree of upstream orientation) or holding station. Although water currents affect the swimming activity of fish, current speed was not correlated with the directionality index and we proposed that the index represents a comparative estimate of the directionality and persistence of upstream movement (Bernatchez and Dodson 1987).

situation where observed mean displacement rate minimises energy cost per unit distance. Index values less than 1 indicate ground speeds below optimal.

For most populations, cost per unit distance is not minimised and observed displacement rates are less than 40% of optimal displacement rates (Fig. 2.2c). The optimum situation was approached only in the case of Stuart Lake sockeye salmon and by Amur River chum salmon. Increasing energy efficiency of migration is associated with distance travelled, as shown by the strong negative relationship between the cost of migration per unit distance travelled and migration distance (Fig. 2.2a). This analysis suggests that migration distance may lead to improved migratory efficiency by selecting for larger body size (Fig. 2.2b), more accurate upstream migration (Fig. 2.2d) and travel speeds that optimise energy efficiency (Fig. 2.2c).

Water currents represent an obvious transport system which marine fishes may exploit for both directional information and energy savings, and which provide another means of examining behavioural adaptations for the efficient use of energy during migration. Weihs (1978) demonstrated mathematically that, in areas where tidal currents are directional, fish may achieve considerable energy savings by swimming in midwater at a constant speed relative to the bottom when the tide flows in the direction of migration and by holding station on the sea floor when the tide flows in other directions. Such behaviour, known as selective tidal stream transport, has been demonstrated in plaice, *Pleuronectes platessa*, and several other marine species (reviewed by Arnold and Cook 1984), and can save up to 40% in energy cost compared with the alternative strategy of swimming continuously in the direction of overall movement. Metcalfe *et al.* (1990) attempted to identify whether plaice use selective tidal stream transport in an energetically effective way by calculating the metabolic cost of transport for individuals tracked at sea. The metabolic cost of transport was 20% less than the cost of continuous swimming at the optimal swimming speed. When in midwater, most fish swam at speeds less than 50% of the optimal speed and headed downtide, but this orientation was not precise. It was, however, sufficient to achieve 88% of the maximum energy saving that could have been made with precise downtide orientation.

These examples demonstrate that the migratory lifestyle is associated with the evolution of a number of behavioural traits improving migratory performance and energetic efficiency. However, it is evident that not all migrants behave to maximise energetic efficiency. Pacific salmon are semelparous and, despite their outstanding energetic efficiency, many populations (but not necessarily all) exhaust their 'fuel' reserves to complete spawning (Groot and Margolis 1991). In such cases, fuel economy is essential for successful reproduction. Iteroparous species could also benefit from minimising costs to improve the probability of survival and/or increased fecundity. However, this does not appear to be a strong selective factor. Overall energy storage may be more important than the energetic efficiency of locomotion; hence the importance of large body size to store sufficient energy reserves for the spawning migration. Only when sufficient energy reserves cannot be stored to successfully migrate and spawn does energetic efficiency approach theoretically optimal levels (Bernatchez and Dodson 1987).

2.3.3 Migration, body size and fecundity

Large body size provides benefits to migrants other than the opportunity to accumulate greater energy stores. Larger body size increases metabolic efficiency, in terms of energy expended per unit body weight, and swimming capacity (Beamish 1978). As the distance moved for one tail beat is the same fraction of a fish's body length, regardless of its size, big fish can migrate farther than small fish and exploit a far greater area during their life cycle (Schmidt-Nielsen 1984). Since fecundity generally increases with body size in fishes (Bagenal 1978), larger size will also contribute to increasing female breeding success through an increase in the number of eggs produced. In addition, larger size will contribute to the fitness of territorial species through success in breeding competition (e.g. van den Berghe and Gross 1989; see also Berglund, Chapter 9 and Dugatkin and FitzGerald, Chapter 10). Larger size at maturity may be achieved by growing faster or staying longer on feeding grounds. Thus, we may predict that the body size of migratory species and populations should be greater than those of non-migratory ones (Roff 1988). In addition, fecundity should also be greater among migratory species because of their larger body size. However, the suggestion has been made that populations allocating a greater proportion of their energy reserves to migration accomplish this by reducing the energy allocated to gonads, thereby reducing relative fecundity (Leggett and Carscadden 1978). Thus, these two hypotheses predict different trends in fecundity with migration distance.

In support of the hypothesised positive relationship between body size, fecundity and migratory life style, a comparison of anadromous and non-anadromous populations of seven salmonid species (*Salmo* and *Salvelinus* spp.) revealed that anadromous fish were significantly bigger and more fecund (number of eggs and total biomass of eggs produced) than non-anadromous ones (Gross 1987). Similarly, migratory marine teleosts are generally bigger than non-migratory species (Roff 1988). The bioenergetic analysis of Bernatchez and Dodson (1987) summarised above revealed that the average adult body size for 15 anadromous populations was positively correlated with migration distance. Positive associations between body size and migration distance and spawning site altitude, with greater growth at sea associated with delayed reproduction, have been observed among Atlantic salmon, *Salmo salar*, (Schaffer and Elson 1975) and anadromous brown trout, *Salmo trutta*, populations (L'Abée-Lund 1991). In a study of Atlantic salmon in Norwegian waters, water discharge was more important than river length in influencing body size (Jonsson *et al.* 1991). A comparison of anadromous and freshwater brook charr, *Salvelinus fontinalis*, inhabiting the same river system revealed that anadromous forms exhibited greater longevity, greater maximal length and a greater age and length at maturity than freshwater forms (Castonguay *et al.*

1982). Female chinook salmon originating in the upper Columbia River (USA) were both larger and more fecund than those sampled in the lower river (Beacham and Murray 1993). Genetically-based variation in growth rate and body size among populations of sticklebacks, *Gasterosteus aculeatus*, that exhibit different migratory strategies is also consistent with the positive body-size/migration-distance relationship (Snyder 1991).

There are, however, intraspecific comparisons that do not support the hypothesised positive relationship between body size, fecundity and migration length. St. John's River (USA) shad suffer complete post-spawning mortality. Despite being the smallest of the Atlantic coast shad studied by Glebe and Leggett (1981b), they displayed the longest and most costly migrations and allocated 1.3 and 3.3 times the absolute energy allocated to migration by Connecticut River and York River shad, respectively (Table 2.1). The St. John's fish also had the greatest fecundity, allocating 1.4 and 1.8 times the absolute energy allocated to the gonad by Connecticut River and York River shad (Leggett and Carscadden 1978). Among Pacific salmon, female chum salmon, *Oncorhynchus keta*, in the long-migrating Yukon River (Canada) populations were smaller and less fecund than those in more coastal populations. In the odd-year broodline of pink salmon, *O. gorbuscha*, upper-Fraser River populations had lower fecundities and were smaller than populations spawning in the lower river. Female sockeye salmon, *O. nerka*, in upper river populations in both the Skeena and Fraser Rivers were less fecund than lower-river populations, but there was no consistent size differential. Upper Fraser River chinook salmon, *O. tshawytscha*, were smaller than those in lower-river populations, but not necessarily less fecund. Female coho salmon in long-distance migrating, upper-river populations were smaller and less fecund than those in more coastal populations (Beacham and Murray 1993). Similarly, Fleming and Gross (1989) reported a negative relationship between female body length and migration distance in coho salmon. Egg production in coho salmon (total biomass of eggs produced) also showed a tendency to decrease with elevation of spawning grounds and migration distance, but the trends were not statistically significant. The majority of these data support the alternative hypothesis mentioned above, namely that fecundity is reduced to pay the costs of longer migrations (Leggett and Carscadden 1978). However, the smaller average body sizes of fish migrating longer distances is not consistent with the positive body-size/migration-distance relationship. Differences in the contribution of body size to breeding success among populations of territorial salmonids may be partially responsible for these trends, although other phenomena may be involved (see Section 2.3.5).

2.3.4 Life-history evolution and reproductive uncertainty

Various authors examining the relationship between reproductive success and the frequency of reproduction have concluded that when reproductive success is unpredictable and the variance of juvenile survival exceeds that of adults, iteroparity significantly increases population stability, lowers the probability of extinction and will thus be selected for (Leggett and Carscadden 1978, and references therein). Such a strategy is generally associated with delayed reproduction, greater size and age at first maturity, lower fecundity and a greater number of lifetime reproductive events. The proximate influence of the energetic requirements of migration on life-history traits cannot be considered independently of the ultimate influence of reproductive uncertainty in the spawning and early rearing habitats. Iteroparity and exhausting migrations are incompatible. The reduction in longevity and lifetime reproductive events associated with increased post-spawning mortality is limited by the probability of extinction associated with high variance in the survival of early life-history stages. Iteroparous populations must maintain a certain level of post-spawning survival by limiting somatic depletion during reproduction. This may be achieved by accumulating sufficient somatic reserves prior to migration and subsequently expending a smaller proportion of these reserves. For example, post-spawning survival in the iteroparous Connecticut River shad population appears to be assured by greater somatic reserves afforded by larger body size and reduction in energy invested in gonads and migration, due to a shorter migration, relative to the semelparous Florida population (Leggett and Carscadden 1978; Table 2.1).

Bell (1980) provided evidence for a negative correlation between migration distance (degree of anadromy) and adult survivorship. Anadromous species that make extensive ocean migrations are semelparous, whereas those exhibiting short oceanic migrations confined to coastal and estuarine waters are all iteroparous. However, semelparity may be selected only if mean juvenile survival is higher, or if its variance is lower, relative to adult survivorship (Bell 1980), regardless of the difficulty of the spawning migration. In fact, complete post-spawning mortality may have little to do with the energetic requirements of migration. Because complete spawning mortality occurs in certain populations of sockeye salmon, pink salmon and American shad that exhibit relatively short freshwater migrations at apparently minimal energetic costs, Glebe and Leggett (1981b) concluded that semelparity was not simply a result of somatic exhaustion due to long migrations. However, the actual energetic cost of these short, semelparous migrations has not been documented. I can think of no reason why such fish would accumulate energy reserves beyond those needed to migrate and spawn, unless the nutrient enrichment of juvenile rearing habitats caused by the death and decomposition of post-spawning adults results in improved juvenile survival (Northcote 1978, and references therein). Intraspecific bioenergetic studies of semelparous populations migrating different distances are required to establish the relationship between body size and energy stores, and the costs of migration and spawning. Considering the potential importance of body size to breeding success in territorial salmonids (van den Berghe and Gross 1989; see also Dugatkin and FitzGerald, Chapter 10), the proportion of energy stores available to compete for mates (and thus the intensity of intrasexual selection) may be far less in salmon populations migrating long distances than in those migrating short distances. For example, the deeper body profile of male sockeye salmon spawning on beaches relative to river spawners within the Kvichak River system of Alaska may be related to differences in the difficulty of their respective migrations (Blair *et al.* 1993). These authors proposed that the relatively short migration distance and the little energy needed to hold position in the lake permit these fish to invest more energy in secondary sexual characteristics (such as body depth) relative to river spawners who must expend more energy to hold position in the river and, in some cases, invest energy in more demanding upstream migrations. In addition, natural selection against deep-bodied forms may be less intense in beach-spawning populations as the lake's depth affords protection from predatory bears and permits the deepest bodied fish access to all spawning areas (Blair *et al.* 1993).

An example of how migration interacts with other life-history traits in iteroparous populations is provided by a comparative study of sympatric populations of anadromous cisco and lake whitefish of James Bay (Canada). Lambert and Dodson (1990a, b) demonstrated that the smaller size and earlier age at first maturity of cisco was associated with higher growth rates and greater adult mortality rates relative to lake whitefish. For the same freshwater migration distance, cisco allocated twice as much energy to migration than lake whitefish did. The resultant higher post-spawning mortality rates in cisco were associated with a less iteroparous lifestyle than that of whitefish, with the expected lifetime fecundity of cisco being achieved in about half the number of spawnings than that of lake whitefish. Thus, for cisco, greater somatic depletion during migration and spawning appears to be associated with smaller size at maturity, greater fecundity and fewer lifetime spawning events. Furthermore, the persistence of anadromous cisco in the geologically younger rivers of Hudson Bay (Canada) is related to migration distances to reach spawning grounds that are much shorter than those in more southern rivers in James Bay. In spite of decreases in pre-migratory energy accumulation in the north of their range due to shorter growing seasons, cisco maintain a similar somatic cost of reproduction without reducing fecundity, which has resulted in similar life-history patterns among populations over a large latitudinal gradient (Kemp *et al.* 1992). Because more difficult migrations in the north of their range would result in greater post-spawning mortality and an almost semelparous lifestyle, the probability of population extinction would be high due to reproductive uncertainty. Anadromous cisco may have succeeded in extending their range to the northernmost rivers because freshwater spawning habitats are closer to marine feeding habitats than in the southern part of their range.

The relative energetic inefficiency of the few iteroparous anadromous migrations studied to date (Table 2.1) may be the result of a strategy to store sufficient energy reserves to insure a level of post-spawning survival appropriate for the uncertainty of the reproductive environment. I speculate that sufficiently high levels of energy accumulated prior to migration may accommodate a wide variety of migration patterns and mechanisms of displacement, with little selective pressure on the evolution of specialised migratory behaviour, physiology or morphology. Sufficient directional bias may be provided by the 'reactive' responses of fish to external stimuli that may simply depend on ortho- and klinokineses (reviewed by Dodson 1988). The observed circuitous routes of many migratory fishes (Leggett 1984, and references therein), with some populations spending from weeks to months residing on spawning grounds prior to spawning (Bernatchez and Dodson 1987), is consistent with this view. This view also implies that adaptations for fuel economy among iteroparous populations are positively correlated with migration distances, and that there exist maximum migration distances beyond which the sedentary lifestyle is favoured or population extinction occurs. The data base presently available to test these predictions is inadequate. Intraspecific comparisons of iteroparous populations that migrate different distances are required to establish how body size and energy stores are related to the difficulty of migration and how the demands of post-spawning survival limit the extent of migration.

2.3.5 Intraspecific migration phenotypes

The greatest opportunity for insight into the evolution of migratory patterns and their relationship with life-history evolution lies in the study of intraspecific migration phenotypes. An evaluation of lifetime fitness ($\Sigma l_x m_x$) among different phenotypes would be an excellent test of the validity of evolutionary models in explaining variations in migratory pattern and life-history traits. When life-history trait optima differ among environments that are inhabited at random within and among generations, selection acts on a trait's norm of reaction (the alteration of a life-history trait in response to environmental change) (Hutchings and Myers 1994, and references therein). Potentially the most important reaction norm with respect to the evolution of migration is a negative correlation between age at maturity and growth rate (Hutchings and Myers 1994). Such adaptive phenotypic plasticity may be responsible for the phenomenon of precocious male maturity among young Atlantic salmon that do not migrate to sea prior to reproduction, but mature in fresh water at a much smaller size (Hutchings and Myers 1994). These authors proposed a simple model incorporating polygenic thresholds of age at maturity (largely environmentally controlled) as a mechanism by which an evolutionary stable continuum of phenotypes could be maintained within a population. Many other examples of alternative migratory and reproductive strategies related to differences in growth rate exist, and these could also be analysed in the same way. Populations of Atlantic salmon may include several migratory phenotypes: the entire life cycle may either occur in fresh water, involve different combinations of freshwater, estuarine and sea residency, or resemble more closely the typical anadromous lifestyle (Power *et al.* 1987). These authors proposed that conditions for growth in the estuary vary from year to year and may determine the proportions of the various migration phenotypes in a population. Smolts may terminate their seaward migration in the estuary when growth exceeds a threshold value at which the reproductive success of non-migrants is equivalent to that of migrants. Equivalent reproductive success may be caused by a greater degree of iteroparity and higher relative fecundity among estuarine salmon relative to anadromous forms. These propositions may be tested using a variety of species exhibiting plasticity of life history and migratory patterns, a situation that appears common among salmonids (Northcote 1978).

The anadromous (sockeye salmon) and non-anadromous (kokanee) forms of *O. nerka* may spawn sympatrically but are genetically distinct in several rivers of British Columbia, Canada (Wood and Foote 1990, and references therein). Kokanee mature at smaller sizes and often at younger ages than sockeye. Genetic differentiation among sockeye and kokanee populations in British Colombia led these authors to suggest that the two forms have diverged in sympatry on numerous, independent occasions. Genetic differentiation is promoted by assortative mating by size and by selection against hybrid progeny (reviewed by Wood and Foote 1990). A possible key to understanding the initial divergence of the two forms may be found in understanding the development of 'residual' sockeye, which do not migrate to sea and are, in part, the progeny of anadromous parents (Burgner 1991). These forms are characterised by faster growth rates among certain age groups (Burgner 1991), suggesting that a negative association between growth rate and age at maturity may be the basis for the evolutionary origin of the non-anadromous form. The relative production rates of anadromous and residual progeny from the spawning of residuals and the genetic component of the phenomenon are unknown. Given that anadromy is considered to be, in phylogenetic terms, a derived character state among salmonines (Stearley 1992), the evolutionary reversal to freshwater residence (the ancestral state) demonstrated by kokanee provides an opportunity to search for the selective factors that have taken precedence over those selecting for anadromy (McLennan 1994).

The different body-size/fecundity relations relative to migration distance among semelparous Pacific salmon species and populations documented by Beacham and Murray (1993) (see above) may be related to different migration phenotypes exhibiting different mortality and fecundity schedules. Among chinook salmon, two such phenotypes exist (Healey 1991). The 'stream-type' phenotype stays longer in fresh water as juveniles, migrates further at sea, matures later and is more fecund, whereas the 'ocean-type' phenotype migrates to sea soon after hatching, remains largely in coastal waters while at sea, matures earlier and is less fecund. Males and females of both phenotypes exhibit a range of ages at first maturity that may span three years. Since the stream-type is distributed further to the north than the ocean-type, it is possible that the two phenotypes evolved as different races in different glacial refugia (Gharret *et al.* 1987) such that differences in life-history traits may be the result of selection in different environments rather than alternative reproductive strategies evolved in a common environment. In the case of coho salmon, two phenotypes are also distinguished on the basis of their marine migrations (Sandercock 1991): an 'ocean' or 'coastal' form that occupies outer coastal or offshore waters, and an 'inshore' or 'interior' form that remains inshore during the marine phase of the life cycle and migrates further upstream during the freshwater phase. If these phenotypes represent evolutionary stable strategies, then their respective lifetime reproductive success ($\Sigma l_x m_x$) should be the same in sympatry, a prediction that requires building life tables for each phenotype to test. Thus, for these semelparous species, smaller body size and/or lower fecundity should be associated with higher pre-spawning survival rates. It is important to note, however, that in cases where the fitness benefits of an increase in fecundity is equalled by its cost in terms of lowered survival, any change in survival or fecundity is selectively neutral (Bell 1980). In such cases, age at maturity may be highly variable within a population.

2.4 Conclusions

If the cost of migration was insignificant, relatively minor benefits to survivorship or breeding success afforded by habitat switching would favour the evolution of migration. As many aquatic organisms drift with oceanic and tidal currents to achieve long-range directed movements, long-distance migration may entail minor costs. In both spring and autumn spawning Atlantic herring of western Newfoundland, reduction in somatic condition due to spawning and overwintering did not attain critically low levels that would reduce survival probability, indicating no survival cost due to reproduction based only on somatic depletion (Y. Lambert, personal communication). It is possible that while somatic costs due to migration and reproduction may be relatively minor, costs due to predation on adults may be major determinants of post-spawning survival. Studies exploring the fitness consequences of the predation-risk component of migrations are needed.

Passive transport is particularly important when considering the displacements of early life-history stages. Given the limited storage capacity and vagility of larvae and the high larval growth requirement associated with survival and reproductive success, most migrations of larvae must be assisted by currents (Miller *et al.* 1985). In the case of immature spot, *Leiostomus xanthurus*, and croaker, *Micropogonias undulatus*, on the east coast of the United States, their 100 km migrations from winter spawning areas near the Gulf Stream to the inlets of Pamlico Sound takes 50 days, during which they grow from 2 to 15 mm (Miller *et al.* 1985). Assuming a net benefit of migration, largely related to the growth of larvae, the migratory lifestyle would have evolved even if there were no additional benefits from switching habitats. Therefore, the evaluation of costs or benefits associated with each leg of a migration becomes important to interpret predicted relationships among life-history traits. As an example, Gross (1987) calculated that egg production by diadromous forms is, on average, three fold greater than that of nondiadromous conspecifics. Therefore, the fitness advantage of migration may be three fold through the enhancement of age-specific fecundity (m_x). Devaluating m_x by the survivorship costs incurred as a consequence of migration, migrants may have as much as a three-fold higher mortality than non-migrants and still be favoured by natural selection. However, recruitment in some pelagic marine fishes can vary by as much as five fold, indicating that placing early life-history stages in favourable habitat through migration may provide at least a five-fold increase in fitness relative to a non-migratory strategy. Thus, adult migrants may have as much as a five-fold higher mortality than non-migrants and still the migratory strategy could be favoured by natural selection. For species whose reproductive success depends largely on the survival of the vulnerable early life-history stages, the passive displacement of larvae and juveniles to favourable rearing habitats hypothetically may provide sufficient benefits in terms of growth and survival to favour the evolution of a migration strategy in the absence of any apparent benefits to switching adult habitats. Variations in the survival of eggs and larvae in the early-rearing habitat may, in many cases, be the most critical variable to consider in the evolution of large-scale migration.

Acknowledgements

I thank G. Arnold, J.-G.J. Godin, J. McNeil, T. Quinn and an anonymous reviewer for constructive criticisms of earlier drafts of this paper, and D. Thivierge and F. Colombani for assistance in the preparation of the manuscript.

References

Andersson, M. (1994). *Sexual selection*. Princeton University Press, Princeton.

Arnold, G.P. and Cook, P.H. (1984). Fish migration by selective tidal stream transport: first results with a computer simulation model for the European continental shelf. In *Mechanisms of migration in fishes* (eds. J.D. McCleave, G.P. Arnold, J.J. Dodson and W.H. Neill), pp. 227–262. Plenum Press, New York.

Bagenal, T.B. (1978). Aspects of fish fecundity. In *Ecology of freshwater fish production* (ed. S.D. Gerking), pp. 75–101. John Wiley and Sons, New York.

Baker, R.R. (1978). *The evolutionary ecology of animal migration*. Hodder and Stoughton, London.

Balon, E.K. (1975). Reproductive guilds of fishes: a proposal and definition. *J. Fish. Res. Board Can.*, **32**, 821–864.

Beacham, T.D. and Murray, C.B. (1993). Fecundity and egg size variation in North American Pacific salmon (*Oncorhynchus*). *J. Fish Biol.*, **42**, 485–508.

Beamish, F.W.H. (1978). Swimming capacity. In *Fish physiology*, Vol. 7 (eds. W.S. Hoar and D.J. Randall), pp. 101–187, Academic Press, New York.

Bell, G. (1980). The costs of reproduction and their consequences. *Am. Nat.*, **116**, 45–76.

Bernatchez, L and Dodson, J.J. (1987). Relationship between bioenergetics and behavior in anadromous fish migrations. *Can. J. Fish. Aquat. Sci.*, **44**, 399–407.

Blair, G.R., Rogers, D.E. and Quinn, T.P. (1993). Variations in life history characteristics and morphology of sockeye salmon in the Kvichak river system, Bristol Bay, Alaska. *Trans. Am. Fish. Soc.*, **122**, 550–559.

Bone, Q. and Marshall, N.B. (1982). *Biology of fishes*. Chapman and Hall, London.

Bowen, B., Meylan, A.B., Ross, J.P., Limpus, C.J., Balazs, G.H. and Avise, J.C. (1992). Global population structure and natural history of the Green Turtle (*Chelonia mydas*) in terms of matriarchal phylogeny. *Evolution*, **46**, 865–881.

Burgner, R.L. (1991). Life history of sockeye salmon (*Oncorhynchus nerka*). In *Pacific salmon life histories* (eds. C. Groot and L. Margolis), pp. 3–117. University of British Columbia Press, Vancouver.

Castonguay, M., FitzGerald, G.J. and Côté, Y. (1982). Life history and movements of anadromous brook charr, *Salvelinus fontinalis*, in the St. Jean River, Gaspé, Québec. *Can. J. Zool.*, **60**, 3084–3091.

Cushing, D.H. (1975). *Marine ecology and fisheries*. Cambridge University Press, Cambridge.

Dingle, H. (1980). Ecology and evolution of migration. In *Animal migration, orientation and navigation* (ed. S.A. Gauthreaux, Jr.), pp. 1–101. Academic Press, New York.

Dodson, J.J. (1988). The nature and role of learning in the orientation and migratory behavior of fishes. *Environ. Biol. Fish.*, **23**, 161–182.

FitzGerald, G.J. (1983). The reproductive ecology and behavior of three sympatric sticklebacks (Gasterosteidae) in a salt marsh. *Biol. Behav.*, **8**, 67–79.

Fleming, I.A. and Gross, M.R. (1989). Evolution of adult female life history and morphology in a Pacific salmon, *Oncorhynchus kisutch*. *Evolution*, **43**, 141–157.

Fortier, L. and Gagné, J.A. (1990). Larval herring dispersion, growth and survival in the St. Lawrence estuary: match/mismatch or membership/vagrancy? *Can. J. Fish. Aquat. Sci.*, **47**, 1898–1912.

Fortier, L., Levasseur, M.E., Drolet, R. and Therriault, J.-C. (1992). Export production and the distribution of fish larvae and their prey in a coastal jet frontal region. *Mar. Ecol. Prog. Ser.*, **85**, 203–218.

Frank, K.T. and Leggett, W.C. (1985). Reciprocal oscillations in densities of larval fish and potential predators: a reflection of present or past predation. *Can. J. Fish. Aquat. Sci.*, **42**, 1841–1849.

Gharret, A.J., Shirley, S.M. and Tromble, G.R. (1987). Genetic relationships among populations of Alaskan chinook salmon (*Oncorhynchus tshawytscha*). *Can. J. Fish. Aquat. Sci.*, **44**, 765–774.

Glebe, B. and Leggett, W.C. (1981a). Temporal, intra-population differences in energy allocation and use by American shad (*Alosa sapidissima*) during the spawning migration. *Can. J. Fish. Aquat. Sci.*, **38**, 795–805.

Glebe, B. and Leggett, W.C. (1981b). Latitudinal differences in energy allocation and use during the freshwater migrations of American shad (*Alosa sapidissima*) and their life history consequences. *Can. J. Fish. Aquat. Sci.*, **38**, 806–820.

Goldman, J.C. (1988). Spatial and temporal discontinuities of biological processes in pelagic surface waters. In *Toward a theory of biological-physical interactions in the world ocean* (ed. B.J. Rothschild), pp. 273–296. Kluwer, Dordrecht.

Goulding, M. (1980). *The fishes and the forest*. University of California Press, Berkeley.

Govoni, J.J. and Pietrafesa, L.J. (1994). Eularian views of layered water currents, vertical distribution of some larval fishes, and inferred advective transport over the continental shelf off North Carolina, USA, in winter. *Fish. Oceanogr.*, **3**, 120–132.

Groot, C and Margolis, L. (eds). (1991). *Pacific salmon life histories*. University of British Columbia Press, Vancouver.

Gross, M.R. (1987). Evolution of diadromy in fishes. *Amer. Fish. Soc. Symp.*, **1**, 14–25.

Gross, M.R., Coleman, R.C. and McDowall, R. (1988). Aquatic productivity and the evolution of diadromous fish migration. *Science*, **239**, 1291–1293.

Harden Jones, F.R. (1968). *Fish migration*. Arnold, London.

Harden Jones, F.R. (1984). A view from the ocean. In *Mechanisms of migration in fishes* (eds. J.D. McCleave, G.P. Arnold, J.J. Dodson and W.H. Neill), pp 1–26. Plenum Press, New York.

Healey, M.C. (1991). The life history of chinook salmon (*Oncorhynchus tshawytscha*). In *Pacific salmon life histories* (eds. C. Groot and L. Margolis), pp. 311–394. University of British Columbia Press, Vancouver.

Heath, M.R. (1992). Field investigations of the early-life stages of marine fish. *Adv. Mar. Biol.*, **28**, 1–174.

Helfman, G.S., Facey, D.E., Hales, L.S. Jr. and Bozeman, E.L. Jr. (1987). Reproductive ecology of the American eel. *Amer. Fish. Soc. Symp.*, **1**, 42–56.

Hutchings, J.A. and Morris, D.W. (1985). The influence of phylogeny, size and behaviour on patterns of covariation in salmonid life histories. *Oikos*, **45**, 118–124.

Hutchings, J.A. and Myers, R.A. (1994). The evolution of alternative mating strategies in variable environments. *Evol. Ecol.*, **8**, 256–268.

Hynes, H.B.N. (1970). *The ecology of running waters*. University of Toronto Press, Toronto.

Johannes, R.E. (1978). Reproductive strategies of coastal marine fishes in the tropics. *Environ. Biol. Fish.*, **3**, 65–84.

Jonsson, N., Hansen, L. and Jonsson, B. (1991). Variation in age, size and repeat spawning of adult Atlantic salmon in relation to river discharge. *J. Anim. Ecol.*, **60**, 937–947.

Kaitala, V. (1990). Evolutionary stable migration in salmon: a simulation study of homing and straying. *Ann. Zool. Fennici*, **27**, 131–138.

Kemp, A., Lambert, Y. and Dodson, J.J. (1992). Relationship between energy accumulation and migration cost and its influence on life-history variations of anadromous cisco (*Coregonus artedii*). *Pol. Arch. Hydrobiol.*, **39**, 443–452.

Kennedy, J.S. (1985). Migration, behavioral and ecological. *Contr. Mar. Sci.*, **27**, 5–26.

Krebs, J.R. and Davies, N.B. (1993). *An introduction to behavioural ecology*, 3rd edn. Blackwell Scientific Publ., Oxford.

L'Abée-Lund, J.H. (1991). Variations within and between rivers in adult size and sea age at maturity of anadromous brown trout, *Salmo trutta. Can. J. Fish. Aquat. Sci.*, **48**, 1015–1021.

Lalli, C.M. and Parsons, T.R. (1993). *Biological oceanography: an introduction*. Pergamon Press, New York.

Lambert, Y. and Dodson, J.J. (1990a). Freshwater migration as a determinant factor in the somatic cost of reproduction of two anadromous coregonines of James Bay. *Can. J. Fish. Aquat. Sci.*, **47**, 318–334.

Lambert, Y. and Dodson, J.J. (1990b). Influence of freshwater migration on the reproductive patterns of anadromous populations of cisco (*Coregonus artedii*) and lake whitefish (*C. clupeaformis*). *Can. J. Fish Aquat. Sci.*, **47**, 335–345.

Landsborough Thompson, A. (1942). *Bird migration*, 2nd edn. Witherby, London.

Laprise, R. and Dodson, J.J. (1989). Ontogenetic changes in the horizontal distribution of two species of larval fish in a turbid, well-mixed estuary. *J. Fish Biol.*, **35** (*Suppl.* A), 39–47.

Legendre, L. and Le Fèvre, J. (1989). Hydrodynamical singularities as controls of recycled versus export production in oceans. In *Productivity of the ocean: present and past* (eds. W.H. Berger, V.S. Smetacek and G. Wefer), pp. 49–63. John Wiley and Sons, New York.

Leggett, W.C. (1984). Fish migrations in coastal and estuarine environments: a call for new approaches to the study of an old problem. In *Mechanisms of migration in fishes* (eds. J.D. McCleave, G.P. Arnold, J.J. Dodson and W.H. Neill), pp. 159–178. Plenum Press, New York.

Leggett, W.C. (1985). The role of migration in the life history evolution of fish. *Contr. Mar. Sci.*, **27**, 277–295.

Leggett, W.C. and Carscadden, J. E. (1978). Latitudinal variation in reproductive characteristics of American shad (*Alosa sapidissima*): evidence for population specific life history strategies in fish. *J. Fish. Res. Board Can.*, **35**, 1469–1478.

McCleave, J.D., Arnold, G.P., Dodson, J.J. and Neill, W.H. (eds.) (1984). *Mechanisms of migration in fishes*. Plenum Press, New York.

McDowall, R.M. (1988). *Diadromy in fishes*. Timber Press, Portland, Oregon.

McDowall, R.M. (1993). A recent marine ancestry for diadromous fishes? Sometimes yes but mostly no! *Environ. Biol. Fish.*, **37**, 329–335.

McKeown, B.A. (1984). *Fish migration*. Timber Press, Portland, Oregon.

McLennan, D.A. (1994). A phylogenetic approach to the evolution of fish behaviour. *Rev. Fish Biol. Fish.*, **4**, 430–460.

Metcalfe, J.D., Arnold, G.P. and Webb, P.W. (1990). The energetics of migration by selective tidal stream transport: an analysis for plaice tracked in the southern North Sea. *J. mar. biol. Ass. UK*, **70**, 149–162.

Miller, J.M., Reed, J.P. and Pietrafesa, L.J. (1984). Patterns, mechanisms and approaches to the study of migrations of estuarine-dependent fish larvae and juveniles. In *Mechanisms of migration in fishes* (eds. J.D. McCleave, G.P. Arnold, J.J. Dodson and W.H. Neill), pp. 209–226. Plenum Press, New York.

Miller, J.M., Crowder, L.B. and Moser, M.L. (1985). Migration and utilization of estuarine nurseries by juvenile fishes: an evolutionary perspective. *Contr. Mar. Sci.*, **27**, 338–352.

Myers, R.A., Mertz, G. and Bishop, C.A. (1993). Cod spawning in relation to physical and biological cycles of the Northern Northwest Atlantic. *Fish. Oceanogr.*, **2**, 154–165.

Nelson, J.S. (1994). *Fishes of the world*, 3rd edn. John Wiley and Sons, New York.

Northcote, T.G. (1978). Migratory strategies and production in freshwater fishes. In *Ecology of freshwater fish production* (ed. S.D. Gerking), pp. 326–359. John Wiley and Sons, New York.

Parrish, R.H., Nelson, C.S. and Bakun, A. (1981). Transport mechanisms and reproductive success of fishes in the California current. *Biol. Oceanogr.*, **1**, 175–203.

Power, G., Power, M.V., Dumas, R. and Gordon, A. (1987). Marine migrations of Atlantic salmon from rivers in Ungava Bay, Québec. *Amer. Fish. Soc. Symp.*, **1**, 364–376.

Quinn, T.P. (1985). Homing and the evolution of Sockeye salmon (*Oncorhynchus nerka*). *Contr. Mar. Sci.*, **27**, 353–366.

Rankin, M.A. and Burchsted, J.C.A. (1992). The cost of migration in insects. *A. Rev. Entomol.*, **37**, 533–559.

Roff, D.A. (1988). The evolution of migration and some life history parameters in marine fishes. *Environ. Biol. Fish.*, **22**, 133–146.

Sandercock, F.K. (1991). The life history of coho salmon (*Oncorhynchus kisutch*). In *Pacific salmon life histories* (eds. C. Groot and L. Margolis), pp. 395–446. University of British Columbia Press, Vancouver.

Schaffer, W.M. and Elson, P.F. (1975). The adaptive significance of variations in life history among local populations of Atlantic salmon in North America. *Ecology*, **56**, 577–590.

Schmidt-Nielson, K. (1984). *Scaling: why is animal size so important?* Cambridge University Press, Cambridge.

Sherman, K., Smith, W., Morse, W., Berman, M., Green, J. and Ejsymont, L. (1984). Spawning strategies of fishes in relation to circulation, phytoplankton production and pulses in zooplankton off the northeastern United States. *Mar. Ecol. Prog. Ser.*, **18**, 1–19.

Sinclair, M.S. (1988). *Marine populations*. University of Washington Press, Seattle.

Smith, R.J.F. (1985). *The control of fish migration*. Springer-Verlag, Berlin.

Snyder, R.J. (1991). Migration and life histories of the threespine stickleback: evidence for adaptive variation in growth rate between populations. *Environ. Biol. Fish.*, **31**, 381–388.

Stearley, R.F. (1992). Historical ecology of the Salmoninae, with special reference to *Oncorhynchus*. In *Systematics, historical ecology and North American freshwater fishes* (ed. R. Mayden), pp. 622–658. Stanford University Press, Stanford.

Taylor, M.H. (1990). Estuarine and intertidal teleosts. In *Reproductive seasonality in teleosts: environmental influences* (eds. A.D. Munro, A.P. Scott and T.J. Lam), pp. 109–124. CRC Press, Boca Raton, Florida.

van den Berghe, E.P. and Gross, M.R. (1989). Natural selection resulting from female breeding competition in a Pacific salmon (Coho: *Oncorhynchus kisutch*). *Evolution*, **43**, 125–140.

Weihs, D. (1978). Tidal stream transport as an efficient method for migration. *J. Cons. Int. Explor. Mer.*, **38**, 92–99.

Wood, C.C. and Foote, C.J. (1990). Genetic differences in the early development and growth of sympatric sockeye salmon and kokanee (*Oncorhynchus nerka*), and their hybrids. *Can. J. Fish. Aquat. Sci.*, **47**, 2250–2260.

Wooster, W.S. and Bailey, K.M. (1989). Recruitment of marine fishes revisited. In *Effects of ocean variability on recruitment and an evaluation of parameters used in stock assessment models* (eds. R.J. Beamish and G.A. McFarlane), *Can. Spec. Publ. Fish. Aquat. Sci.*, **108**, 153–159.

3 *Habitat selection: patterns of spatial distribution from behavioural decisions*

Donald L. Kramer, Robert W. Rangeley and Lauren J. Chapman

'There should be a trout of at least eighteen inches in that notch,' I said. 'There are roots just downstream a couple of feet for security. The current moves well along that bank, which is what caused the roots to wash out and the tree to fall. So a good supply of food will be coming to the fish without his having to work for it. Let's go see if he's there.'

(Brooks 1974, p. 26)

3.1 Introduction

Anyone who observes or tries to capture wild fishes soon becomes aware of patterns in their distribution. Each species is reliably found in some kinds of locality and not in others. Even within the confines of an aquarium, some species typically remain near the bottom, others near the surface; some stay in shelters, while others avoid such enclosed spaces. With experience comes a sense that one can even predict the size or sex of an individual from the characteristics of the site.

Habitat selection is a key process in these distribution patterns of fishes. In this chapter we provide an overview of habitat selection: what it is, why it is important and how it is studied. We introduce the theoretical approaches used to explain and predict distribution patterns arising from behaviour, and we indicate where specific models have been tested on fishes. Finally, we review briefly the environmental variables important to habitat selection by teleost fishes in a number of the principal habitats in which they have been studied. Our focus is primarily on intraspecific patterns of distribution among the habitats used for growth and survival; space limitations prevent a more complete treatment of reproductive habitat selection, which is partly considered by Dodson (Chapter 2), and of interspecific interactions reviewed by Persson *et al.* (Chapter 12). Huntingford (1993) recently reviewed aspects of the behavioural ecology of habitat selection in fishes, and a symposium volume on habitat selection in birds (Cody 1985) illustrates the diversity of approaches to the topic of habitat selection.

3.2   What is habitat selection?

'Habitat' is a widely used term that has been inconsistently defined in ecology (Whittaker *et al.* 1973). In the broadest sense, *habitat* and *habitat type* have meanings similar to ecosystem and refer to areas of the physical environment more or less distinct from other areas in a broad range of abiotic and biotic variables. Examples are running water (lotic) habitat and coral reef habitat. Habitat is also applied at a smaller scale within an ecosystem for areas that differ in enough ways to appear qualitatively distinct. Riffle and pool habitats within a stream and reef crest and backreef habitats within a coral reef are examples. At a still finer scale, the term *microhabitat* is usually used for subdivisions of habitats that are relatively homogeneous and differ primarily in quantitative values of a small number of variables. Examples are deep and shallow water microhabitats within a pool and live branched coral and rubble substrate microhabitats within a backreef. There is no absolute distinction between habitats and microhabitats, but the term habitat is often used for the general type of place an animal lives and therefore applies to a scale larger than the animal's normal daily range. In contrast, microhabitat is typically used for finer subdivisions of space use that occur within the normal daily range.

Habitats and microhabitats may be isolated or continuous relative to the range of the animal. Behavioural ecologists usually use the term *patch* to describe delimited, relatively homogeneous parts of habitats or microhabitats that differ, especially in resource availability, from the rest of the habitat (Stephens and Krebs 1986). However, in other subdisciplines of ecology, patchiness can refer to environmental heterogeneity on any scale (Wiens 1976), and patch can therefore have a meaning almost synonymous with habitat. Habitats also differ in temporal variability and predictability; they may be constant, ephemeral, seasonal or unpredictable (Southwood 1977).

The density of fishes can vary dramatically among habitats and microhabitats. These patterns are the result of many processes, including differential reproduction and mortality, colonisation and extinction, involuntary transport and voluntary movements. We use the term *habitat selection* to refer to the non-random use of space resulting from voluntary movements. These movements can be simple locomotor or settlement responses to key stimuli or behaviourally sophisticated 'decisions' (*sensu* Dill 1987; Krebs and Kacelnik 1991) concerning the allocation of time to different parts of a familiar home range. The spatial scale over which habitat selection takes place can vary from a few centimetres, as for poeciliid fish trapped in a pool of an ephemeral stream (Chapman and Kramer 1991), to thousands of kilometres, as exemplified by bluefin tuna, *Thunnus thynnus*, on seasonal transatlantic movements (Mather 1962; see also Dodson, Chapter 2). Habitat selection occurs over a broad range of environmental complexity, from very similar, adjacent microhabitats (e.g. a salmonid choosing the current speed of a foraging station in a riffle (Everest and Chapman 1972)) to highly distinct habitat types (e.g. a newly metamorphosed pomacentrid choosing between the water column and proximity to solid substratum or between coral reef and adjacent sea grass and bare sand habitats (Doherty 1991; Victor 1991)). For animals that move between different environments at different stages of their life cycle, spatial heterogeneity at many different scales (e.g. Williams 1991) means that habitat selection may be a hierarchical process in which a general region is selected first (e.g. nearshore), then a major habitat type (e.g. reef), followed by successively smaller scale decisions about the local habitat features it encounters (see Orians and Wittenberger 1991). When habitat selection changes in a regular pattern with state of tide or time of day or year, it is considered *migration* (Neill 1984; Dodson, Chapter 2). We consider *habitat choice* to be synonymous with habitat selection and *habitat preference* to be a measure of the degree to which one habitat is selected over another (Johnson 1980). Note that habitat selection has been used occasionally in a very different sense to refer to natural selection exerted by a habitat on the organisms that disperse to it (Bazzaz 1991).

Studies of habitat selection differ from most other phenomena of interest in behavioural ecology in that the behaviour is almost never observed directly. Rather, it is inferred from non-random distributions that cannot be attributed to passive transport, to differential reproduction and mortality rates or to colonisation events. These alternative explanations for distribution patterns are easily controlled in laboratory or field enclosure studies, but for field observations assessing their role is not always straightforward. We can be quite sure that the absence of large fish from the shallow waters of a lake or stream is a result of behavioural avoidance of these areas if large fish are present in nearby deeper waters, but we can be less certain about the origin of differences in density between opposite ends of a large lake. Thus, field studies of habitat selection depend on knowledge of the mobility of the organism in relation to the strength and direction of physical forces and demographic processes in the environment. Some distribution patterns may be the result of combined effects of habitat selection and other ecological processes. When the role of active habitat selection in relation to other processes is unknown, the terms *habitat use* (Johnson 1980) or *habitat correlation* (Wiens 1976) are preferable to habitat selection.

3.3 Why is fish habitat selection studied?

Habitat characteristics are so fundamental to the evolution and ecology of the organisms that occupy them that habitat use has been proposed as the basis for a variety of ecological classifications of organisms, including fishes (see Balon 1975). Habitat selection is one of the processes that maintains this association. As the behavioural component of animal distributions, habitat selection provides an important link between behavioural ecology and population ecology. Distributions resulting from habitat selection influence the partitioning of resources in fish communities (e.g. Ross 1986; Persson *et al.*, Chapter 12). In addition, microhabitat selection is relevant to many aspects of physiology, morphology, life history and behaviour because it is involved in functional processes such as the regulation of body temperature and ionic balance, feeding, respiration, reproduction and avoidance of mortality from predators, parasites and extreme physical and chemical conditions.

Many researchers study habitat selection out of sheer fascination with the patterns of fish distribution and curiosity about the processes leading to these patterns. However, numerous applied problems in the exploitation and management of commercial stocks and the preservation of endangered species require an understanding of habitat selection. These problems include locating and enumerating mobile populations, differentiating separate populations within species, identifying appropriate modifications to the environment to increase population productivity, reducing the negative impact of dams and other development projects, and designing protective regulations and reserve areas.

3.4 Approaches to the study of habitat selection

Researchers approach habitat selection in fishes from both empirical and theoretical directions. The empirical approach starts with the distribution of particular species or communities in nature. It asks which habitat variables are correlated with the presence or abundance of fishes and how these patterns change with temporal and spatial variation in the environment and with the ontogenetic stage, physiological condition and genotype of the individuals involved. Having established a distribution pattern, the next question concerns the relative importance of habitat selection in producing these patterns. Confirmation of a role for habitat selection raises challenging questions about the sensory, motor and integrative mechanisms involved and about the evolutionary processes that led to these mechanisms. The ultimate goals for behavioural ecologists are understanding the adaptive significance of habitat selection decisions and the constraints which limit those decisions.

While empirical approaches end with adaptive significance and constraints, theoretical approaches start with these as assumptions to develop models predicting space use by individuals and the consequent distribution of populations. The models can be used to explore the implications of factors, such as conflicting demands, variable environments and changing population density, for spatial patterns. These predictions can be tested in the field or in controlled laboratory systems. Some aquatic habitats, especially small streams, ponds and coral reefs, are so well suited for experimental manipulation that fishes have been widely used to test general theories of spatial distribution. The small size, short life cycle and ease of adjustment to captivity of some species have made fishes excellent model systems for controlled laboratory tests. Thus, fishes have played an important role in the development and testing of habitat selection theory.

Empirical and theoretical approaches both have insights to offer, and both will be needed before habitat selection in nature can be predicted. Unfortunately, the two approaches are usually adopted by different researchers with different training and interests who have not always recognised their common goals. Field studies of habitat distributions often take a multivariate approach to identifying which of several easily measured physical and chemical habitat variables are most closely correlated with fish population density, but variables such as food supply or predator abundance are frequently ignored. Furthermore, it is difficult to recognise conspecific density as a variable potentially influencing habitat selection when it is also the measure of habitat use. In many field studies, several species are considered at the same time, but habitat distributions for each species are reduced to mean values that eliminate important distribution data or to abstract scores from a multivariate analysis, such as a canonical analysis of discriminants, that make comparisons between studies impossible. Sometimes, such studies examine only differences between species and not which habitat variables are used by the different species in a non-random manner. In contrast, much of behavioural ecological theory considers only a single species and does not attempt to predict which environmental variables are important; instead the focus is on how populations should be distributed in relation to one or two complex and often density-dependent variables, assumed *a priori* to be important. This division has been so great that it is difficult to integrate the presentation of empirical patterns and tests of theory. We have therefore divided this review into separate sections on the patterns of habitat use of fishes in nature and on the theoretical framework that should explain and predict those patterns. Recently, there are encouraging signs that the historical dichotomisation of this field is breaking down. In providing examples of tests of theory, we will attempt to highlight studies that are leading the way toward a fuller integration between a theoretical framework and actual field distributions.

3.5 Identifying the important habitat variables

Patterns of habitat use are documented by relating the abundance of different components of a fish population to specific environmental variables under particular conditions. Potentially important subdivisions of the population include genotype, age/size class, sex, reproductive condition, parasite load and physiological condition. Environmental variables important for habitat selection in fishes can be broadly categorised into physical characteristics of the water, chemical characteristics of the water, characteristics of the adjacent substratum and other solid objects, position in relation to surface, bottom and other objects, and the occurrence of individuals of the same and other species (Table 3.1). Environmental variables

Table 3.1 Environmental variables associated with fish distributions on a behavioural scale

1. *Physical characteristics of the water:*
    Presence of water
    Permanence of water
    Temperature
    Depth/pressure
    Current speed
    Turbulence
    Suspended material/turbidity
    Light intensity and spectral composition

2. *Chemical characteristics of the water:*
    Dissolved gases (oxygen, carbon dioxide, hydrogen sulfide)
    pH
    Salinity/conductivity
    Ionic composition
    Other natural dissolved substances (e.g. homing cues, conspecific odours)
    Chemical pollutants

3. *Characteristics of adjacent substratum and structure:*
    Composition of substratum
    Amount, form and composition of other structure (e.g. coral, algae, macrophytes,
    woody debris)
    Availability of caves and holes

4. *Position relative to:*
    Substratum and other structure
    Surface
    Shore and depth of water column
    Permanent refuges for individuals invading temporarily available habitat (e.g. intertidal zone,
    flood plains, intermittent streams)

5. *Presence or absence of other species:*
    Prey
    Predators and parasites
    Competitors and mutualists

may interact, resulting in spatial and temporal changes in preferences. Thus, the researcher has to be alert for changes in habitat selection in relation to environmental gradients, as well as to daily, tidal, and seasonal cycles and less predictable variation in weather.

When fish density and a series of environmental variables are measured at a set of predetermined locations, the correlation between abundance and the quantitative level of each variable can be examined (e.g. Orth and Maughan 1982; Kwak *et al.* 1992; Perry and Smith 1994). This approach is widely used to define 'habitat suitability' for the management of aquatic ecosystems (Reiser *et al.* 1989). A common variation of this approach is to define qualitative habitat types *a priori* and use transects or stratified sampling to estimate the density of fish in each habitat type (e.g. Keast *et al.* 1978; Hallacher and Roberts 1985; Urban and Brandt 1993). Alternatively, the researcher can first locate the fish, by underwater observations or radiotracking, for example, then measure the environmental variables or record the habitat type at these specific locations. These measures of habitat use are compared to the availability of the habitats to test for non-randomness. Availability is estimated by measures of the distribution of habitat characteristics at randomly selected locations (e.g. Grossman and Freeman 1987; Kinzie 1988) or the proportion of different habitat categories such as depth strata or vegetation type estimated from maps (e.g. Margenau 1986; Chapman and Mackay 1984). In such studies, it is important to distinguish the magnitude of preference from the magnitude of use. For example, it is possible that highly preferred habitats are relatively unimportant because they comprise such a small area that only a small proportion of the population occupies them. Conversely, a common error is to conclude that the most preferred habitat is the one in which the greatest proportion of the population occurs. Furthermore, because many habitat features are correlated in nature, the features which the researcher associates with preferential use may not necessarily be the features selected by the fish. Multivariate analyses (e.g. Sale *et al.* 1984; Grossman and Freeman 1987; Rakocinski 1988; Baltz *et al.* 1993) can help to sort out this complexity, but experimentation is necessary to determine which features provide the cues for the fish.

The influence of specific environmental variables on habitat selection can be determined experimentally with relative ease in the laboratory using an apparatus with two or more areas identical in all but the variable under consideration. A non-random distribution of fish or of time spent among areas provides evidence of habitat selection. Such tests are very useful for examining the ability of fish to distinguish among levels of a particular habitat variable such as temperature, pH, dissolved oxygen, or food availability (e.g. Sale 1969; Ingersoll and Claussen 1984; Suthers and Gee 1986). In many cases, laboratory results are related to field distributions. However, field and laboratory measures of preferences do not always agree (e.g. Sale 1968; Ross *et al.* 1992). Such discrepancies are likely to be the result of interactions between environmental variables, a subject in need of considerably more research.

Although laboratory studies have contributed much to our knowledge of the influence of specific environmental variables on habitat selection, they are less suitable for examining selection of distinct habitat types. Even when attempting to look at the effects of single factors, problems can arise from unnoticed environmental variables that are correlated with treatments. For example, a gradient of light intensity in the room or associations between the extremes of an experimental gradient and the corners of the test apparatus, which certain species tend to prefer and others to avoid, can confound experimental treatments. Within the confined area of most laboratory testing chambers, social behaviour of the species can also result in complicating patterns. For example, dominant individuals may exclude others from the most preferred area (Beitinger and Magnuson 1975). The solution to these problems is not necessarily to test individuals alone because many social species show atypical behaviour in the absence of conspecifics, and even species which do sometimes occur alone alter habitat selection as a function of number of conspecifics in the vicinity (Rangeley 1994). However, it is important to understand the social behaviour of the test species and take it into account in the experimental design and interpretation of the results. In many situations the experimental approach can be extended to the field, either in the open environment or in large enclosures that allow control of the test population, thus permitting tests of specific hypotheses on natural populations and on a larger scale than possible in the laboratory. This approach has been particularly important in the investigation of substratum and cover selection, such as in studies of stream modification and artificial reefs (Bohnsack *et al.* 1991).

Studies of habitat selection frequently fail to acknowledge the statistical complexity underlying even simple questions of distribution. Non-linear associations between habitat variables create problems in defining valid null distributions. A simple example involves the interplay between two-dimensional and three-dimensional measures of habitat availability. If fish swimming above a sloping bottom are distributed randomly with respect to the water volume, they will be non-random with respect to the substratum and surface area (appearing to avoid the shallower water). If they are distributed randomly with respect to substratum and surface area, they will be non-random with respect to the water volume (appearing to prefer the shallower water). With most tests, statistical significance applies to the difference between the entire distribution and a specified null hypothesis and can be strongly influenced by inclusion of a single strongly underutilised habitat in the set (Johnson 1980; Perry and Smith 1994). Subsequent assessment of differences in selection among the frequently used habitats is complicated by the non-independence of the observations among the different habitats; presence in one habitat necessitates absence in the others. Furthermore, measures of distribution frequently involve repeated measurements of the same individuals, and statistical approaches appropriate for examining frequency distributions involving multiple measures of multiple individuals are not widely available. Audet *et al.* (1986) discuss statistical issues involved in the analysis of preferences in laboratory gradients, and Manly *et al.* (1993) provide a recent discussion of the statistics of preference and avoidance.

3.6 Optimal habitat selection

3.6.1 Theory

Behavioural ecologists usually take an adaptationist approach to predicting habitat selection. This approach assumes that the decision rules underlying the relationship between movement patterns and environmental variables have been shaped by the fitness consequences of those movements. Therefore, from the relationship between available environmental conditions and correlates or components of fitness, such as survival, rate of food intake or growth, mating or reproductive success, one should be able to predict habitat selection by assuming that animals will choose the option yielding the highest fitness. For example, in Fig. 3.1A the solid line shows a hypothetical 'fitness function', the expected relationship between a fitness-related benefit and an environmental variable such as temperature, water depth or density of aquatic vegetation. If all else is equal, and in particular there are no differences among habitats in other fitness components, the highest fitness should be obtained by animals selecting habitat 7. Therefore, animals should be selected to prefer that habitat. If the entire population has the same fitness function, if there are no negative effects of increasing density on the benefits obtained and if animals can select habitats perfectly, the population is predicted to occur only in the habitat offering the highest benefits (Fig. 3.1B). If that habitat is not available, the population should occur in the available habitat providing the next highest benefit level. If different portions of the population have different fitness functions, each should occur in the habitat resulting in the highest benefit level for it. The basic structure and approaches of optimality theory as applied to behaviour are presented by Krebs and McCleery (1984), Stephens and Krebs (1986) and Krebs and Kacelnik (1991).

Although the idea that animals will choose the best available option appears almost trivially self evident, optimality theory has proven its value in making subtle and non-intuitive predictions in other areas of behavioural ecology, especially in foraging behaviour. Indeed, one important aspect of optimal foraging theory is the prediction of foraging habitat selection and use (Pyke 1984). Many of the subtleties arise from situations which complicate the definition of the best option. One major complication is that it is often not possible to maximise all components of fitness with the same choice. This is known as a trade-off.

In habitat selection, trade-offs occur because many different components of fitness are affected by the place in which the animal spends its time. The habitat that maximises one component of fitness is not necessarily the habitat that maximises another. Trade-offs occur between closely related components of fitness. For example, the maximum availability of two prey types may occur in different habitats or the habitat with greater prey availability may also require higher energy expenditure while foraging. Trade-offs can also occur between very different components of fitness. For example, the habitat that provides the best conditions for developing offspring may result in higher risk of predation for the adults caring for those offspring. The habitat in which fitness is maximised is the one in which the combined effect of the affected components of fitness is maximised. A hypothetical example is provided in Fig. 3.2. In this case benefit to one component of fitness (y) is maximised at level 7 of the habitat variable as in Fig. 3.1. However, a second component of fitness (x) is also affected by the same habitat variable; for this component, the maximum benefit is at level 1.4 (Fig. 3.2A). When the effect of both components of fitness is taken into account by combining them, the resulting fitness function has a different shape, which depends on whether the functions are combined additively (Fig. 3.2B) or multiplicatively (Fig. 3.2C). In the

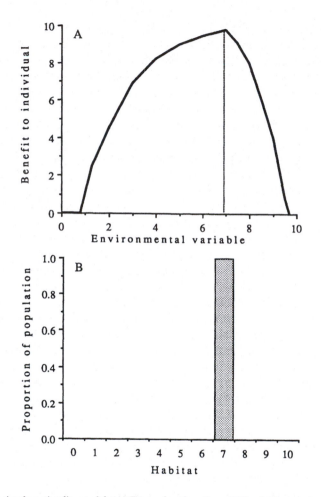

Fig. 3.1   A simple optimality model. (A) Fitness function: the solid line indicates a hypothetical relationship between an environmental variable characterising a habitat (e.g. temperature, depth, density
of vegetation) and a fitness-related benefit of occupying that habitat (e.g. growth rate, survival, fecundity). If all else is equal, the highest fitness is obtained by individuals selecting the habitat associated with the maximum benefit, indicated by the vertical dotted line. (B) Expected distribution of
the population among a series of habitats characterised by their level of the environmental variable
in A: if individuals are capable of perfect habitat selection and if there is no effect of density on the
benefit, all should be found within the available habitat that provides the highest benefit.

example provided, the optimal habitat level for the combined benefit is slightly
higher than that which would have been predicted considering only component
x and much lower than that which would have been predicted considering only
component *y*.

    In practice, it is sometimes straightforward to see how to combine different
components of fitness. For example, trade-offs between habitats that differ in the

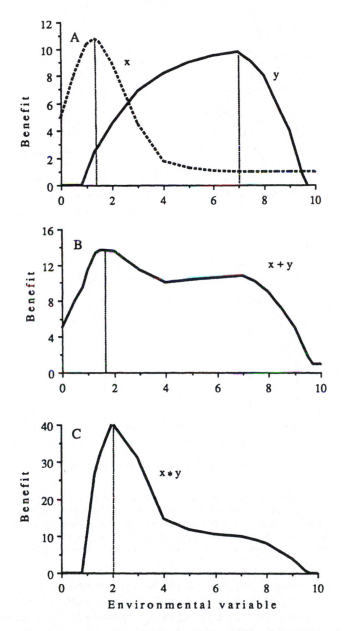

Fig. 3.2   A simple optimality model with trade-offs. (A) Separate fitness functions: an environmental variable significantly affects two components of fitness, *x* and *y*. (B) Combined fitness function resulting from adding the benefits of components *x* and *y*. (C) Combined fitness function resulting from multiplying the benefits of components *x* and *y*. Vertical dotted lines indicate the optimal level of the environmental variable for each component of fitness considered separately and combined.

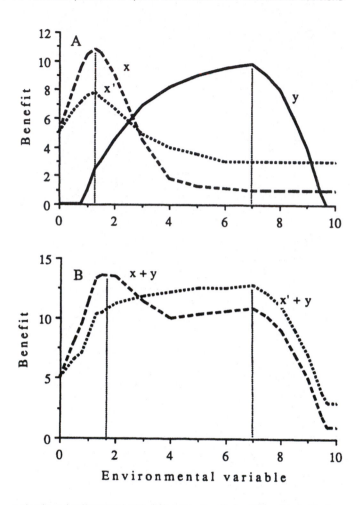

Fig. 3.3   A simple optimality model with different trade-offs for different individuals or conditions. (A) The environmental variable affects components of fitness *x* and *y*, but for some members of the population component *x* is more affected than for other members. This is illustrated by the lines marked *x* and *x'*, respectively. (B) Combined fitness functions for *x* + *y* and *x'* + *y*. Optimal levels of the environmental variable for the combined fitness functions are shown by vertical dotted lines.

abundance of individual prey types might be resolved by selecting the habitat with sum of prey types. Trade-offs between habitats that differ in encounter rate with predators and in the probability of being captured per encounter might be resolved by selecting the habitat with the lowest product of encounter rate and capture probability. In other cases, a more complex life history or bioenergetic model is required to determine the appropriate function (see below). However, it is not rare to have situations in which too little information is available to determine how to combine quantitatively disparate components of fitness such as predation risk, energy intake and reproductive success. To predict habitat distribution using optimality models, it is necessary to be able to have a fitness function that incorporates all the major effects of the habitat on fitness for the part of the population being considered and at the temporal scale at which the fish are selecting habitats.

Even when it is not clear how different components of fitness should be combined to generate a quantitative prediction, the recognition of a trade-off permits qualitative predictions. This is because the component of fitness that is more affected by the environmental variable should have a greater effect on habitat selection as illustrated in Fig. 3.3. As in Fig. 3.2, the environmental variable affects two components of fitness x and y. However, for one part of the population, component x is less affected than for other part (Fig. 3.3A). When the benefits are combined additively (Fig. 3.3B), the optimal level of the environmental variable is 7 for the less affected individuals but 1.7 for the more affected individuals. As the effect of an environmental variable on a fitness component increases in magnitude, its effect on habitat choice should increase. However, the size of a fitness benefit effect that should produce a change in habitat choice will depend on the quantitative values of the fitness functions. With the proviso that one cannot be certain of the required magnitude of effect, one can therefore make qualitative predictions relating a difference in expected fitness benefits to a difference in expected habitat choice, even when the fitness functions are not known.

Fitness benefits for a given component can vary between different parts of the population and between different situations. For example, if there is a trade-off between two habitats in foraging profitability versus predation risk, then the part of the population that is relatively more vulnerable (e.g. smaller individuals) should use the dangerous habitat less, and individuals for whom energy gain is relatively more important (e.g. animals in poor condition) should use the dangerous habitat more. Similarly, an increase in predation risk (due, for example, to an influx of predators or to lower light levels) should decrease use of the dangerous habitat. Qualitative predictions also can be derived from optimality models with trade-offs when a directional shift in the position of the peak of one variable can be identified.

When the selected habitat itself influences the animal's state, static optimality models cannot readily predict the expected distribution. By taking into account the changing state of the animal, the recently developed technique of dynamic modelling may be able to predict the final state and how habitat selection will change over time. The relevance of this approach to habitat selection was indicated in Mangel and Clark's (1988) summary of the method and illustrated by Clark and Levy's (1988) analysis of vertical migration in sockeye salmon, *Oncorhynchus nerka*, and Burrows' (1994) investigation of habitat selection by plaice, *Pleuronectes platessa*, over the diel and tidal cycles.

Limitations in animals' abilities to sample all available habitats, to assess habitat quality accurately and to track environmental fluctuations form a set of constraints which may prevent animals from occupying their optimal habitat. Differences within and among species in the magnitude of these constraints may be useful for predicting final distributions, but this approach does not appear to have been taken. When animals do not possess the sensory capacity to discriminate levels of all environmental variables relevant to their fitness, selection may favour simple 'rules of thumb' (Stephens and Krebs 1986; Bouskila and Blumstein 1992). These are responses to variables which have little importance in themselves but which are normally correlated with relevant variables in nature. For example, many of the preferences for shallow water by small fishes and for deep water by large fishes are probably related to the risk of aquatic and terrestrial predation, respectively (Goodyear 1973; Power 1987; see also Godin, Chapter 8). Water depth is likely to be much easier than predation risk for fish to assess directly, especially when the assessment of predation risk must be carried out before the actual appearance of a predator.

Assessment of habitats is difficult for the researcher and is likely to be at least as difficult for the fish. Failure by either one to include important variables in habitat assessment will result in a mismatch between theory and observation. For the fish, the problems include perceiving relevant habitat characteristics, sampling a sufficient number of the habitats available and integrating the information. Sometimes, the selection decision may precede the appearance of key habitat characteristics. For example, choice of a tide pool or stream pool may take place during a high-water period when characteristics at low water are difficult to predict. Furthermore, in a spatially variable and temporally dynamic environment, habitat decision rules favoured by selection at one time and place may not be favoured at another, resulting in a mismatch between selection and local conditions. Finally, density dependence and despotic behaviour may prevent simple optima from being achieved, as discussed in the next section. Thus, optimality models will not always predict habitat selection successfully. Failures of the models can provide powerful insights into additional variables or constraints that need to be considered. It is important, then, to determine objectively the predictive power of various models and how this power varies with the situation, rather than approach optimal habitat selection as if it were a logical necessity (see Townsend and Calow (1981) for further discussion).

3.6.2 Examples

Changes in water chemistry can have important effects on the survival and growth rates of fishes. Laboratory studies have documented the capacity of fishes to change their distributions appropriately in response to several important natural chemical variables, including oxygen (e.g. Stott and Cross 1973; Stott and Buckley 1979), pH (e.g. Pedder and Maly 1986; Peterson *et al.* 1989) and salinity (e.g. Audet *et al.* 1985). Chemical pollutants introduced into the aquatic environment by humans provide an interesting contrast to natural variables, because fishes have little or no evolutionary history of exposure to these substances which can have major effects on fitness. Reviews by Giattina and Garton (1983) and Beitinger and Free-man (1983) have documented the wide diversity of response patterns. For some substances and test situations, avoidance responses are strong, but in other cases avoidance is lacking or ineffective. For example, golden shiners, *Notemigonus crysoleucas*, avoided chromium, copper and arsenic at levels two orders of mag-nitude below acute lethal concentrations while they failed to avoid cadmium and selenium at nearly lethal levels (Hartwell *et al.* 1989). Sometimes, fishes may even show preferences for toxicants (Beitinger and Freeman 1983).

A predictive approach to the responses of ectotherms to temperature variation has been developed using optimality theory (Huey and Slatkin 1976; Huey 1991). In some situations, animals must avoid exposure to lethal extremes; in other situations the critical problem is selecting a temperature at which one or more physiological processes are maximised. For fishes, there is evidence for a close correspondence between temperature preference and several fitness correlates, es-pecially growth rate (Brett 1971; Magnuson *et al.* 1979; Crowder and Magnuson 1983; see Fig. 3.4). Because temperature and food supply can show independent spatial variation in freshwater lakes and marine systems, there is likely to be a trade-off between the effects of temperature and food availability on selection for habitat variables such as water depth. Both are likely to have a major impact on fitness through their effect on growth rate. Thus, bioenergetic studies of the effects of interactions between food supply and temperature on fish growth may be able to predict the solution to the trade-off. A laboratory test found that bluegill sunfish, *Lepomis macrochirus*, failed to choose the combination of temperatures and food supply that would have maximised their growth rates according to a detailed bioenergetic model (Wildhaber and Crowder 1990). Field studies have also indicated that temperate-zone freshwater fish show a preference for thermal effluents in the winter, even though they lose mass at a higher rate than conspe-cifics in adjacent cooler water (Merriman and Thorpe 1976, cited by Crowder and Magnuson 1983). However, there is qualitative support for bioenergetic-based predictions. Because the temperature that maximises growth declines as food supply is reduced (e.g. Brett *et al.* 1969), preferred temperatures should decline when food supply is reduced (Crowder and Magnuson 1983). This directional change in preference was found in Wildhaber and Crowder's (1990) laboratory study of bluegills. Similarly, cod, *Gadus morhua*, in the Gulf of St. Lawrence, Canada were found to occupy lower temperatures during years of higher total population size when food intake was probably reduced (Swain and Kramer 1995).

The choice of current speed by stream-dwelling fishes provides another example of a trade-off for which a combined fitness function can be calcu-lated. Fish choosing higher current speeds experience higher encounter rates with drifting prey but capture a lower proportion of encountered prey and pay a higher energetic cost of swimming to remain in position in the stream (Godin and Rangeley 1989; Hughes and Dill 1990; Hill and Grossman 1993;

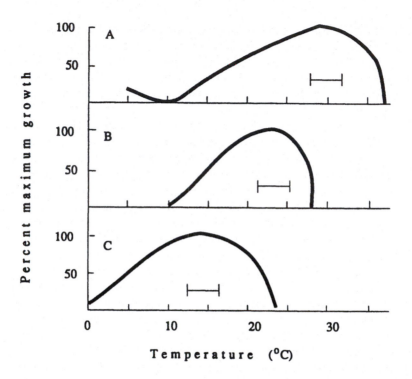

Fig. 3.4   The relationship between preferred temperature and growth rate on excess ration as a function of temperature for three species of North American freshwater fishes (A) bluegill sunfish, *Lepomis macrochirus*, (B) yellow perch, *Perca flavescens*, and (C) sockeye salmon, *Oncorhynchus nerka*. The horizontal bar represents median temperature preference ± 33% as determined by laboratory experiments. Growth rate is expressed as a percentage of the maximum. Data sources are bluegill temperature preference: Magnuson and Beitinger (1978), bluegill growth: Lemke (1977), Beitinger, Stuntz and Magnuson (unpublished data), yellow perch temperature preference: McCauley and Read (1973), yellow perch growth: Huh (1975), Huh *et al.* (1976), Kitchell *et al.* (1977), sockeye growth and temperature preference: Brett (1971). Redrawn from Magnuson *et al.* (1979, Fig. 9) with permission of the author and Kluwer Academic Publishers.

Tyler 1993). Quantitative analysis of this trade-off has permitted quite successful predictions of selected current speeds in cyprinids and salmonids in natural streams (Hughes and Dill 1990; Hill and Grossman 1993) and in a laboratory experiment also involving intraspecific competition for prey (Tyler and Gilliam 1995).

Determining a combined fitness function to make a quantitative prediction of habitat selection is much more difficult when habitat trade-offs involve survival as well as resource gain. The appropriate way to combine components of fitness with different types of units is not obvious. Werner and Gilliam (1984) concluded that minimisation of the ratio of predation rate to growth rate would be an appropriate combined fitness function for juvenile fish. Gilliam and Fraser (1988) extended this conclusion with a set of graphical models for habitat choice based on minimising the ratio of predation rate to foraging rate rather than growth rate. Clark and Levy (1988) used this approach in their dynamic optimisation model which showed how the interaction of feeding and predation risk could explain vertical migration patterns in juvenile sockeye salmon. A similar perspective was taken by Burrows (1994) to explain variation between localities in patterns of diel and tidal onshore–offshore movements by juvenile plaice. In many situations, the estimation of absolute rather than only relative mortality rates, as required for quantitative predictions, is a daunting task. Therefore, researchers may be restricted to qualitative predictions. Examples of such predictions are that more vulnerable size classes will use dangerous habitats less than other size classes will or that exposed habitats will be used less in the presence of predators than in their absence (Werner *et al.* 1983; Power 1984a; Fraser and Gilliam 1992).

3.7 Frequency-dependent habitat selection

3.7.1 Theory

Simple optimality models of habitat selection assume that the fitness consequences of selecting a particular habitat are independent of the number of animals in that habitat. There are many situations, however, in which this assumption is not likely to be valid. Feeding rate, predation risk and reproductive success are all known to be affected in a variety of ways by conspecific numbers or density. Although models that take density dependence into account still assume that animals make adaptive individual decisions, they are very different in structure from simple optimality models. This is because the optimal decision depends on the decisions of other individuals. With negative density dependence, an intrinsically high-quality habitat becomes poor if too many individuals choose it; conversely, with positive density dependence, the value of an intrinsically poor-quality habitat can be improved if it is occupied by an appropriate number of conspecifics. Negative density dependence may therefore cause the population to spread out over a range of habitat types, while positive density dependence may promote aggregation. As a result, predictions cannot be based on maximal individual fitness. Instead, they are based on the stability of the distribution: the final distribution should be one in which no individual can raise its fitness by moving to another habitat. No single habitat can be considered 'optimal' because the best choice of habitat depends on the frequency distribution of choices among the other members of the population. Thus, density dependence in habitat quality should lead to frequency dependence in habitat selection.

An important model of frequency-dependent habitat selection is known as the Ideal Free Distribution (IFD), following Fretwell and Lucas (1970; Fretwell 1972) who first described it. It has been reviewed recently by Milinski and Parker (1991), Kacelnik *et al.* (1992) and Kennedy and Gray (1993). (See Milinski (1994) and Gray and Kennedy (1994a) for further discussion). If animals can assess realised habitat quality (i.e. intrinsic quality discounted by the effect of competitors) and are free to enter any habitat and to share equally the available resources, they should move to any habitat that offers a higher fitness benefit. If fitness benefits are negatively density dependent, net movement between habitats should cease when expected fitness is equal in all occupied habitats and no unoccupied habitat offers a greater benefit to a settling individual. In this stable distribution, density will be positively correlated with intrinsic habitat quality (e.g. total resource availability). At larger total population sizes, average fitness will be lower and habitats of lower intrinsic quality will be included in the occupied range.

An example may help to make these patterns clearer. Fig. 3.5A shows fitness functions for different population densities. Unlike the situation modelled in Fig. 3.1, benefits of each habitat decline as density increases. In this hypothetical example, the level of the environmental variable that offers the highest benefit is the same for all population densities. In addition, small increases in density lower benefits to a greater extent in habitats of intermediate quality than in very good habitats. We assume that benefits do not increase at densities below one individual per unit area. A horizontal line cutting across the fitness functions can be used to represent expected benefits in each habitat at the IFDs where fitness in all occupied habitats is equal; three examples (lines a, b and c) are illustrated in Fig. 3.5A. The population density in each habitat at the IFD is determined by which fitness function crosses the horizontal line at that level of the environmental variable. For example, line b represents an IFD in which all individuals would have a benefit level of 5. In the best habitat (level 7 of the environmental variable), a benefit level of 5 is achieved at a density of about 34. For habitat 8 the density would be about 13, while habitats 9 and 10 would be unoccupied because even at densities ≤ 1 the benefits would be less than 5. The same process can be used to determine the IFD densities in habitats 1–6. These are shown for IFDs a, b and c in Fig. 3.5B. Note that at lower benefit levels, densities are higher and a broader range of the available habitats is occupied. Fig. 3.5C shows the proportion of the population that would occur in each habitat, if each habitat occupied the same area. The population is aggregated in a smaller proportion of the total potential range when the population size is lower.

A special case of the IFD occurs when the benefit in any habitat equals the intrinsic quality of the habitat divided by the number of competitors present. In Fig. 3.6A, for example, the environmental variable represents a divisible characteristic such as the rate of arrival or production of food resources per unit area. The sloping lines represent individual benefits for different population densities when the benefits are equal to the value of the environmental variable divided by the number of competitors. Thus, at a density of 1 individual per unit area, each individual obtains all the resources from that area. At a density of 2, each individual obtains half

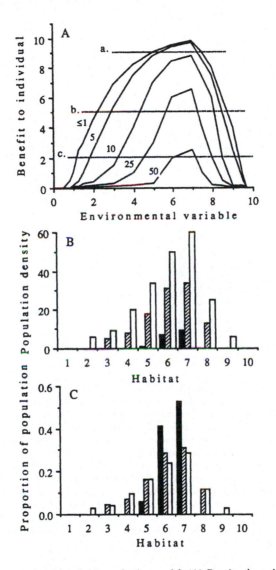

Fig. 3.5 A frequency-dependent habitat selection model. (A) Density dependence of the fitness function: the solid lines indicate fitness-related benefits obtained by individual animals at densities of 50, 25, 10, 5 and ≤ 1 animals per unit area in relation to the level of an environmental variable. The dashed horizontal lines (a, b and c) represent examples of possible Ideal Free Distributions in which equivalent fitness is obtained in better habitats at high densities and in poor habitats at low densities. (B) Population densities expected in a series of habitats defined by the level of the environmental variable in panel A. Expected patterns are shown for the three Ideal Free Distributions represented by horizontal lines in panel A: a (filled bars), b (hatched bars) and c (open bars). (C) Proportions of the population expected to occur in each of a series of habitats at the Ideal Free Distributions a, b and c. (Symbols as in panel B.) Proportions were calculated assuming an equal area for each habitat.

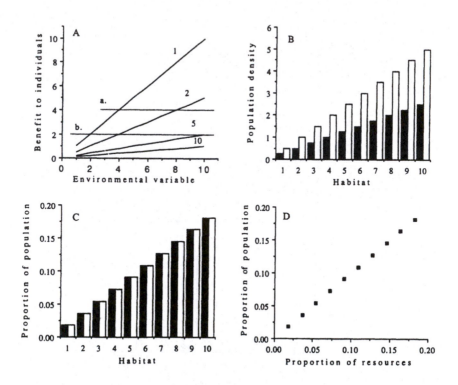

Fig. 3.6   An Ideal Free Distribution model of habitat selection when fitness benefits are directly related to the amount of resources available divided by the number of competitors. (A) Solid lines show the hypothetical relationship between individual acquisition of fitness-related benefits and level of a resource-related environmental variable, such as food arrival or production rate, for population densities of 1, 2, 5 and 10 individuals per unit area. Dashed horizontal lines (a and b) indicate two possible Ideal Free Distributions at benefit levels 4 and 2, respectively. (B) Expected population density in each of a series of habitats, defined by the level of the environmental variable, at the Ideal Free Distributions represented by horizontal lines a (black bars) and b (open bars) from panel A. (C) Expected proportions of the population occurring in each habitat at Ideal Free Distributions a (black bars) and b (open bars), if all habitats are equal in area. (D) Expected proportion of the population in each habitat in relation to the proportion of the total resources occurring in that habitat. Proportion of resources was calculated by dividing the resource level in each habitat by the sum of resource levels in all habitats. Values for benefit levels 4 and 2 fall on the same line.

the resources, at a density of 5, one fifth, and so on. Such a situation might occur when there is scramble competition for a resource such as food which is consumed immediately upon its arrival or production. Dashed horizontal lines indicate two possible IFDs. Fig. 3.6B shows the expected population density at two IFDs in each of a series of habitats defined by their resource productivity. Density is higher for the IFD with the lower benefit level, but all habitats are occupied. Fig. 3.6C shows that, in each habitat, the proportion of the population expected to occur there is the same at both IFDs (if all habitats are equal in area). This proportion is equal to the proportion of the total resources arriving in each habitat. The prediction that the proportion of the total population occupying a given habitat should equal the proportion of total resources arriving or produced in that habitat, whatever the population size, is known as 'input matching' (Parker 1978), and is one of the few explicit quantitative predictions of distributions in the habitat selection literature. Recently, Lessells (1995) showed that input matching can occur even when the resource is not consumed immediately upon its arrival, provided that consumption rate increases when the standing crop of the accumulating resource increases.

It is quite unlikely that major fitness correlates would be inversely proportional to population density over a broad range of densities in nature. For example, when food availability is high relative to density, gain rates will be limited by ability to handle or digest food rather than by its availability. Even when animals are limited by resource availability, it is unlikely that individuals using the standing crop of a non-renewing resource would experience a halving of their foraging rate each time the population density doubles. Negative effects of competitors may include active interference such as aggressive behaviour and passive interference such as disturbance of prey or the added foraging costs when several individuals pursue the same prey item, but these are not likely to be linearly related to resource availability and density. Furthermore, many fish probably experience positive effects of the presence of other individuals in reducing predation risk (Pitcher and Parrish 1993; Smith, Chapter 7; Godin, Chapter 8). Therefore, proportional resource availability and proportional population density are not predicted to be equal (Sutherland 1983; Tregenza 1994). Fagen (1987) proposed an alternative general model for the relationship between resource availability and population distribution. Unless the relationship between major components of fitness and population density can be determined *a priori*, the IFD can only predict that density should be correlated with intrinsic quality and that average fitness should be equivalent in all occupied habitats. In addition, if habitats differ in what Fretwell (1972) called their 'basic suitability' (i.e. potential fitness benefits as density approaches zero), the range of habitats occupied should increase with population density, and the order of occupation should follow the order of their basic suitabilities. Although qualitative, these are predictions of considerable value to the study of habitat selection.

The potential importance of the IFD for populations under exploitation by fisheries was recognised by MacCall (1990), who developed a graphical version called the 'basin model'. He focussed on systems in which intrinsic habitat quality might vary continuously over the geographical range of a species, for example, in pelagic or demersal marine fish populations. Then density dependence of realised habitat quality should result in a positive correlation between a population's size and its geographical range. A population in decline as a result of fishing pressure would gradually contract to the region of highest intrinsic suitability. Thus, the decreasing population size would not be strongly reflected by decreased catch rates in the highest quality part of the range. Fishermen in prime habitat could continue to make good catches until the population was nearly eliminated. MacCall proposed that this could explain the sudden collapse of some clupeid and engraulid populations. If this model were to prove to be broadly applicable, it would raise important concerns about the use of catch-per-unit-effort data for monitoring commercial fish populations.

MacCall's (1990) model also has important implications for the location of reserves. In a situation of frequency-dependent habitat selection and strong spatial variation in intrinsic habitat quality, reserves in suboptimal habitat could be depleted by heavy fishing in prime habitat, whereas the reverse would not be the case. This is because at lower population sizes fish would be expected to move from intrinsically less suitable to intrinsically more suitable habitats. An example of such a response is provided by Wellington and Victor (1988) in a study of a species considered relatively sedentary, the Acapulco damselfish, *Stegastes acapulcoensis*, on an eastern Pacific reef in Panama. Following removals, high-quality shallow habitat was quickly re-populated by adults moving from other areas, whereas poorer-quality but still suitable deeper habitat was not. Responses to population reduction in part of the range will also be important for reserves intended to supply harvestable fish to adjacent fishing grounds. Such applications of frequency-dependent models also clearly depend on the spatial scale over which fish redistribute in response to local perturbations. At the present time, researchers are only beginning to examine the implications of frequency-dependent habitat selection for the potential of reserves to retain fish populations (Roberts and Polunin 1991; DeMartini 1993; Rakitin and Kramer 1996).

Frequency-dependent habitat selection need not necessarily result in IFDs. In Fretwell's (1972; Fretwell and Lucas 1970) original analysis, the IFD was one of three alternative models for the distribution of territories of passerine birds settling onto summer breeding habitats. The other two models were the 'Allee-type Ideal Free Distribution', an IFD model appropriate when fitness increases with population density over at least some of the range of population densities, and the 'Ideal Despotic Distribution', a non-IFD model appropriate when aggressive behaviour prevents the free entry or equal sharing of resources between new immigrants and settled residents of a habitat. Fretwell presented his models not to predict the patterns of settlement or final distributions but rather to discriminate among the underlying processes on the basis of the observed patterns. It is now clear that all three types of distribution can occur in nature. A large body of work on aggregation in fishes makes it clear that fitness can increase with local density under some circumstances (Pitcher and Parrish 1993), suggesting that Allee-type models will sometimes be appropriate. On the other hand, the despotic model certainly applies to other cases such as the feeding territories of stream-dwelling salmonids (Grant and Kramer 1990). But Freeman and Grossman (1992) found little direct or indirect evidence of fitness costs or benefits to aggregation in the rosyside dace, *Clinostomus funduloides*.

Fretwell (1972; Fretwell and Lucas 1970) proposed that the different distributions could be distinguished by the relationship between fitness and density: fitness should be equal in all habitats in an IFD, but positively correlated with density in an Ideal Despotic Distribution. However, one could imagine the opposite, a Despotic Distribution in which poorer competitors were forced into poor habitat at higher density than territorial individuals in good habitat. Recognition of distribution types remains a difficulty, but has been the subject of some research. Fretwell's approach was extended by Morris (1994, and references therein), who considered how changing density should affect the relative density in habitat pairs under different distributions. In IFDs, plots relating density in one habitat to density in another (called 'isodars') should reveal the underlying patterns of density dependence. For situations in which fairly large density variation occurs over time or among sites, the isodar approach might be extended to permit qualitative recognition of different forms of Ideal Free, Ideal Despotic and Allee-type Ideal Free Distributions. Rodríguez (1995) recently applied the isodar framework to selection of pools versus riffles by sympatric Atlantic salmon, *Salmo salar*, and brook charr, *Salvelinus fontinalis*.

Which ecological situations and spatial scales are appropriate for the expectation of an IFD is a question of fundamental importance that has received too little attention. Tests of the IFD so far have selected spatial and temporal scales and resource distributions for which positive density dependence and despotism were not likely to have strong effects. However, to develop a more broadly applicable theory of habitat selection, the body of theory relating environmental distribution of risks and resources to aggregation and resource defence must be developed to permit researchers to predict when to expect IFD, Allee-type and Despotic distributions. Promising starts have been made (e.g. Pulliam and Caraco 1984; Grand and Grant 1994a, b; Oksanen *et al.* 1995), but the task has only just begun.

There is currently considerable interest in extending frequency-dependent habitat selection models to a wider range of situations. Important issues of continuing interest include the potential habitat segregation of individuals of different competitive ability (Sutherland and Parker 1985; Parker and Sutherland 1986; Milinski and Parker 1991), the interaction between density-dependent and density-independent habitat features (Abrahams and Dill 1989; Kennedy *et al.* 1994; Tyler and Gilliam 1995; Swain and Kramer 1995), the effect of constraints on the ability of organisms to distinguish between habitats of different quality (Abrahams 1986; Gray and Kennedy 1994b) and the tracking of environmental change (Stephens 1987; Bernstein *et al.* 1988, 1991). Another important concern is the development of models to make predictions over more than two trophic levels, when a population is being tracked by its predators at the same time as it is tracking its food resource (Sih 1980; Hugie and Dill 1994; Persson *et al.*, Chapter 12).

3.7.2 Examples

Tests of the input-matching prediction of the IFD have been restricted to small and highly simplified laboratory environments, perhaps because no completely appropriate field application of this form of the model has yet been recognised. Input matching was first examined using the distribution of a small number of threespine sticklebacks, *Gasterosteus aculeatus*, between two renewing food patches at opposite ends of an aquarium (Milinski 1979). The proportion of fish at each site conformed closely to the proportion of food items arriving there. This pattern has since been confirmed for a variety of other species in a number of similar situations (Godin and Keenleyside 1984; Milinski 1984, 1987; Talbot and Kramer 1986; Gillis and Kramer 1987; Sutherland *et al.* 1988; Abrahams 1989; van Duren and Glass 1992; Utne *et al.* 1993, Kennedy *et al.* 1994). Fishes, like other taxa in which the model has been tested, usually show a tendency for the richer food patch to be slightly underused and the poorer patch to be slightly overused in comparison to the IFD prediction (Kennedy and Gray 1993).

The most thorough field examination of an IFD in fishes so far is Power's (1984b) study of the distribution of loricariid catfishes in a Panamanian stream (Fig. 3.7). Power determined that the loricariids were limited by their algal food supply, which was controlled by the amount of light reaching the water surface through the canopy of trees along the stream. Loricariid population densities were positively correlated with light levels and therefore food production rates. The algae, however, were grazed down to similar levels in pools with different light levels, probably equalising intake rates, and the growth and survivorship of pre-reproductive individuals of the most abundant species, *Ancistrus spinosus*, were similar among pools. Although Power's data did not permit her to make an *a priori* prediction of population distribution, the differences in density and the similarities in fitness measures among habitats of different resource availability strongly supported the hypothesis that the catfishes were distributed in an ideal free fashion. More recently, a similar approach has been taken to the distribution of effort by commercial fishermen (Gillis *et al.* 1993; see also Abrahams and Healey 1990). The authors found evidence for competition among fishermen and for equalisation of catch success. This would imply that when fishermen have access to a number of different stocks or fishing areas, spatial variation in fishing effort would be a better measure of spatial variation in population size than would spatial variation in catch rate.

The IFD prediction that the range of habitats occupied should increase with population density was examined by Fraser and Sise (1980), who found that the juveniles of two species of cyprinids and the adults of one of these species tended to be more evenly distributed among pools in a stream as population density increased. Recently, this prediction has attracted attention in the marine fisheries

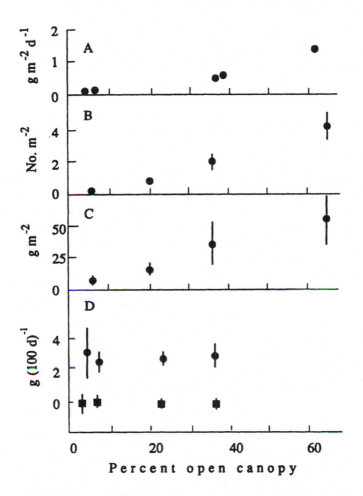

Fig. 3.7 Algal productivity (A) and the density (B), biomass (C) and growth rate (D) of loricariid catfish in relation to light levels, as estimated by percent open canopy, in pools of the Rio Frijoles, Panama. Algal productivity, as well as the density and biomass of four loricariid species, increased with light levels. Growth rates of pre-reproductive individuals of *Ancistrus spinosus*, the most common species, were higher in the rainy season (circles) than in the dry season (squares), but were not related to light levels in different pools. Vertical lines represent two standard errors. Redrawn from Fig. 6 in Power (1983) with permission of the author and *American Zoologist*.

literature. Marshall and Frank (1994) reviewed evidence for changes in geographical range associated with changes in population size and discussed some statistical problems with the indices of spatial distribution. Swain and Sinclair (1994) proposed an alternative index of spatial spread of a population and provided evidence for increases in the area occupied by cod in the Gulf of St. Lawrence when the population size was larger. Swain (1993) suggested that shifts in age-specific depth distributions of cod with changes in population density could be the result of frequency-dependent habitat selection, resulting from density-dependent benefits associated with food resources and density-independent costs associated with temperature.

3.8   Fish habitat selection in nature

3.8.1   Overview

The taxonomic and morphological diversity of fishes is parallelled by the tremendous diversity of environments in which they live (Davenport and Sayer 1993; Nelson 1994). Fish are found in almost every type of aquatic habitat from high altitude streams and lakes to 7000 meters below the ocean surface in the dark waters of the abyssal plains, from almost pure fresh water to hypersaline lakes, from Antarctic waters to soda lakes with temperatures reaching 41°C, and from torrential mountain streams and wave-swept shores to stagnant swamps and cave pools. They occur in relatively stable, unlimited and nearly structureless habitats of the open ocean, in tiny ephemeral pools, in nearly impenetrable mats of submerged vegetation and buried in the substratum. Fishes are not even restricted to aquatic habitats; more than 60 species regularly spend some time out of water (Liem 1987; Sayer and Davenport 1991).

Habitat types differ both in the environmental variables that exhibit sufficient spatial variation to permit habitat selection to occur and the variables that have the greatest potential impact on fitness. These differences could be the basis of a comparative study of habitat use and habitat selection. However, students of fish habitat use often seem to be habitat specialists. While numerous studies contrast species within habitats, there has been surprisingly little comparative analysis of habitat use and preference patterns among the major habitat types. Furthermore, variables considered by researchers in one habitat are often ignored by those working in other habitats. In addition, habitat use has not received equal attention in all the major habitats in which fish occur. As easier systems in which to observe fish, coral reefs, clear temperate-zone streams and small lakes have been the subject of most behavioural research. Large lakes and marine systems have been investigated at a larger scale using nets and hydroacoustics. But other major habitats in which habitat selection is likely to be important, especially swamps and marshes, large rivers, and marine shorelines other than tidepools, are greatly underrepresented in habitat selection studies. In this section we review the associations between environmental variables and fish distributions, that is, habitat use in a variety of environments. Although necessarily brief and incomplete, we hope that the accounts and examples will illustrate the diversity of ways that fish, and fish behavioural ecologists, have approached the problems of habitat selection in very different habitats and will stimulate more comparative analysis of differences and similarities between habitats.

3.8.2 Lotic freshwater habitats

The running waters of rivers and streams (lotic habitats) are usually sufficiently mixed that vertical gradients of temperature and water chemistry do not develop, but they often exhibit a great deal of spatial variation in current speed, water depth and substratum type (Hynes 1970). Qualitative differences often exist between shallow, fast-moving sections (riffles) and deeper, slow-moving sections (pools). Within each of these habitat types, quantitative variation exists in current speed, depth and substratum particle size. This variation is often well correlated with habitat use. Some species or size classes can be characterised as pool or riffle dwellers (e.g. Angermeier and Karr 1983), although such specificity is not always observed (e.g. Grossman and Freeman 1987). In two streams in Idaho, USA, Everest and Chapman (1972) found that variation in the densities of juvenile steelhead trout, *Oncorhynchus mykiss,* and chinook salmon, *Oncorhynchus tshawytscha,* could be successfully accounted for by a set of related physical variables: substratum particle size, water depth and current velocity. Newly emerged steelhead live in shallow water that is flowing at low velocity and move into deeper and faster water as they grow; thus the abundance of age 0+ fish is negatively correlated with velocity and depth, but the abundance of age 1+ fish is positively correlated with these variables. Chinook, on the other hand, tend to be found in pools or eddies, so their abundance is negatively correlated with velocity but positively correlated with depth. However, for both species, the depth and current velocity of the specific location at which a fish is observed correlate positively with the individual's size. In other studies, distance from the substratum is an important microhabitat variable (Baker and Ross 1981; Grossman *et al.* 1987). For some darters (*Etheostoma* spp.), experiments demonstrate that substratum particle size is a selected variable (Schlosser and Toth 1984). The functional significance of habitat selection in stream-dwelling fish seems to be related to trade-offs among swimming costs, the supply rate and capture success of food items, and safety from predators, both aquatic and terrestrial (Hughes and Dill 1990).

Many lotic habitats are subject to strong temporal variation in flow rate. As water levels rise and fall, conditions change at each specific location; a site with a gentle current may dry up completely or turn into a raging torrent. Under these unstable conditions habitat selection must be dynamic, with fish changing location in response to changing conditions. Matthews and Hill (1980) provide an extreme example for cyprinid fishes in a highly seasonal shallow river in the southwestern USA. If fish maintain their microhabitat preference as conditions change, the effect of flow changes on habitat suitability can be predicted. This is the basis for using 'Instream Flow Incremental Methodology' (Reiser *et al.* 1989) to make decisions on stream flow regulation from the perspective of its impact on the fish. Shirvell (1994) showed that individually marked juvenile coho salmon, *Oncorhynchus kisutch,* in the Canadian northwest responded to variation in stream discharge by changes in horizontal position to maintain most of the same microhabitat characteristics. However, he also noted some significant changes in microhabitat preference under experimentally controlled drought or flood conditions. Evidence for changes in microhabitat selection with flow rate, season and habitat type (riffles versus pools) was provided by Vondracek and Longanecker (1993).

3.8.3 Lentic freshwater habitats

Lakes, ponds and stagnant pools in intermittent streams and rivers (lentic habitats) develop spatial heterogeneity in water quality much more readily than running waters. Vertical gradients in light and exposure to the atmosphere combined with reduced vertical mixing can create large vertical gradients in temperature, oxygen and other variables (Wetzel 1975). Even within a mixed layer, vertical gradients in light can result in important variation in food abundance or detectability and in predation risk. Thus, vertical distribution is often an important part of habitat selection in lentic habitats. For example, striped bass, *Morone saxatilis*, an estuarine species introduced into reservoirs in the USA, avoid deep water when it becomes low in dissolved oxygen during the summer. At the same time, high temperature surface waters are also avoided, and, under extreme conditions, most of the fish may be concentrated into a narrow band just above the poorly oxygenated hypolimnion (Matthews *et al.* 1985). Physiological tolerances and habitat preferences of striped bass and the spatial distributions that result from them may be important determinants of the potential of particular reservoirs to sustain populations (Coutant 1985). Similar constraints imposed by dissolved oxygen level on the expression of temperature preferences may apply to a number of other species (e.g. Rudstam and Magnuson 1985). However, even poorly oxygenated deep water zones may be used to some extent. For example, mudminnows, *Umbra limi*, routinely venture into the the hypolimnion of a northern Wisconsin (USA) lake to forage on phantom midge larvae (*Chaoborus* sp.), even though the oxygen is so low that the fish cannot survive for an extended period (Rahel and Nutzman 1994). A potentially widespread use of vertical heterogeneity in the water column is the regulation of metabolic rate through habitat selection of temperature strata (Brett 1971, 1983; Crowder and Magnuson 1983; Wurtsbaugh and Neverman 1988).

Along the shores of lentic waters, physical habitat structure in the form of aquatic vegetation, rocks or submerged terrestrial vegetation can be very important to fish distributions. In North American lakes, fish density is often highest in areas with abundant aquatic vegetation, and different species may associate with different types of vegetation (Werner *et al.* 1977; Keast *et al.* 1978; Aboul Hosn and Downing 1994). Often, vegetation is used to a greater extent by smaller species and size classes, apparently because it provides greater protection from predation (see Smith, Chapter 7; Godin, Chapter 8; Persson *et al.*, Chapter 12). Using the habitat of origin of prey in the stomach as evidence of habitat use in experimental ponds, Werner *et al.* (1983) found that larger size classes of bluegill sunfish select habitats on the basis of foraging profitability, but that the smallest size class uses the less profitable vegetated habitat to a greater extent when predatory largemouth bass, *Micropterus salmoides*, are present. Using a combination of observational and experimental approaches, Foster *et al.* (1988) concluded that the greater use of vegetated habitats by small juvenile threespine sticklebacks was a consequence of their greater vulnerability to predation by conspecific adults.

3.8.4 Freshwater wetlands

Shoreline environments of both lotic and lentic systems grade into shallow wetlands, such as swamps and marshes, in which restricted water movement and variable shading of the surface by vegetation produce complex horizontal as well as vertical variation in oxygen, temperature and other water quality variables. These shallow lentic habitats frequently experience dramatic diel variations in response to changes in insolation and air temperature. Wetland habitats are often subject to large seasonal fluctuations as well. In larger floodplain rivers, such as the Amazon and the Niger, enormous areas of forest and savanna are inundated annually, producing temporary lentic conditions in which many species feed and reproduce before returning to the main channel or isolated ponds for the dry season (Lowe-McConnell 1964, 1975; Welcomme 1979; Goulding 1980). Similar phenomena take place on a smaller scale in floodplain rivers around the world.

The difficulties of working in shallow, heavily vegetated waters with soft substrata have resulted in few detailed studies of habitat selection in wetland environments. But some information is available. For example, yellow perch, *Perca flavescens*, feeding and sheltering in a large marsh adjacent to a lake in central Canada show avoidance of heavily vegetated areas at night during mid-summer when oxygen levels are very low (Suthers and Gee 1986). *Barbus neumayeri*, a species of cyprinid that is very tolerant of hypoxia, lives in many habitats including the dense interior of African papyrus swamps. Large gills, the use of aquatic respiration at the air-water interface and selection of microhabitats with higher levels of dissolved oxygen permit survival during the dry season when conditions become extreme (Chapman and Liem 1995). In contrast, the distribution of *Clarias liocephalus*, an air-breathing catfish, in the same swamp is independent of dissolved oxygen levels (Chapman 1995). Both species are more widely distributed throughout the swamp during the rainy season than during the dry season.

Little is known about how safe refuges from dry-season desiccation are selected or what determines the distance moved from such refuges during the wet season. Although many fish are trapped in pools which eventually dry up or become so shallow that fish are very vulnerable to avian and crocodylian predators (Johnels 1954; Lowe-McConnell 1964; Welcomme 1979), many others escape to more secure locations. In the River Sokoto, Nigeria, regularities in the occurrence of species in pools from year to year and the recapture of marked individuals in the same pools in different years suggest habitat selection for refuge pools during the dry season (Holden 1963; Chapman and Chapman 1993).

3.8.5 Intertidal marine habitats

The intertidal and near subtidal zones are dominated by the strong physical forces of waves and tides, producing heterogeneity in substratum type and exposure to currents but restricting the development of gradients in water quality. Tidal cycles produce a dynamic environment in which conditions can change from deep water with heavy wave action to stagnant tide pools or waterless flats within a few hours. In such environments, habitat selection is often a continuous process. In the outer Bay of Fundy, Canada, an area with a tidal range of over 8 m, juvenile pollock, *Pollachius virens*, make extensive use of intertidal habitats for feeding and probably predator avoidance, moving up to 200 m inshore with the tides (Rangeley and Kramer 1995a). These movements do not involve simply tracking the changing location of particular microhabitat characteristics; habitat preferences also change over the tidal cycle. On rising tides, the pollock select relatively open habitat at lower tidal stages before switching to dense patches of macroalgae at higher tidal stages, then moving rapidly downshore as the tide starts to fall (Rangeley and Kramer 1995b). Distribution in relation to conspecifics also changes: the pollock disperse when using dense algae but aggregate strongly in open habitat.

Many fishes remain in the intertidal zone throughout the tidal cycle, in tide pools, under stones or algae or buried in the substratum. Habitat selection is critical to achieving a favourable position during the low tide period. Species using the intertidal zone commonly have a strong homing capability (Gibson 1986). Tide pools provide a permanent aquatic habitat within the intertidal zone, but water quality over the low tide period can vary from relatively stable to highly variable in temperature, dissolved oxygen and other characteristics, depending on pool position and size and local weather conditions. In intertidal salt marshes of the eastern USA, juveniles of the cyprinodont, *Fundulus heteroclitus*, remain in residual pools over the tidal cycle, while adults move in and out with the tide (Kneib 1987). Fish remaining in pools are subjected to more extreme physical and chemical conditions and to a lower food supply, but avoid the risk of fish predation associated with moving into the tidal channels at low tide (McIvor and Odum 1988; Rountree and Able 1992; Kneib 1993). In salt marshes on the Gulf of St. Lawrence, Canada, threespine sticklebacks breed in pools at the upper limit of the intertidal zone which are only accessible for a few hours at high tide on only a few days each month (FitzGerald *et al.* 1992). In selecting breeding habitat, the fish are capable of avoiding shallow pools that would eventually dry out (Whoriskey and FitzGerald 1989). A similar pattern has been suggested for spotted kelpfish, *Gibbonsia elegans*, avoiding tidepools at times of the year in which nocturnal isolation from the sea allows them to become hypoxic (Congleton 1980).

3.8.6 Subtidal marine habitats

Most inshore marine environments have sufficient water movement that important gradients in water chemistry or temperature are unlikely to occur. However, depth of the water column and type of substratum are often highly variable over small distances and correlate with other important physical and biological variables. Many shallow marine environments are characterised by physical structure in the form of coral reefs, rocky outcrops and boulders, macroalgae and seagrass beds, or mangroves. Greater abundances of fishes near such structure than in adjacent more open habitats provides indirect evidence for habitat selection and for the influence of such structured habitats on the productivity of fish populations in inshore areas. Increasing habitat structure in relatively unstructured habitats is a fisheries management technique of growing importance (Bohnsack *et al.* 1991; Seaman and Sprague 1991).

Coral reefs provide a particularly complex environment in terms of biodiversity and physical structure (Sale 1991a). Substratum type and depth appear to be very important determinants of habitat selection (e.g. Robertson and Lassig 1980; Waldner and Robertson 1980; Wellington 1992). Luckhurst and Luckhurst (1978) indicate the highly specific types of structure (e.g. sponges, worm tubes, small crevices or ledges) that may be occupied. Extremely narrow substratum preferences are illustrated by anemonefish, *Amphiprion* spp., that are closely associated with particular species of sea anemones (Fautin 1986).

Like the majority of marine fish, most species inhabiting coral reefs have a pelagic larval stage (Sale 1980). Thus, there is the opportunity for an important habitat selection decision at the time the juvenile settles from the plankton onto the reef. Clearly, both active selection and stochastic events are involved in this process (Doherty 1991; Jones 1991; Sale 1991b; Wellington 1992). Wellington (1992), for example, demonstrated that the differences in adult habitats of two closely related and ecologically similar damselfish species, *Stegastes leucostictus* (found in backreef habitats at 1–2 m depth) and *S. variabilis* (found on forereef at 10–15 m depth), were the consequence of differences in settlement location. Elliott *et al.* (1995) showed that close association between particular species of anemonefish and their hosts can be explained by chemical-mediated attraction at the time of settlement.

Sometimes, there are post-settlement changes in habitat by the growing juveniles. For example, the three-spot damselfish, *Stegastes planifrons,* exhibits very specific ontogenetic habitat shifts on a Honduras coral reef (Lirman 1994). Adults prefer to establish territories on live foliose coral heads where algal mats that are important feeding resources are maintained. Post-settlement juveniles are found within the territories of adults, but as the juveniles grow, adult aggression eventually forces them to leave. They then prefer dead foliose coral heads where they presumably remain until they mature and move back to establish feeding and reproductive territories on live coral heads.

Distinctly different habitat preferences by day and night occur in a number of species. For example, French grunts, *Haemulon flavolineatum*, remain in groups at specific sites on reefs during the day, migrating to adjacent sand flats to feed at night; the bluestriped grunt, *H. sciurus*, in contrast moves to sea grass beds and other structured habitat for noctural foraging (McFarland *et al.* 1979; Burke 1995).

Many species associated with coral reefs remain on or very close to the substratum (e.g. Sale 1969). This may be so obvious to researchers that it is rarely specifically recorded, but Hobson (1991, Fig. 3) provides an example of interspecific differences in distance from substratum. Distance above the bottom has been recognised as an important microhabitat variable more often for temperate marine species. For example, among the rockfishes, *Sebastes* spp., inhabiting kelp forests on the coast of California (USA), some species remain near the substratum while others stay well off the bottom (Hallacher and Roberts 1985). Predation and food availability are potentially important selective pressures influencing distance from the substratum in temperate marine fishes. An unusual case where parasitism or predation plays a role has been described off southern California. At night, adult fish remaining near the substratum are attacked by swarms of ostracods and isopods. Many fish swim higher in the water column or hide in sand or crevices. Adult kelpfish, *Heterostichus rostratus*, and other species held in cages on the seafloor were killed by swarms of the crustaceans which invaded the gills and entered the body cavity through the anus (Stepien and Brusca 1985).

3.8.7 Offshore marine habitats

Many species of fish inhabit the pelagic zone which provides little or no physical structure. These include the early stages of many species whose adults use more structured habitats. Even in the apparently simple environment of the pelagic zone, habitat selection may be very important. A few species use the limited structure available—floating macroalgae and debris or animals such as larger fishes and gelatinous zooplankton (Gooding and Magnuson 1967; Safran and Omori 1990). But for others, position in the water column is the primary form of habitat selection. As with lentic freshwater habitats, marine systems show important vertical variation in temperature and water chemistry, and depth therefore can strongly affect survival and net rates of energy gain. For example, Fortier and Harris (1989) documented differences in vertical distribution of pre-feeding and feeding fish larvae that could be explained by avoiding predators at the pre-feeding stage and preference for greatest prey abundance once feeding started. Neilson and Perry (1990) summarise a wide range of studies that document changes in such vertical distributions over the diel cycle in fish ranging in size from newly hatched larvae only a few mm long to adult swordfish, *Xiphias gladius*, reaching over 4 m. Their review illustrates the complexity of vertical habitat selection and its modulation by numerous environmental factors.

For highly mobile fish, differences in water conditions or food supply can also influence horizontal distribution, sometimes over hundreds of kilometres. For example, Atlantic cod distributions in the northern Gulf of St. Lawrence are influenced by both temperature and the abundance of their major prey, capelin, *Mallotus villosus* (Rose and Leggett 1989). Some oceanic and coastal fish migrations are associated with the movement of temperature isotherms (Neill 1984). When ocean currents move in different directions at different depths, vertical position is very important to the direction and velocity of 'passive' transport (e.g. Fortier and Leggett 1982, 1983). Thus, vertical habitat selection can be an important component of final horizontal spatial distributions.

For benthic species, depth and substratum characteristics can be important. For example, adult red hake, *Urophycis chuss,* in the North Atlantic are benthic fish found only over sand or mud substrata. They are broadly dispersed over the continental shelf in the fall, moving farther offshore in winter to the edge of the shelf where they are most abundant at temperatures between 8–10°C. Adults migrate inshore in the spring and then offshore again to spawn. At a length of 3.5 cm, the pelagic larvae become benthic and enter live sea scallops, *Placopecten magellanicus,* through the exhalant opening. When they outgrow the scallops at about 13 cm or when winter temperatures on the scallop beds reach 4°C, the juveniles move to deeper water (Musick 1974; Steiner *et al.* 1982).

3.9 Conclusions

Most field studies of habitat selection in fishes simply document associations between abundance and various abiotic and biotic variables, rarely considering the underlying processes and frequently even failing to distinguish active selection from passive distribution and demographic processes. In contrast, laboratory studies usually simplify the environment to a small number of sites within centimetres of each other and differing only in the quantitative level of a single variable. There is a large and rapidly growing body of habitat selection theory, but models that consider more than one or two variables have yet to be developed. Tests of theoretical models have taken place in laboratory aquaria far more often than in the natural environment. Despite these problems, field workers are more often acknowledging the value of predictions from habitat selection theory and theoreticians are more often developing models that can be tested in nature. To fishery biologists and conservation scientists, habitat selection is extremely important as one of the processes influencing the distribution and abundance of populations, and how they change with age, reproductive status, population size, community composition and natural and man-made environmental perturbation. To behavioural ecologists, the diversity of fish behaviour and of fish habitats is a source of inspiration and intellectual challenge. The amenability of some species and habitats to experimental manipulation have made them preferred systems for testing new theoretical approaches to habitat selection. We hope that recent trends toward productive interactions among the many approaches to habitat selection will continue, but we must not underestimate the task of understanding fish distributions. In the words of a Canadian writer, academic and avid sports fisherman:

'So long as water moves, so long as fins press against it, as long as weather changes and man is fallible, fish will remain in some measure unpredictable.'

(Haig-Brown 1975, p. 252)

Acknowledgements

Our studies of fish habitat selection have been supported by NSERC research grants to DLK, NSERC/Canada Department of Fisheries and Oceans grants to DLK and RWR, an FCAR graduate fellowship to RWR, and NSERC postgraduate and postdoctoral fellowships and USAID and NSF research grants to LJC. We are grateful for this continuing support. The manuscript benefited from criticisms and comments by Jean-Guy Godin and an anonymous referee, Joel Elliott, Luc-Alain Giraldeau, James Grant, David Lee, Andrew McAdam, Mary Power, Derek Roff, Gray Stirling, Doug Swain and Fred Whoriskey.

References

Aboul Hosn, W. and Downing, J.A. (1994). Influence of cover on the spatial distribution of littoral-zone fishes. *Can. J. Fish. Aquat. Sci.*, **51**, 1832–1838.

Abrahams, M.V. (1986). Patch choice under perceptual constraints: a cause for departures from an ideal free distribution. *Behav. Ecol. Sociobiol.*, **19**, 409–415.

Abrahams, M.V. (1989). Foraging guppies and the ideal free distribution: influence of information on patch choice. *Ethology*, **82**, 116–126.

Abrahams, M.V. and Dill, L.M. (1989). A determination of the energetic equivalence of the risk of predation. *Ecology*, **70**, 999–1007.

Abrahams, M.V. and Healey, M.C. (1990). Variation in the competitive abilities of fishermen and its influence on the spatial distribution of the British Columbia salmon troll fleet. *Can. J. Fish. Aquat. Sci.*, **47**, 1116–1121.

Angermeier, P.L. and Karr, J.R. (1983). Fish communities along environmental gradients in a system of tropical streams. *Environ. Biol. Fish.*, **9**, 117–135.

Audet, C., FitzGerald, G.J. and Guderley, H. (1985). Salinity preferences of four sympatric species of sticklebacks (Pisces: Gasterosteidae) during their reproductive season. *Copeia*, **1985**, 209–213.

Audet, C., FitzGerald, G.J. and Guderley, H. (1986). Environmental control of salinity preferences in four sympatric species of sticklebacks: *Gasterosteus aculeatus*, *Gasterosteus wheatlandi*, *Pungitius pungitius* and *Apeltes quadracus*. *J. Fish Biol.*, **28**, 725–739.

Baker, J.A. and Ross, S.T. (1981). Spatial and temporal resource utilization by southeastern cyprinids. *Copeia*, **1981**, 178–189.

Balon, E.K. (1975). Reproductive guilds of fishes: a proposal and definition. *J. Fish. Res. Board Can.*, **32**, 821–864.

Baltz, D.M., Rakocinski, C. and Fleeger, J.W. (1993). Microhabitat use by marsh-edge fishes in a Louisiana estuary. *Environ. Biol. Fish.*, **36**, 109–126.

Bazzaz, F.A. (1991). Habitat selection in plants. *Am. Nat.*, **137**, S116–S130.

Beitinger, T.L. and Freeman, L. (1983). Behavioral avoidance and selection responses of fishes to chemicals. *Residue Rev.*, **90**, 35–55.

Beitinger, T.L. and Magnuson, J.J. (1975). Influence of social rank and size on thermoselection behavior of bluegill (*Lepomis macrochirus*). *J. Fish. Res. Board Can.*, **32**, 2133–2136.

Bernstein, C., Kacelnik, A. and Krebs, J.R. (1988). Individual decisions and the distribution of predators in a patchy environment. *J. Anim. Ecol.*, **57**, 1007–1026.

Bernstein, C., Kacelnik, A. and Krebs, J.R. (1991). Individual decisions and the distribution of predators in a patchy environment. II. The influence of travel costs and structure of the environment. *J. Anim. Ecol.*, **60**, 205–225.

Bohnsack, J.A., Johnson, D.L. and Ambrose, R.F. (1991). Ecology of artificial reef habitats and fishes. In *Artificial habitats for marine and freshwater fishes* (eds. W. Seaman, Jr. and L.M. Sprague), pp. 61–107. Academic Press, New York.

Bouskila, A. and Blumstein, D.T. (1992). Rules of thumb for predation hazard assessment: predictions from a dynamic model. *Am. Nat.*, **139**, 161–176.

Brett, J.R. (1971). Energetic responses of salmon to temperature. A study of some thermal relations in the physiology and freshwater ecology of sockeye salmon (*Oncorhynchus nerka*). *Amer. Zool.*, **11**, 99–113.

Brett, J.R. (1983). Life energetics of sockeye salmon, *Oncorhynchus nerka*. In *Behavioral energetics: the cost of survival in vertebrates* (eds. W.P. Aspey and S.I. Lustick), pp. 29–63. Ohio State University Press, Columbus.

Brett, J.R., Shelbourn, J.E. and Shoop, C.T. (1969). Growth rate and body composition of fingerling sockeye salmon, *Oncorhynchus nerka*, in relation to temperature and ration size. *J. Fish. Res. Board Can.*, **26**, 2363–2394.

Brooks, C.E. (1974). *The trout and the stream*. Crown Publishers Inc., New York.

Burke, N.C. (1995). Nocturnal foraging habitats of French and bluestriped grunts, *Haemulon flavolineatum* and *H. sciurus*, at Tobacco Caye, Belize. *Environ. Biol. Fish.*, **42**, 365–374.

Burrows, M.T. (1994). An optimal foraging and migration model for juvenile plaice. *Evol. Ecol.*, **8**, 125–149.

Chapman, C.A. and Mackay, W.C. (1984). Versatility in habitat use by a top aquatic predator, *Esox lucius* L. *J. Fish Biol.*, **25**, 109–115.

Chapman, L.J. (1995). Seasonal dynamics of habitat use by an air-breathing catfish (*Clarias liocephalus*) in a papyrus swamp. *Ecol. Freshwat. Fish*, **4**, 113–123.

Chapman, L.J. and Chapman, C.A. (1993). Fish populations in tropical floodplain pools: a re-evaluation of Holden's data on the River Sokoto. *Ecol. Freshwat. Fish*, **2**, 23–30.

Chapman, L.J. and Kramer, D.L. (1991). The consequences of flooding for the dispersal and fate of poeciliid fish in an intermittent tropical stream. *Oecologia*, **87**, 299–306.

Chapman, L.J. and Liem, K.F. (1995). Papyrus swamps and the respiratory ecology of *Barbus neumayeri*. *Environ. Biol. Fish*, **44**, 183–197.

Clark, C.W. and Levy, D.A. (1988). Diel vertical migrations by juvenile sockeye salmon and the antipredation window. *Am. Nat.*, **131**, 271–290.

Cody, M.L. (ed.) (1985). *Habitat selection in birds.* Academic Press, New York.

Congleton, J.L. (1980). Observations on the responses of some southern California tidepool fishes to noctural hypoxic stress. *Comp. Biochem. Physiol.*, **66A**, 719–722.

Coutant, C.C. (1985). Striped bass, temperature, and dissolved oxygen: a speculative hypothesis for environmental risk. *Trans. Am. Fish. Soc.*, **114**, 31–61.

Crowder, L.B. and Magnuson, J.J. (1983). Cost-benefit analysis of temperature and food resource use: a synthesis with examples from the fishes. In *Behavioral energetics: the cost of survival in vertebrates* (eds. W.P. Aspey and S.I. Lustick), pp. 189–221. Ohio State University Press, Columbus.

Davenport, J. and Sayer, M.D.J. (1993). Physiological determinants of distribution in fish. *J. Fish Biol.*, **43** *(Suppl.* A), 121–145.

DeMartini, E.E. (1993). Modeling the potential of fishery reserves for managing Pacific coral reef fishes. *Fish. Bull.*, 91, 414–427.

Dill, L.M. (1987). Animal decision making and its ecological consequences: the future of aquatic ecology and behaviour. *Can. J. Zool.*, **65**, 803–811.

Doherty, P.J. (1991). Spatial and temporal patterns in recruitment. In *The ecology of fishes on coral reefs* (ed. P.F. Sale), pp. 261–293. Academic Press, New York.

Elliott, J.K., Elliott, J.M. and Mariscal, R.N. (1995). Host selection, location, and association behaviors of anemonefishes in field settlement experiments. *Mar. Biol.*, **122**, 377–389.

Everest, F.H. and Chapman, D.W. (1972). Habitat selection and spatial interaction by juvenile chinook salmon and steelhead trout in two Idaho streams. *J. Fish. Res. Board Can.*, **29**, 91–100.

Fagen, R. (1987). A generalized habitat matching rule. *Evol. Ecol.*, **1**, 5–10.

Fautin, D.G. (1986). Why do anemone fishes inhabit only some host actinians? *Environ. Biol. Fish.*, **15**, 171–180.

FitzGerald, G.J., Whoriskey, F.G., Morrissette, J. and Harding, M. (1992). Habitat scale, female cannibalism, and male reproductive success in three-spined sticklebacks (*Gasterosteus aculeatus*). *Behav. Ecol.*, **3**, 141–147.

Fortier, L. and Harris, R.P. (1989). Optimal foraging and density-dependent competition in marine fish larvae. *Mar. Ecol. Prog. Ser.*, **51**, 19–33.

Fortier, L. and Leggett, W.C. (1982). Fickian transport and the dispersal of fish larvae in estuaries. *Can. J. Fish. Aquat. Sci.*, **39**, 1150–1163.

Fortier, L. and Leggett, W.C. (1983). Vertical migrations and transport of larval fish in a partially mixed estuary. *Can. J. Fish. Aquat. Sci.*, **40**, 1543–1555.

Foster, S.A., Garcia, V.B. and Town, M.Y. (1988). Cannibalism as the cause of an ontogenetic shift in habitat use by fry of the threespine stickleback. *Oecologia*, **74**, 577–585.

Fraser, D.F. and Gilliam, J.F. (1992). Nonlethal impacts of predator invasion: facultative suppression of growth and reproduction. *Ecology*, **73**, 959–970.

Fraser, D.F. and Sise, T.E. (1980). Observations on stream minnows in a patchy environment: a test of a theory of habitat distribution. *Ecology*, **61**, 790–797.

Freeman, M.C. and Grossman, G.D. (1992). Group foraging by a stream minnow: shoals or aggregations? *Anim. Behav.*, **44**, 393–403.

Fretwell, S.D. (1972). *Populations in a seasonal environment.* Princeton University Press, Princeton.

Fretwell, S.D. and Lucas, H.L., Jr. (1970). On territorial behavior and other factors influencing habitat distribution in birds. I. Theoretical development. *Acta Biotheoretica*, **19**, 16–36.

Giattina, J.D. and Garton, R.R. (1983). A review of the preference-avoidance responses of fishes to aquatic contaminants. *Residue Rev.*, **87**, 43–90.

Gibson, R.N. (1986). Intertidal teleosts: life in a fluctuating environment. In *The behaviour of teleost fishes* (ed. T.J. Pitcher), pp. 388–408. Croom Helm, London.

Gilliam, J.F. and Fraser, D.F. (1988). Resource depletion and habitat segregation by competitors under predation hazard. In *Size-structured populations* (eds. B. Ebenman and L. Persson), pp. 173–184. Springer-Verlag, Berlin.

Gillis, D.M. and Kramer, D.L. (1987). Ideal interference distributions: population density and patch use by zebrafish. *Anim. Behav.*, **35**, 1875–1882.

Gillis, D.M., Peterman, R.M. and Tyler, A.V. (1993). Movement dynamics in a fishery: application of the ideal free distribution to spatial allocation of effort. *Can. J. Fish. Aquat. Sci.*, **50**, 323 -333.

Godin, J.-G.J. and Keenleyside, M.H.A. (1984). Foraging on patchily distributed prey by a cichlid fish (Teleostei, Cichlidae): a test of the ideal free distribution theory. *Anim. Behav.*, **32**, 120–131.

Godin, J.-G.J. and Rangeley, R.W. (1989). Living in the fast lane: effects of cost of locomotion on foraging behaviour in juvenile Atlantic salmon. *Anim. Behav.*, **37**, 943–954.

Gooding, R.M. and Magnuson, J.J. (1967). Ecological significance of a drifting object to pelagic fishes. *Pacific Sci.*, **21**, 486–497.

Goodyear, C.P. (1973). Learned orientation in the predator avoidance behaviour of mosquitofish, *Gambusia affinis*. *Behaviour*, **45**, 191–224.

Goulding, M. (1980). *The fishes and the forest*. University of California Press, Berkeley.

Grand, T.C. and Grant, J.W.A. (1994a). Spatial predictability of food influences its monopolization and defence by juvenile convict cichlids. *Anim. Behav.*, **47**, 91–100.

Grand, T.C. and Grant, J.W.A. (1994b). Spatial predictability of resources and the ideal free distribution in convict cichlids, *Cichlasoma nigrofasciatum*. *Anim. Behav.*, **48**, 909–919.

Grant, J.W.A. and Kramer, D.L. (1990) Territory size as a predictor of the upper limit to population density in stream-dwelling salmonids. *Can. J. Fish. Aquat. Sci.*, **47**, 1724–1737.

Gray, R.D. and Kennedy, M. (1994a). Misconceptions or misreadings? Missing the real issues about the IFD. *Oikos*, **71**, 167–170.

Gray, R.D. and Kennedy, M. (1994b). Perceptual constraints on optimal foraging: a reason for departures from the ideal free distribution? *Anim.Behav.*, **47**, 469–471.

Grossman, G.D., de Sostoa, A., Freeman, M.C. and Lobon-Cerviá, J. (1987). Microhabitat use in a Mediterranean riverine fish assemblage. Fishes of the lower Matarraña. *Oecologia*, **73**, 490–500.

Grossman, G.D. and Freeman, M.C. (1987). Microhabitat use in a stream fish assemblage. *J. Zool., Lond.*, **212**, 151–176.

Haig-Brown, R.L. (1975). *Fisherman's summer*. Totem Books, Don Mills, Ontario.

Hallacher, L.E. and Roberts, D.A. (1985). Differential utilization of space and food by the inshore rockfishes (Scorpaenidae: *Sebastes*) of Carmel Bay, California. *Environ. Biol. Fish.*, **12**, 91–110.

Hartwell, S.I., Jin, J.H., Cherry, D.S. and Cairns, J., Jr. (1989). Toxicity versus avoidance response of golden shiner, *Notemigonus crysoleucas*, to five metals. *J. Fish Biol.*, **35**, 447–456.

Hill, J. and Grossman, G.D. (1993). An energetic model of microhabitat use for rainbow trout and rosyside dace. *Ecology*, **74**, 685–698.

Hobson, E.S.. (1991). Trophic relationships of fishes specialized to feed on zooplankters above coral reefs. In *The ecology of fishes on coral reefs* (ed. P.F. Sale), pp. 69–95. Academic Press, New York.

Holden, M.J. (1963). The populations of fish in dry season pools of the River Sokoto. *Colonial Office Fishery Publications*, **19**, 1–65.

Huey, R.B. (1991). Physiological consequences of habitat selection. *Am. Nat.*, **137**, S91–S115.

Huey, R.B. and Slatkin, M. (1976). Costs and benefits of lizard thermoregulation. *Q. Rev. Biol.*, **51**, 363–384.

Hughes, N.F. and Dill, L.M. (1990). Position choice by drift-feeding salmonids: model and test for arctic grayling (*Thymallus arcticus*) in subarctic mountain streams, interior Alaska. *Can. J. Fish. Aquat. Sci.*, **47**, 2039–2048.

Hugie, D.M. and Dill, L.M. (1994). Fish and game: a game theoretic approach to habitat selection by predators and prey. *J. Fish Biol.*, **45** (Suppl. A), 151–169.

Huh, H.T. (1975). *Bioenergetics of food conversion and growth of yellow perch (Perca flavescens) and walleye (Stizostedion vitreum vitreum) using formulated diets*. Ph.D. thesis, University of Wisconsin, Madison.

Huh, H.T., Calbert, H.E. and Stuiber, D.A. (1976). Effects of temperature and light on growth of yellow perch and walleye using formulated feed. *Trans. Am. Fish. Soc.*, **105**, 254–258.

Huntingford, F.A. (1993). Can cost–benefit analysis explain fish distribution patterns? *J. Fish Biol.*, **43** (Suppl. A), 289–308.

Hynes, H.B.N. (1970). *The ecology of running waters*. University of Toronto Press, Toronto.

Ingersoll, C.G. and Claussen, D.L. (1984). Temperature selection and critical thermal maxima of the fantail darter, *Etheostoma flabellare*, and johnny darter, *E. nigrum*, related to habitat and season. *Environ. Biol. Fish.*, **11**, 131–138.

Johnels, A.G. (1954). Notes on fishes from the Gambia River. *Arkiv för Zoologi*, 6, 327–411.

Johnson, D.H. (1980). The comparison of usage and availability measurements for evaluating resource preference. *Ecology*, **61**, 65–71.

Jones, G.P. (1991). Postrecruitment processes in the ecology of coral reef fish populations: a multifactorial perspective. In *The ecology of fishes on coral reefs* (ed. P.F. Sale), pp. 294–328. Academic Press, New York.

Kacelnik, A., Krebs, J.R. and Bernstein, C. (1992). The ideal free distribution and predator–prey populations. *Trends Ecol. Evol.*, **7**, 50–55.

Keast, A., Harker, J. and Turnbull, D. (1978). Nearshore fish habitat utilization and species associations in Lake Opinicon (Ontario, Canada). *Environ. Biol. Fish.*, **3**, 173–184.

Kennedy, M. and Gray, R.D. (1993). Can ecological theory predict the distribution of foraging animals? A critical analysis of experiments on the ideal free distribution. *Oikos*, **68**, 158–166.

Kennedy, M., Shave, C.R., Spencer, H.G. and Gray, R.D. (1994). Quantifying the effect of predation risk on foraging bullies: no need to assume an IFD. *Ecology*, **75**, 2220–2226.

Kinzie, R.A., III (1988). Habitat utilization by Hawaiian stream fishes with reference to community structure in oceanic island streams. *Environ. Biol. Fish.*, **22**, 179–192.

Kitchell, J.F. Stewart, D.J. and Weininger, D. (1977). Applications of a bioenergetics model to yellow perch (*Perca flavescens*) and walleye (*Stizostedion vitreum vitreum*). *J. Fish. Res. Board Can.*, **34**, 1922–1935.

Kneib, R.T. (1987). Predation risk and use of intertidal habitats by young fishes and shrimp. *Ecology*, **68**, 379–386.

Kneib, R.T. (1993). Growth and mortality in successive cohorts of fish larvae within an estuarine nursery. *Mar. Ecol. Prog. Ser.*, **94**, 115–127.

Krebs, J.R. and Kacelnik, A. (1991). Decision-making. In *Behavioural ecology: an evolutionary approach*, 3rd edn. (eds. J.R. Krebs and N.B. Davies), pp. 105–136. Blackwell Scientific Publ., Oxford.

Krebs, J.R. and McCleery, R.H. (1984). Optimisation in behavioural ecology. In *Behavioural ecology: an evolutionary approach*, 2nd edn. (eds. J.R. Krebs and N.B. Davies), pp. 91–121. Blackwell Scientific Publ., Oxford.

Kwak, T.J., Wiley, M.J., Osborne, L.L. and Larimore, R.W. (1992). Application of diel feeding chronology to habitat suitability analysis of warmwater stream fishes. *Can. J. Fish. Aquat. Sci.*, **49**, 1417–1430.

Lemke, A.E. (1977). Optimum temperature for growth of juvenile bluegills. *Prog. Fish. Cult.*, **39**, 55 -57.

Lessells, C.M. (1995). Putting resource dynamics into continuous input ideal free distribution models. *Anim. Behav.*, **49**, 487–494.

Liem, K.F. (1987). Functional design of the air ventilation apparatus and overland excursions by teleosts. *Fieldiana, Zoology*, **37**, 1–29.

Lirman, D. (1994). Ontogenetic shifts in habitat preferences in the 3-spot damselfish, *Stegastes planifrons* (Cuvier), in Roatan Island, Honduras. *J. Exp. Mar. Biol. Ecol.*, **180**, 71–81.

Lowe-McConnell, R.H. (1964). The fishes of the Rupununi savanna district of British Guiana, South America. Part 1. Ecological groupings of fish species and effects of the seasonal cycle on the fish. *J. Linn. Soc. (Zool.)*, **45**, 103–144.

Lowe-McConnell, R.H. (1975). *Fish communities in tropical freshwaters*. Longman, London.

Luckhurst, B.E. and Luckhurst, K. (1978). Diurnal space utilization in coral reef fish communities. *Mar. Biol.*, **49**, 325–332.

MacCall, A.D. (1990). *Dynamic geography of marine fish populations*. University of Washington Press, Seattle.

McCauley, R.W. and Read, L.A.A. (1973). Temperature selection by juvenile and adult yellow perch (*Perca flavescens*) acclimated to 24C. *J. Fish. Res. Board. Can.*, **30**, 1253–1255.

McFarland, W.N., Ogden, J.C. and Lythgoe, J.N. (1979). The influence of light on the twilight migrations of grunts. *Environ. Biol. Fish.*, **4**, 9–22.

McIvor, C.C. and Odum, W.E. (1988). Food, predation risk, and microhabitat selection in a marsh fish assemblage. *Ecology*, **69**, 1341–1351.

Magnuson, J.J. and Beitinger, T.L. (1978). Stability of temperatures preferred by centrarchid fishes and terrestrial reptiles. In *Contrasts in behavior* (eds. E.S. Reese and F.J. Lighter), pp. 181–216. Wiley-Interscience, New York.

Magnuson, J.J., Crowder, L.B. and Medvick, P.A. (1979). Temperature as an ecological resource. *Amer. Zool.*, **19**, 331–343.

Mangel, M. and Clark, C.W. (1988). *Dynamic modeling in behavioral ecology*. Princeton University Press, Princeton.

Manly, B.F.J., McDonald, L.L. and Thomas, D.L. (1993). *Resource selection by animals. Statistical design and analysis for field studies*. Chapman and Hall, London.

Margenau, T. (1986). Habitat preferences and movement of northern pike during fall and early winter in Potato Lake, Washburn County. Department of Natural Resources Research Report 139. Madison, WI.

Marshall, C.T. and Frank, K.T. (1994). Geographic responses of groundfish to variation in abundance: methods of detection and their interpretation. *Can. J. Fish. Aquat. Sci.*, **51**, 808–816.

Mather, F.J. (1962). Transatlantic migration of two large bluefin tuna. *J. Conseil*, **27**, 325–327.

Matthews, W.J. and Hill, L.G. (1980). Habitat partitioning in the fish community of a southwestern river. *Southwest. Nat.*, **25**, 51–66.

Matthews, W.J., Hill, L.G. and Schellhaass, S.C.. (1985). Depth distribution of striped bass and other fish in Lake Texoma (Oklahoma-Texas) during summer stratification. *Trans. Am. Fish. Soc.*, **114**, 84–91.

Merriman, D. and Thorpe, L.M. (1976). The Connecticut River ecological study. *Am. Fish. Soc. Monogr.*, **1**, 1–252.

Milinski, M. (1979). An evolutionarily stable feeding strategy in sticklebacks. *Z. Tierpyschol.*, **51**, 36–40.

Milinski, M. (1984). Competitive resource sharing: an experimental test of a learning rule for ESSs. *Anim. Behav.*, **32**, 233–242.

Milinski, M. (1987). Competition for non-depleting resources: the ideal free distribution in sticklebacks. In *Foraging behavior* (eds. A.C. Kamil, J.R. Krebs and H.R. Pulliam), pp. 363–388. Plenum Press, New York.

Milinski, M. (1994). Ideal free theory predicts more than only input matching—a critique of Kennedy and Gray's review. *Oikos*, **71**, 163–166.

Milinski, M. and Parker, G.A. (1991). Competition for resources. In *Behavioural ecology: an evolutionary approach*, 3rd edn. (eds. J.R. Krebs and N.B. Davies), pp. 137–168. Blackwell Scientific Publ., Oxford.

Morris, D.W. (1994). Habitat matching: alternatives and implications to populations and communities. *Evol. Ecol.*, **8**, 387–406.

Musick, J.A. (1974). Seasonal distribution of sibling hakes, *Urophycis chuss* and *U. tenuis* (Pisces, Gadidae) in New England. *Fish. Bull.*, **72**, 481–495.

Neill, W.H. (1984). Behavioral enviroregulation's role in fish migration. In *Mechanisms of migration in fishes* (eds. J.D. McCleave, G.P. Arnold, J.J. Dodson and W.H. Neill), pp. 61–66. Plenum Press, New York.

Neilson, J.D. and Perry, R.I. (1990). Diel vertical migrations of marine fishes: an obligate or facultative process? *Adv. Mar. Biol.*, **26**, 115–168.

Nelson, J.S. (1994). *Fishes of the world*, 3rd edn. John Wiley and Sons, New York.

Oksanen, T., Power, M.E. and Oksanen, L. (1995). Ideal free habitat selection and consumer-resource dynamics. *Am. Nat.*, **146**, 565–583.

Orians, G.H. and Wittenberger, J.F. (1991). Spatial and temporal scales in habitat selection. *Am. Nat.*, **137**, S29-S49.

Orth, D.J. and Maughan, O.E. (1982). Evaluation of the incremental methodology for recommending instream flows for fishes. *Trans. Am. Fish. Soc.*, **111**, 413–445.

Parker, G.A. (1978). Searching for mates. In *Behavioural ecology: an evolutionary approach* (eds. J.R. Krebs and N.B. Davies), pp. 214–244. Sinauer Associates, Inc., Sunderland.

Parker, G.A. and Sutherland, W.J. (1986). Ideal free distributions when individuals differ in competitive ability: phenotype-limited ideal free models. *Anim. Behav.*, **34**, 1222–1242.

Pedder, S.C.J. and Maly, E.J. (1986). The avoidance response of groups of juvenile brook trout, *Salvelinus fontinalis*, to varying levels of acidity. *Aquatic Toxicology*, **8**, 111–119.

Perry, R.I. and Smith, S.J. (1994). Identifying habitat associations of marine fishes using survey data: an application to the northwest Atlantic. *Can. J. Fish. Aquat. Sci.*, **51**, 589–602.

Peterson, R.H., Coombs, K., Power, J. and Paim, U. (1989). Responses of several fish species to pH gradients. *Can. J. Zool.*, **67**, 1566–1572.

Pitcher, T.J. and Parrish, J.K. (1993). Functions of shoaling behaviour in teleosts. In *Behaviour of teleost fishes*, 2nd edn (ed. T.J. Pitcher), pp. 363–439. Chapman and Hall, London.

Power, M.E. (1983). Grazing responses of tropical freshwater fishes to different scales of variation in their food. *Environ. Biol. Fish.*, **9**, 103–115.

Power, M.E. (1984a). Depth distributions of armored catfish: predator-induced resource avoidance? *Ecology*, **65**, 523–528.

Power, M.E. (1984b). Habitat quality and the distribution of algae-grazing catfish in a Panamanian stream. *J. Anim. Ecol.*, **53**, 357–374.

Power, M.E. (1987). Predator avoidance by grazing fishes in temperate and tropical streams: importance of stream depth and prey size. In *Predation: direct and indirect impacts in aquatic communities* (eds. W.C. Kerfoot and A. Sih), pp. 333–351. University Press of New England, Hanover, NH.

Pulliam, H.R. and Caraco, T. (1984). Living in groups: is there an optimal group size? In *Behavioural ecology: an evolutionary approach*, 2nd edn. (eds. J.R. Krebs and N.B. Davies), pp. 122–147. Blackwell Scientific Publ., Oxford.

Pyke, G.H. (1984). Optimal foraging theory: a critical review. *A. Rev. Ecol. Syst.*, **15**, 523–575.

Rahel, F.J. and Nutzman, J.W. (1994). Foraging in a lethal environment: fish predation in hypoxic waters of a stratified lake. *Ecology*, **75**, 1246–1253.

Rakitin, A. and Kramer, D.L. (1996). The effect of a marine reserve on the distribution of coral reef fishes in Barbados. *Mar. Ecol. Prog. Ser.*, **131**, 97–113.

Rakocinski, C. (1988). Population structure of stream-dwelling darters: correspondence with habitat structure. *Environ. Biol. Fish.*, **23**, 215–224.

Rangeley, R.W. (1994). *Habitat selection in juvenile pollock, Pollachius virens: behavioural responses to changing habitat availability.* Ph.D. thesis, McGill University, Montreal.

Rangeley, R.W. and Kramer, D.L. (1995a). Use of rocky intertidal habitats by juvenile pollock *Pollachius virens. Mar. Ecol. Prog. Ser.*, **126**, 9–17.

Rangeley, R.W. and Kramer, D.L. (1995b). Tidal effects on habitat selection and aggregation by juvenile pollock *Pollachius virens* in the rocky intertidal zone. *Mar. Ecol. Prog. Ser.*, **126**, 19–29.

Reiser, D.W., Wesche, T.A. and Estes, C. (1989). Status of instream flow legislation and practices in North America. *Fisheries*, **14**, 22–29.

Roberts, C.M. and Polunin, N.V.C. (1991). Are marine reserves effective in management of reef fisheries? *Rev. Fish Biol. Fish.*, **1**, 65–91.

Robertson, D.R. and Lassig, B. (1980). Spatial distribution patterns and coexistence of a group of territorial damselfishes from the Great Barrier Reef. *Bull. Mar. Sci.*, **30**, 187–203.

Rodríguez, M.A. (1995). Habitat-specific estimates of competition in stream salmonids: a field test of the isodar model of habitat selection. *Evol. Ecol.*, **9**, 169–184.

Rose, G.A. and Leggett, W.C. (1989). Interactive effects of geophysically-forced sea temperatures and prey abundance on mesoscale coastal distributions of a marine predator, Atlantic cod (*Gadus morhua*). *Can. J. Fish. Aquat. Sci.*, **46**, 1904–1913.

Ross, S.T. (1986). Resource partitioning in fish assemblages: a review of field studies. *Copeia*, **1986**, 352–388.

Ross, S.T., Knight, J.G. and Wilkins, S.D. (1992). Distribution and microhabitat dynamics of the threatened bayou darter, *Etheostoma rubrum*. *Copeia*, **1992**, 658–671.

Rountree, R.A. and Able, K.W. (1992). Fauna of polyhaline subtidal marsh creeks in southern New Jersey: composition, abundance and biomass. *Estuaries*, **15**, 171–185.

Rudstam, L.G. and Magnuson, J.J. (1985). Predicting the vertical distribution of fish populations: analysis of cisco, *Coregonus artedii*, and yellow perch, *Perca flavescens*. *Can. J. Fish. Aquat. Sci.*, **42**, 1178–1188.

Safran, P. and Omori, M. (1990). Some ecological observations on fishes associated with drifting seaweed off Tohoku coast, Japan. *Mar. Biol.*, **105**, 395–402.

Sale, P.F. (1968). Influence of cover availability on depth preference of the juvenile manini, *Acanthurus triostegus sandvicensis*. *Copeia*, **1968**, 802–807.

Sale, P.F. (1969). Pertinent stimuli for habitat selection by the juvenile manini, *Acanthurus triostegus sandvicensis*. *Ecology*, **50**, 616–623.

Sale, P.F. (1980). The ecology of fishes on coral reefs. *A. Rev. Oceanogr. Mar. Biol.*, **18**, 367–421.

Sale, P.F. (ed.) (1991a). *The ecology of fishes on coral reefs*. Academic Press, New York.

Sale, P.F. (1991b). Introduction. In *The ecology of fishes on coral reefs* (ed. P.F. Sale), pp. 3–15. Academic Press, New York.

Sale, P.F., Douglas, W.A. and Doherty, P.J. (1984). Choice of microhabitats by coral reef fishes at settlement. *Coral Reefs*, **3**, 91–99.

Sayer, M.D.J. and Davenport, J. (1991). Amphibious fish: why do they leave water? *Rev. Fish Biol. Fish.*, **1**, 159–181.

Seaman, W., Jr. and Sprague, L.M. (1991). Artificial habitat practices in aquatic systems. In *Artificial habitats for marine and freshwater fisheries* (eds. W. Seaman, Jr. and L.M. Sprague), pp. 1–29. Academic Press, New York.

Schlosser, I.J. and Toth, L.A. (1984). Niche relationships and population ecology of rainbow (*Etheostoma caeruleum*) and fantail (*E. flabellare*) darters in a temporally variable environment. *Oikos*, **42**, 229–238.

Shirvell, C.S. (1994). Effect of changes in streamflow on the microhabitat use and movements of sympatric juvenile coho salmon (*Oncorhynchus kisutch*) and chinook salmon (*O. tshawytscha*) in a natural stream. *Can. J. Fish. Aquat. Sci.*, **51**, 1644–1652.

Sih, A. (1980). Optimal behavior: can foragers balance two conflicting demands? *Science*, **210**, 1041–1043.

Southwood, T.R.E. (1977). Habitat, the templet for ecological strategies? *J. Anim. Ecol.*, **46**, 337–365.

Steiner, W.W., Luczkowich, J.J. and Olla, B.L. (1982). Activity, shelter usage, growth and recruitment of juvenile red hake *Urophycis chuss*. *Mar. Ecol. Prog. Ser.*, **7**, 125–135.

Stephens, D.W. (1987). On economically tracking a variable environment. *Theor. Popul. Biol.*, **32**, 15–25.

Stephens, D.W. and Krebs, J.R. (1986). *Foraging theory*. Princeton University Press, Princeton.

Stepien, C.A. and Brusca, R.C. (1985). Nocturnal attacks on nearshore fishes in southern California by crustacean zooplankton. *Mar. Ecol. Prog. Ser.*, **25**, 91–105.

Stott, B. and Buckley, B.R. (1979). Avoidance experiments with homing shoals of minnows, *Phoxinus phoxinus*, in a laboratory stream channel. *J. Fish Biol.*, **14**, 135–146.

Stott, B. and Cross, D.G. (1973). The reactions of the roach [*Rutilus rutilus* (L.)] to changes in the concentration of dissolved oxygen and free carbon dioxide in a laboratory channel. *Water Res.*, **7**, 793–805.

Sutherland, W.J. (1983). Aggregation and the 'ideal free' distribution. *J. Anim. Ecol.*, **52**, 821–828.

Sutherland, W.J. and Parker, G.A. (1985). Distribution of unequal competitors. In *Behavioural ecology* (eds. R.M. Sibly and R.H. Smith), pp. 255–273. Blackwell Scientific Publ., Oxford.

Sutherland, W.J., Townsend, C.R. and Patmore, J.M. (1988). A test of the ideal free distribution with unequal competitors. *Behav. Ecol. Sociobiol.*, **23**, 51–53.

Suthers, I.M. and Gee, J.H. (1986). Role of hypoxia in limiting diel spring and summer distribution of juvenile yellow perch (*Perca flavescens*) in a prairie marsh. *Can. J. Fish. Aquat. Sci.*, **43**, 1562–1570.

Swain, D.P. (1993). Age- and density-dependent bathymetric pattern of Atlantic cod (*Gadus morhua*) in the southern Gulf of St. Lawrence. *Can. J. Fish. Aquat. Sci.*, **50**, 1255–1264.

Swain, D.P. and Kramer, D.L. (1995). Annual variation in temperature selection by Atlantic cod (*Gadus morhua*) in the southern Gulf of St. Lawrence and its relation to population size. *Mar. Ecol. Prog. Ser.*, **116**, 11–23.

Swain, D.P. and Sinclair, A.F. (1994). Fish distribution and catchability—what is the appropriate measure of distribution? *Can. J. Fish. Aquat. Sci.*, **51**, 1046–1054.

Talbot, A.J. and Kramer, D.L. (1986). Effects of food and oxygen availability on habitat selection by guppies in a laboratory environment. *Can. J. Zool.*, **64**, 88–93.

Townsend, C.R. and Calow, P. (1981). *Physiological ecology: an evolutionary approach to resource use.* Sinauer Associates, Inc., Sunderland, MA.

Tregenza, T. (1994). Common misconceptions in applying the ideal free distribution. *Anim. Behav.*, **47**, 485–487.

Tyler, J.A. (1993). Effects of water velocity, group size and prey availability on the stream-drift capture efficiency of blacknose dace, *Rhinichthys atratulus*. *Can. J. Fish. Aquat. Sci.*, **50**, 1055–1061.

Tyler, J.A. and Gilliam, J.F. (1995). Ideal free distributions of stream fish: a model and test with minnows, *Rhinichthys atratulus*. *Ecology*, **76**, 580–592.

Urban, T.P. and Brandt, S.B.. (1993). Food and habitat partitioning between young-of-the year alewives and rainbow smelt in southeastern Lake Ontario. *Environ. Biol. Fish.*, **36**, 359–372.

Utne, A.C.W., Aksnes, D.L. and Giske, J. (1993). Food, predation risk and shelter: an experimental study on the distribution of adult two-spotted goby *Gobiusculus flavescens* (Fabricius). *J. Exp. Mar. Biol. Ecol.*, **166**, 203–216.

van Duren, L.A. and Glass, C.W. (1992). Choosing where to feed: the influence of competition on feeding behaviour of cod, *Gadus morhua* L. *J. Fish. Biol.*, **41**, 463–471.

Victor, B.C. (1991). Settlement strategies and biogeography of reef fishes. In *The ecology of fishes on coral reefs* (ed. P.F. Sale), pp. 231–260. Academic Press, New York.

Vondracek, B. and Longanecker, D.R. (1993). Habitat selection by rainbow trout *Oncorhynchus mykiss* in a California stream: implications for Instream Flow Incremental Methodology. *Ecol. Freshwat. Fish.*, **2**, 173–186.

Waldner, R.E. and Robertson, D.R. (1980). Patterns of habitat partitioning by eight species of territorial Caribbean damselfishes (Pisces: Pomacentridae). *Bull. Mar. Sci.*, **30**, 171–186.

Welcomme, R.L. (1979). *Fisheries ecology of floodplain rivers*. Longman, London.

Wellington, G.M. (1992). Habitat selection and juvenile persistence control the distribution of two closely related Caribbean damselfishes. *Oecologia*, **90**, 500–508.

Wellington, G.M. and Victor, B.C. (1988). Variation in components of reproductive success in an undersaturated population of coral-reef damselfish: a field perspective. *Am. Nat.*, **131**, 588–601.

Werner, E.E. and Gilliam, J.F. (1984). The ontogenetic niche and species interactions in size-structured populations. *A. Rev. Ecol. Syst.*, **15**, 393–425.

Werner, E.E., Hall, D.J., Laughlin, D.R., Wagner, D.J., Wilsmann, L.A. and Funk, F.C. (1977). Habitat partitioning in a freshwater fish community. *J. Fish. Res. Board Can.*, **34**, 360–370.

Werner, E.E., Gilliam, J.F., Hall, D.J. and Mittelbach, G.G. (1983). An experimental test of the effects of predation risk on habitat use in fish. *Ecology*, **64**, 1540–1548.

Wetzel, R.G. (1975). *Limnology*. W.B. Saunders, Philadelphia.

Whittaker, R.H., Levin, S.A. and Root, R.B. (1973). Niche, habitat, and ecotope. *Am. Nat.*, **107**, 321–338.

Whoriskey, F.G. and FitzGerald, G.J. (1989). Breeding-season habitat use by sticklebacks (Pisces: Gasterosteidae) at Isle Verte, Quebec. *Can. J. Zool.*, **67**, 2126–2130.

Wiens, J.A. (1976). Population responses to patchy environments. *A. Rev. Ecol. Syst.*, **7**, 81–120.

Wildhaber, M.L. and Crowder, L.B. (1990). Testing a bioenergetics-based habitat choice model: bluegill (*Lepomis macrochirus*) responses to food availability and temperature. *Can. J. Fish. Aquat. Sci.*, **47**, 1664–1671.

Williams, D.McB. (1991). Pattern and processes in the distribution of coral reef fishes. In *The ecology of fishes on coral reefs* (ed. P.F. Sale), pp. 437–474. Academic Press, New York.

Wurtsbaugh, W.A. and Neverman, D. (1988). Post-feeding thermotaxis and daily vertical migration in a larval fish. *Nature, Lond.*, **333**, 846–848.

4 *Territoriality*

James W.A. Grant

4.1 Introduction

When competing for resources, animals commonly defend access to the resource
or its means of production by attempting to exclude others from specific areas.
Like other animals, fish often defend food, shelter, mates, nest sites, spawning
sites and offspring (Noakes 1978). The importance of such territorial behaviour
in the lives of many fishes is obvious to anyone that has snorkelled on a coral reef
or kept cichlids in an aquarium. One of the classical model systems of early ethology was the male stickleback, *Gasterosteus aculeatus*, pugnaciously defending
its nest (Lorenz 1952). Nevertheless, this rich natural history is often overwhelmed
in general reviews of territoriality by the extensive bird literature, which seems to
have dominated this field ever since Howard's (1920) book on 'Territory in bird
life'.

I have two primary goals in writing this chapter. First, I review territoriality
in fishes as an important component of their behavioural ecology. To accomplish
this, I (1) describe the frequency of occurrence of territoriality in three different
faunas, use a cost/benefit approach to (2) ask whether an individual should defend a territory at all and, if so, (3) how large an area should be defended, and (4)
discuss the ecological consequences of territoriality both to individuals and populations. My second goal is to see what new insights can be derived about territoriality in general by examining the phenomenon from a fishes' perspective. To do
so, I compare fishes with birds, the best studied group with respect to territoriality. Several unique characteristics about the behaviour of fishes emerge from this
comparison (e.g. concentric territoriality) that deserve broader recognition in the
general literature on territoriality.

Definitions of territoriality usually focus on two attributes: the defence behaviour of the owner and the relatively exclusive area defended (Wilson 1975;
Wittenberger 1981). I will concentrate on the first attribute because it is easier to apply to a wide variety of species and situations. Hence, my working
definition of a territory is 'any defended area' (Noble 1939), the oldest and
most common definition used in the vertebrate literature (Maher and Lott
1995).

Table 4.1 The occurrence of territorial behaviour in families of fishes from three habitats[1]

Behaviour	Coral reef	Marine fishes of eastern Canada	Freshwater fishes of Canada	Total
(a) Defence of food?				
% defending (n)[2]	67.6 (37)	6.3 (16)	9.1 (11)	42.2 (64)
% unknown (n)[3]	33.9 (56)	70.4 (54)	52.2 (23)	51.9 (133)
(b) Defence of spawning sites, mates or offspring?				
% defending (n)	84.2 (38)	73.3 (15)	68.4 (19)	77.8 (72)
% unknown (n)	32.1 (56)	72.2 (54)	17.4 (23)	45.9 (133)
(c) Defence of any kind? (including a and b)				
% defending (n)	79.1 (43)	45.8 (24)	68.4 (19)	67.4 (86)
% unknown (n)	23.2 (56)	55.6 (54)	17.4 (23)	35.3 (133)

[1] Coral reefs, Thresher (1984); Marine fishes of eastern Canada, Scott and Scott (1988); Freshwater fishes of Canada, Scott and Crossman (1979).
[2] The number of families with at least one species defending a territory of a particular type × 100/the total number of families with behaviour described sufficiently well to score territoriality (n).
[3] The number of families where the descriptions of behaviour were too poor to score territoriality × 100/the total number of families (n).

4.2 Incidence of territoriality

A comprehensive survey of the frequency of occurrence of territoriality in the more than 24 000 known species of fishes, many of whose behaviour has not been well described, is beyond the scope of this chapter. As a first attempt to estimate the incidence of territoriality, three faunas that have been moderately well studied were selected for examination: freshwater fishes of Canada as described by Scott and Crossman (1979), marine fishes of eastern Canada as described by Scott and Scott (1988), and coral reef fishes as described by Thresher (1984). I hope this selective treatment will encourage a more comprehensive review of the phenomenon in fishes.

For each species, I recorded whether it defended a territory or not and, if so, whether the apparent function was feeding, reproduction, or both. The descriptions of behaviour in the three books were supplemented by more detailed information in Breder and Rosen (1966) and by a data set summarising the territory sizes of 134 species of fishes representing 36 families (T.C. Grand and J.W.A. Grant, unpublished data). Related species share some behavioural characteristics because of descent from a common ancestor (Harvey and Pagel 1991). Hence, I use families of fishes as independent data in any given comparative analysis.

The most striking result of the survey was how little is known about the behaviour of marine fishes in habitats other than coral reefs. For the marine fishes of eastern Canada, a family was included in the survey only if at least one species was sufficiently well known to merit a full description in Scott and Scott (1988).

Hence, 97 families of fishes were immediately excluded from the analysis; most were pelagic or deep-sea fishes. Of the remaining 54 families, the behaviour of 56% was too poorly described to score territoriality, compared with 23% of coral reef families and 17% of freshwater families (Table 4.1; $G = 16.64$, $df = 2$, $P < 0.005$).

The incidence of feeding territoriality differed among the three faunas (Table 4.1; $G = 26.34$, $df = 2$, $P < 0.005$). Of the families with well described behaviour, the defence of food occurred in 68% of coral-reef families, but in only 9% of the freshwater families of Canada and in 6% of the marine families of eastern Canada. The paucity of feeding territoriality in freshwater fishes was first noted in a qualitative analysis by Barlow (1993). Juvenile salmonids were the only fish to defend feeding territories in the freshwater fauna of Canada, whereas the Arctic shanny, *Stichaeus punctatus* (Stichaeidae) (Brown and Green 1976) was the only marine example from eastern Canada.

Defence of spawning sites, mates or offspring occurred in 78% of the families of fishes whose behaviour was well described, and did not differ significantly among the three faunas (Table 4.1; $G = 2.03$, $df = 2$, $P > 0.10$). Territories for reproduction were more common than territories for feeding in all three faunas ($\chi^2 = 25.80$, $P < 0.005$; Fisher's combined probability test; Sokal and Rohlf 1981), suggesting that spawning sites, mates or offspring are more often economically defendable than food (see Section 4.3).

Territorial behaviour of any kind occurred in only 67% of the families of fishes whose behaviour was well described. In contrast, virtually every family, if not species, of bird defends some kind of reproductive territory (Oring 1982). The prevalence of territoriality in birds may be related to the ubiquity of parental care within the group. Offspring are valuable entities that are almost always worth defending by an attending parent (see Section 4.3 and Sargent, Chapter 11).

4.3 Whether or not to defend?

Resource defence theory attempts to predict whether an animal will defend a resource at all. According to the theory of economic defendability (Brown 1964), an animal will only defend a territory if the fitness benefits of defence exceed the costs, or if the net benefits of defence exceed the net benefits of not defending. Predictions about when resources should be economically defendable come from the literatures on mating (Emlen and Oring 1977), foraging (Milinski and Parker 1991) and social (Warner 1980; Lott 1991) systems (for a review, see Grant 1993). Six key ecological variables are thought to influence the economic defendability of a resource (Fig. 4.1). An idealised description of an economically defendable resource is one that is moderately dense, spatially clumped, temporally dispersed, spatially predictable and temporally predictable, with low intruder density.

Many of the key tests of resource defence theory have involved fishes, primarily because they are convenient experimental animals and exhibit considerable behavioural plasticity. Some of these tests are summarised in Table 4.2 (also see Grant

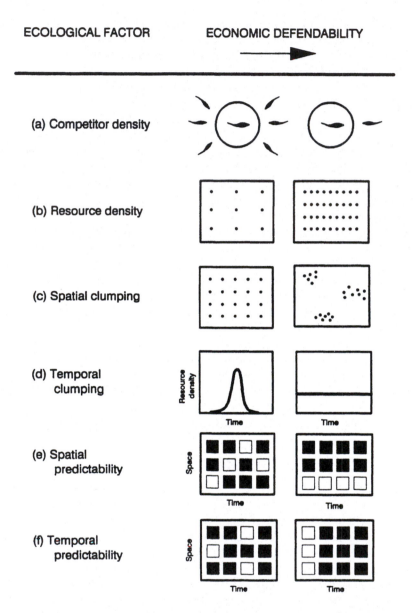

Fig. 4.1 The economic defendability of a resource is predicted to increase with: (a) decreasing competitor density (circles represent territory owners defending against intruders); (b) increasing resource density (dots represent units of a renewing resource); (c) increasing spatial clumping of resources (dots as in b); (d) decreasing temporal clumping of resources; (e) increasing spatial predictability of resources (intensity of shading represents resource density); or (f) temporal predictability of resources (shading as in e). Economic defendability will eventually decrease at extremely low levels of competitor density or extremely high levels of resource density or spatial clumping.

Table 4.2 Effects of five ecological factors on whether or not territories are defended or on the effectiveness of territorial defence

Species	Description	Resource[1]/ Type[2]	References
(a) Competitor density			
Plecoglossus altivelis	Territorial at low density, schooling at high density	F/FO	Kawanabe (1969)
Poecilia reticulata	Dominant males monopolise a patch longer at intermediate densities	F/LE	Magurran and Seghers (1991)
Pseudolabrus celiodotus	% of interactions with conspecifics that are aggressive decreases with density	F/FO	Jones (1983)
Thalassoma bifasciatum	Success of large, territorial males declines with increasing population size	M/FO	Warner and Hoffman (1980a,b)
Trichogaster trichopterus	Aggression per fish decreases with increasing density	F/LE	Syarifuddin and Kramer (1996)
Reef fishes	Schooling fish successfully intrude on territories, whereas solitary fish are excluded	F/FO	Barlow (1974b), Robertson et al. (1976), Foster (1985)
(b) Resource density			
Oryzias latipes	Cease defence when food is superabundant.	F/LE	Magnuson (1962)
Oryzias latipes	Non-aggressive fish grow faster than aggressive fish when food is superabundant and spatially clumped	F/LE	Ruzzante and Doyle (1993)
Cichlid fishes	Territorial defence is most intense at intermediate densities	F/LE	Wyman and Hotaling (1988)
(c) Spatial clumping of resources			
Cichlasoma nigrofasciatum	Defence and monopolisation are more effective when food is clumped than dispersed	F/LE	Grant and Guha (1993)
Elassoma evergladei	Territorial when food is clumped, nonterritorial when dispersed	F/LE	Rubenstein (1981)
Oryzias latipes	Territorial when food is clumped, nonterritorial when dispersed	F/LE	Magnuson (1962)
Oncorhynchus kisutch	Solitary and aggressive when food is clumped, otherwise schooling	F/LE	Ryer and Olla (1991)
(d) Temporal clumping of resources			
Brachydanio rerio	Effectiveness of defence and resource monopolisation decrease as food arrives more synchronously	F/LE	Grant and Kramer (1992)
Oryzias latipes	Frequency of aggression and variance of male mating success are higher when females arrive asynchronously rather than synchronously	M/LE	Grant et al. (1995)
Oryzias latipes	Frequency of aggression, resource monopolisation, and variance in fitness are higher when food arrives asynchronously rather than synchronously	F/LE	Bryant and Grant (1995)
(e) Spatial predictability of resources			
Cichlasoma nigrofasciatum	Aggressiveness and resource monopolisation increase with increasing predictability of food	F/LE	Grand and Grant (1994)

[1] F = food, M = mates

[2] F = field study, L = laboratory study, O = observational study, E = experimental

1993). The studies listed in Table 4.2 collectively suggest broad support for the qualitative predictions of resource defence theory, but also point out gaps in this empirical literature. First, there have been no tests of the influence of temporal predictability of resources. Second, there is no clear evidence to support the prediction that aggressiveness initially increases with food density; fish in aquaria are aggressive even when food is scarce. Third, there have been relatively few experimental manipulations of the dispersion of mates, so it is premature to conclude that resource defence theory applies equally well to foraging and mating systems. Fourth, there have been relatively few field experiments testing resource defence theory.

Resource defence theory and tests of its predictions may help explain some of the broader patterns of territoriality that were noted in Section 4.2. Mates, spawning sites and offspring were defended more often than food (Table 4.1), perhaps because of the spatial-clumping effect (Fig. 4.1c). Mates, spawning sites and offspring are spatially clumped resources in the sense that a valuable unit of reproductive success comes in a small package that can be guarded for a short duration (see also Berglund, Chapter 9; Sargent, Chapter 11). The equivalent value of food, if it could ever be measured in fitness units (e.g. Abrahams 1993), would likely be more dispersed in space and need to be defended for much longer.

Fish defend food in coral reef habitats more often than in other habitats (Table 4.1a), probably because of the resource-density effect (Fig. 4.1b). Coral reefs are amongst the most productive habitats in the world; on average, rates of primary production are 10 times higher on reefs than in lakes or streams (Begon *et al.* 1990). Productive habitats allow animals to occupy small home ranges, which are more easily defended than large home ranges (Grant *et al.* 1992). But not all coral reef fishes defend food; diet seems to play an important role. For example, butterflyfishes that feed on coral, a spatially predictable resource, are territorial whereas those feeding on plankton, a spatially unpredictable resource, are non-territorial, shoaling species (Roberts and Ormond 1992).

Barlow (1993) observed that the defence of food is rare among freshwater fishes in the wild. Yet, these same species readily defend feeding territories when placed in aquaria with a spatially clumped and predictable food supply. Barlow concluded that food must rarely be economically defendable in the wild. Observations of guppies, *Poecilia reticulata*, support this idea. Guppies forage on benthic invertebrates and algae and are rarely aggressive in the wild. However, they readily defend small, dense patches of food in aquaria (Magurran and Seghers 1991) or when these artificial patches are placed in natural streams (A.E. Magurran and B.H. Seghers, personal communication).

So far, the decision of whether or not to defend a territory has been examined only from an environmental perspective. If a population is exposed to consistent environmental conditions over evolutionary time, then selection may either promote or diminish aggressive behaviour (e.g. Ruzzante and Doyle 1993). For instance, populations or species of salmonids living in streams are inherently more aggressive than those occupying lentic habitats (Ferguson and Noakes 1982; Taylor 1988). Presumably, drifting aquatic invertebrates in streams are denser, more spatially predictable resources than plankton in lentic habitats. In guppies, individuals from populations with low predation risk are inherently more aggressive and have weaker schooling tendencies than those from high-risk populations (Magurran and Seghers 1991). Hence, selection may restrict the behavioural options that are available to individuals.

4.4 How large an area to defend?

Given that an animal is defending a territory, an important 'decision' may be how large an area to defend. Models of optimal territory size assume that an individual is free to make this decision. In practice, this means that there must be unoccupied space between neighbouring territories (i.e. non-contiguous territories) to allow individuals to expand their territories.

4.4.1 Non-contiguous territories

Optimal territory size models attracted attention when two papers in the same issue of *The American Naturalist* (Hixon 1980; Ebersole 1980) made contradictory predictions about how energy maximisers should respond to an increase in food density (also see Dill 1978). Hixon predicted a decrease in optimal territory size, whereas Ebersole predicted an increase. Coincidentally, both authors were influenced by their observations of reef fishes defending feeding territories.

Schoener (1983) reconciled the differences between the contrasting predictions of the models of Hixon, Ebersole and many others by recognising that the predictions of an optimal territory-size model depend on the 'goal' of the territory holder (time minimisation versus energy maximisation), possible constraints on energy maximisation (time versus processing capability), and the relationship between food density and intruder pressure. Luckily, much of this complexity can be avoided for most fishes. Fish are generally energy maximisers (but see Metcalfe *et al.* 1986), because they are indeterminate growers and because fitness generally increases with body size (Ware 1982). In addition, fish rarely grow at maximum rates in the wild (but see Elliott 1984), so food intake will likely be limited by time spent foraging rather than processing capabilities. Therefore, optimal territory size for fishes should generally decrease with either an increase in food density or intruder pressure (Hixon 1980; Schoener 1983).

Of the few experimental tests of optimal territory-size models using fishes, two were by Hixon (1981) and Ebersole (1980). Hixon manipulated food availability on the territories of male surfperch, *Embiotoca jacksoni*, by covering portions of territories with nylon netting or by adding food-rich substrates to unproductive parts of other territories. Despite small sample sizes, the results were convincing

(Fig. 4.2). When food abundance was reduced, focal fish increased their territories as if trying to maintain a constant food supply. Their response to an increase in food was more conservative, perhaps hedging against unpredictable supplies in the future.

Like his model, Ebersole's (1980) study produced results opposite to those of Hixon (1980). In response to food augmentation, female damselfish, *Eupomacentrus leucostictus*, expanded their territories. Norman and Jones (1984) have questioned these results because Ebersole added high quality food in a grid pattern both inside and outside the original territory. Hence, females may have moved more widely to exploit this new and valuable food source.

4.4.2 Contiguous territories

As in other taxa, many territorial fish occupy habitats that are subdivided into a mosaic of contiguous territories (Keenleyside 1979). When a fish is removed from the mosaic, its neighbours often expand their territories to absorb the free space (Sale 1975; Nursall 1977; Larson 1980; Norman and Jones 1984). These data suggest that fish often occupy territories smaller than their non-contiguous optimum because of the territorial 'pressure' exerted by their contiguous neighbours. At extremely high densities, territories become so compressed that they form a mosaic of polygons (Barlow 1974a; Clayton 1987). Though elegant, simple models of optimal territory size are not appropriate for contiguous territories if residents are not free to expand their territories. Hence, most optimal territory-size models may not be applicable for most territorial fishes in the wild.

Only Hixon (1980) has modelled the case of the contiguous territory owner (Fig. 4.3). In his model, food density only influences territory size when it is sufficiently abundant ($> F'$, Fig. 4.3) that the optimal territory size for non-contiguous territories is smaller than that forced upon the focal fish by its neighbours.

Two elegant field studies of reef fishes have investigated the size of contiguous territories. In a factorial experiment, Norman and Jones (1984) manipulated the presence of neighbours and food density while measuring the territories of a pomacentrid fish, *Parma victoriae*. Territory size was not significantly influenced by either a 34% increase or a 50% decrease in food density. However, focal fish expanded their territories by 64% when two or more of their nearest neighbours were removed. Their experiment suggests that food density, even though increased by 34%, was less than F' (Fig. 4.3).

The territory size of pairs of butterflyfish, *Chaetodon multicinctus*, was also influenced primarily by the presence of neighbours (Tricas 1989). When adjacent contiguous pairs were removed, focal pairs expanded their territories by 160%. Tricas cleverly simulated an increase in intruder pressure by neighbours by placing live, bottled conspecifics inside the territories of focal fish. Focal pairs responded to the intruders with increased rates of aggression, but eventually decreased their

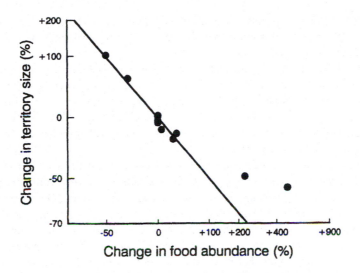

Fig. 4.2 Changes in territory size of male *Embiotoca jacksoni* were inversely related ($r = -0.942$, $n = 11$, $P < 0.01$) to changes in food abundance. The line indicates the territory sizes that would include a constant amount of food. Data are modified from Hixon (1981).

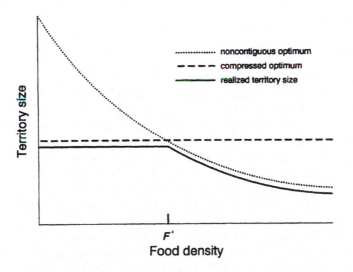

Fig. 4.3 The predicted effects of food density on the size of contiguous territories. At food densities below F', the realised territory size equals the compressed territory size (the size forced upon the occupant by its neighbours). At food densities greater than F', the realised territory size equals the noncontiguous optimum. Modified from Hixon (1980).

territories by 37%. Food alone had only modest effects on territory size, whereas a large increase in food accompanied by an increase in intruder pressure caused a 70% decrease in territory size.

Salmonid fish occupying contiguous territories show similar responses to food and intruders. Territory size decreased when intruder pressure increased (Dill *et al.* 1981; McNicol and Noakes 1984) or when an increase in food was accompanied by an increase in intruder pressure (Slaney and Northcote 1974). Observational data suggest that territory size is inversely related to food abundance (Dill *et al.* 1981; Keeley and Grant 1995).

The importance of intruder pressure to contiguous territory owners is beautifully illustrated by the mudskipper, *Boleophthalmus boddarti*. Individual male or female mudskippers occupy permanent, pentagonal territories on the intertidal mudflats of Kuwait (Clayton 1987). By depositing mouthfuls of mud at the territory boundary, they surround their territory with walls 30–40 mm in height. At low population densities, territories are non-contiguous, randomly dispersed and unwalled. At intermediate densities, walls are built only between contiguous territory owners. At high densities, territories are completely walled, contiguous, overdispersed and compressed into a pentagonal shape. Experimental removal of walls suggest that they reduce the costs of territorial defence by acting as a visual barrier between neighbouring territorial fish (Clayton 1987).

In summary, the size of contiguous territories are influenced more by the territorial pressure exerted by neighbours than by food density, as predicted by Hixon's (1980) model.

4.4.3 Concentric territories

Models of optimal territory size assume that residents defend a single territory boundary against a single type of intruder, presumably conspecifics. This simplifying assumption has been known to be false ever since Howard (1920) devoted a chapter of his book to interspecific territoriality in birds. Yet the idea that 'territoriality is primarily an intraspecific affair' (Brown and Orians 1970) has persisted, perhaps because the study of territorial behaviour has been dominated by bird behavioural ecologists. Most bird territories at least partly serve a reproductive function, which is by definition primarily intraspecific. As late as 1971, Murray argued that interspecific territoriality in birds is non-adaptive, resulting from errors in species recognition.

Reef-fish biologists bring a dramatically different perspective to the study of territorial behaviour. For example, damselfish regularly defend their territories against 20 species of intruders, with conspecifics accounting for less than 10% of territorial encounters (Low 1971; Thresher 1976). In addition, the resident defends a characteristic territory radius for each species of intruder (Myrberg and Thresher 1974; Thresher 1976). This phenomenon only became obvious when Myrberg and Thresher (1974) measured the maximum distance of attack towards live intruders

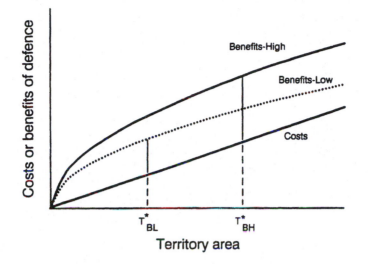

Fig. 4.4 A basic optimal territory-size model modified to explain concentric territories. When the benefits of excluding an intruder are higher (e.g. higher diet overlap with the resident), the optimal territory size increases from T^*_{BL} to T^*_{BH}.

in glass bottles that were slowly moved toward the centre of the territory. This 'intruder-in-a-bottle' technique permits the accurate measurement of the territory radius for a wide range of intruding species. Concentric or serial (*sensu* Myrberg and Thresher 1974) territories (a series of concentric territory boundaries against different intruders) seem to be ubiquitous among reef fishes.

A simple modification of a basic optimal territory-size model can explain concentric territoriality (Fig. 4.4). Optimal territory size increases as the benefits of excluding an intruder increase. For example, a positive correlation between territory radius and dietary overlap between the intruder and defender (Fig. 4.5a; also see Thresher 1976) is consistent with the model.

Reef-fish biologists more commonly measure the percentage of intruders that are chased rather than territory radius. The percentage of intruders chased increases with increasing diet overlap up to about 100% for conspecifics (Low 1971; Ebersole 1977; Hixon 1981). These observations can be explained by the concentric territory concept, because a fish is usually scored as an intruder when it crosses the intraspecific boundary, usually the outermost boundary. Hence, a positive correlation between diet overlap and territory radius produces a positive correlation between diet overlap and percentage of intruders chased.

Diet overlap was not a good predictor of the size of concentric territories in *Parma microlepis* (Moran and Sale 1977). Instead, territory radius was positively correlated with the similarity of microhabitat use between species (Fig. 4.5b). *P. microlepis* spends most of its time within 20 cm of the substratum, often within a

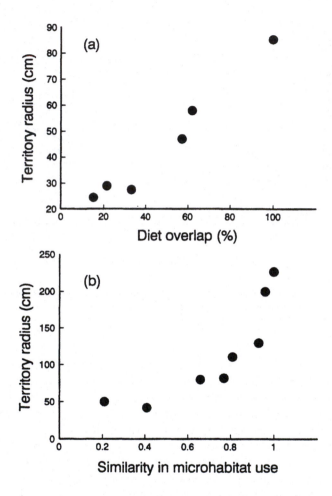

Fig. 4.5   The radii of all-purpose territories defended by (a) dusky damselfish, *Eupomacentrus dorsopunicans*, increase as its diet overlap with intruding species increases ($r_s = 0.94$, $P = 0.005$; data modified from Mahoney 1981), and (b) *Parma microlepis* increase as its overlap in microhabitat use with intruding species increases ($r_s = 0.98$, $P < 0.001$; from Moran and Sale 1977).

rock crevice. It defends a large territory against other crevice-dwelling fishes, but has a smaller territory radius towards fish that are consistently off the substratum.

The concentric territory-size model also applies to the defence of other resources, such as offspring. Optimal territory size increases with the threat that a predator poses to the brood. The biparental cichlid fish *Cichlasoma maculicauda* defends a series of concentric territories in a manner consistent with the model (Perrone 1978). During the egg stage of parental care, they defend a larger territory against potential egg predators than against *Cichla ocellaris*, a cichlid that attacks only fry. Parents increase their territory radius towards *C. ocellaris* during the fry stage.

Furthermore, larger *C. ocellaris* attack fry from greater distances, and parents adjust their defence radii accordingly (Perrone 1978).

Concentric territoriality has not been widely reported in other taxonomic groups. However, the defence of dominions (a decrease in the exclusiveness of defence with increasing distance from the centre of the territory) is common in birds (Wittenberger 1981). I suspect some of these dominions are really cases of concentric territoriality. A definitive answer may require a modification of the intruder-in-a-bottle technique that is suitable for birds.

4.4.4 Allometry of territory size

Ecologists have traditionally approached the question of how large an area to defend from an interspecific, allometric perspective. Unlike for birds and mammals, there has been no broad comparative study of space use in fishes. T.C. Grand and J.W.A. Grant (unpublished data) have completed a preliminary study for fishes. For each of seven families, they calculated mean body length (L) and territory area (A) for territories that have either a feeding function (feeding or all-purpose) or are exclusively for reproduction (spawning or parental). Territory size increased with body length (Fig. 4.6a; least squares regression: $\log_{10}A = 4.30 \log_{10} L - 8.43$, $r^2 = 0.581$, $P = 0.0015$; model II regression: $\log_{10} A = 5.64 \log_{10}L - 11.19$). Assuming that body mass (M) scales as $L^{3.09}$ (Roff 1988), territory size scales as $M^{1.83}$ (5.64/3.09). Hence, space use apparently scales with body mass at a faster rate than does metabolic rate ($\approx M^{0.75}$) in fishes, as it does in birds (Schoener 1968) and mammals (Harestad and Bunnell 1979).

To compare the sizes of feeding versus reproductive territories, Grand and Grant used an analysis of covariance (Fig. 4.6a). On average, feeding territories were 11.0 times larger than territories exclusively for reproduction ($F = 19.24$, $P = 0.001$). Feeding territories may be larger than reproductive territories for two reasons. First, as suggested earlier, food may be less spatially clumped than mates when measured in fitness units. Second, feeding territories are often permanent, whereas reproductive territories are usually seasonal. Permanent territories should be larger than seasonal territories to guard against fluctuations in resource abundance over time.

Because of indeterminate growth in fishes, the allometry of territory size can also be studied intraspecifically. Stream-dwelling salmonids are the best studied group. Slopes of the territory size regressions vary between 2.24 and 3.91 when plotted against body length (Fig. 4.6b), and between 0.82 and 1.12 when plotted against body mass (Keeley and Grant 1995). Breaks in the Atlantic salmon, *Salmo salar*, and brown trout, *S. trutta*, regressions occur because of ontogenetic shifts in habitat use and morphology, respectively (Keeley and Grant 1995). Metabolic rate in juvenile salmonids scales to body mass by exponents between 0.78 and 0.97 (Brett 1965). Hence, unlike other taxonomic groups, data for salmonid fishes are consistent with the hypothesis that the allometric increase in territory size is to meet their increasing energetic requirements.

Salmonid territories are five and four orders of magnitude smaller than the territories of birds and the home ranges of lizards of a comparable mass, respectively (Grant and Kramer 1990). Similarly, coral reef fishes occupy home ranges that are at least one order of magnitude smaller than those occupied by birds, mammals

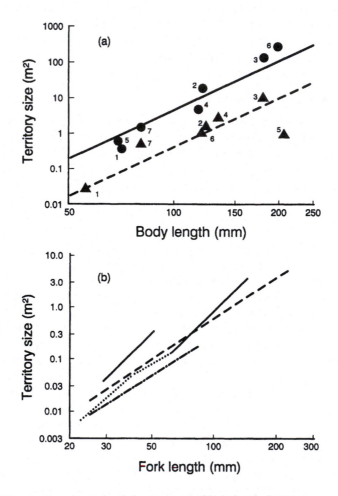

Fig. 4.6   Mean territory size, A, in relation to mean total body length, L, (a) among seven families of fishes when the territory is used for feeding (●, solid line: $\log_{10}A = 4.56 \log_{10}L - 8.44$) or exclusively for reproduction (▲, dotted line: $\log_{10}A = 4.56 \log_{10}L - 9.49$) (1 = Blennidae, 2 = Cichlidae, 3 = Labridae, 4 = Pomacentridae, 5 = Salmonidae, 6 = Serranidae, 7 = Tripterygiidae) (data from T.C. Grand and J.W.A. Grant, unpublished), and (b) among stream salmonids: solid line = Atlantic salmon (Keeley and Grant 1995); dotted line = brown trout (Elliott 1990); dashed line = interspecific regression (Grant and Kramer 1990); dash-dotted line = brook trout (Grant *et al.* 1989). Figure modified from Keeley and Grant (1995).

or lizards (Sale 1978). These results suggest that benthic fishes occupy less space than most other vertebrates of similar mass. Alternatively, the unique ecological conditions experienced by salmonid and coral-reef fishes may account for their small space requirements. Salmonids feed on drifting insects that flow over their territories, thus permitting small home ranges in running water. Coral reef fishes may also be able to occupy small home ranges because of the productivity of their environment (see Section 4.3). Whether the small space requirements of salmonid and coral-reef fishes can be generalised for all fishes awaits a thorough review of space use in fishes and other ectotherms.

4.5 Ecological consequences of territoriality

4.5.1 For individuals

In Sections 4.3 and 4.4, I have implicitly assumed that fish gain benefits and incur costs by defending territories. Here, I briefly review the evidence supporting these assumptions.

There is evidence that some territorial fish have higher foraging success, mating success, survivorship and offspring survival, and suffer lower rates of predation than non-territorial fish (Table 4.3a). Unfortunately, some of these comparisons are not ideal because territorial fish are often larger, older or more dominant than non-territorial fish. For example, territorial male bluegill sunfish, *Lepomis macrochirus*, experience higher mating success but mature later in life than non-territorial males, so the overall fitness of the two strategies appears to be equal (Gross 1991). Table 4.3a shows only that the short-term benefits gained by territorial individuals exceed that of non-territorial individuals.

Territorial fish also pay higher short-term time, energy, injury and opportunity costs than non-territorial individuals (Table 4.3b). I could find no clear example that an increased risk of predation is a cost of territoriality. However, the reduction or absence of red nuptial coloration in breeding male threespine sticklebacks in some populations is thought to be a response to intense predation by fish (Semler 1971).

Territoriality also promotes resource monopolisation, the uneven distribution of resources among individuals. When the resource is food, monopolisation leads to variation in body mass that typically increases with time as the cohort grows (e.g. Thorpe *et al.* 1992). Social behaviour apparently plays a major causal role in this phenomenon (Magnuson 1962; Thorpe *et al.* 1992), which Ricker (1975) called growth depensation.

Aquaculturists are interested in maximising the efficient production of fish of a relatively uniform size. Hence, they try to minimise aggressive or territorial behaviour that wastes energy, causes damage to subordinates, and promotes growth depensation. Indeed, there is a fear that artificial selection for growth in aquaculture conditions will also select for aggressive fish (Doyle and Talbot 1986). An obvious solution is to provide conditions that render territoriality uneconomical

Table 4.3 Benefits (a) and costs (b) of territoriality in fishes

Description of benefit or cost	References
(a) *Benefits*	
Foraging	
The biomass and production of algae was higher inside than outside damselfish (Pomacentridae) territories	Vine (1974), Brawley and Adey (1977)
Territorial salmonids (Salmonidae) fed more frequently than nonterritorial fish	Puckett and Dill (1985), Grant (1990)
Mating	
Territorial bluehead wrasse,*Thalassoma bifasciatum*, pupfish, *Cyprinodon pecosensis*, and bluegill sunfish, *Lepomis macrochirus*, spawned more frequently and successfully than their nonterritorial competitors	Warner *et al.* (1975), Kodric-Brown (1986), Gross (1991)
Offspring survival	
Predation of offspring was higher when guarding male mottled sculpins, *Cottus bairdi*, or one or both cichlid (Cichlidae) parents were removed	Downhower and Yost (1976) Keenleyside (1978), Perrone (1978)
Survival	
Only brown trout, *Salmo trutta*, with territories survived the first 1–2 months after emergence from gravel nests	Elliott (1990)
The number of spinyhead blennies, *Acanthemblemaria spinosa*, occupying a coral head was limited by the number of shelter holes, which were defended vigorously	Buchheim and Hixon (1992)
Antipredator	
Only 8% of territorial Atlantic salmon, *Salmo salar*, were eaten by a predator compared to 26% of nonterritorial fish	Symons (1974)
(b) *Costs*	
Time	
Time spent on agonistic activity by territorial coho salmon, *Oncorhynchus kisutch*, was higher than nonterritorial fish but lower than floaters	Puckett and Dill (1985)
Energy	
Liver glycogen levels decrease with the duration of stickleback, *Gasterosteus aculeatus*, fights	Chellappa and Huntingford (1989)
Agonistic interactions were the most costly activity (energy/time) in the daily energy budget of a juvenile coho salmon	Puckett and Dill (1985)
Injury	
Aggression caused dorsal fin damage in hatchery-reared salmonids (Salmonidae)	Abbott and Dill (1985)
Amount of injury to scales and fins increased with the degree of escalation of fights between cichlid fish, *Nannacara anomala*	Enquist *et al.* (1990)
Opportunity	
Male fathead minnows, *Pimephales promelas*, and rockbass, *Ambloplites rupestris*, lost weight while defending broods	Unger (1983), Sabat (1994)

(e.g. Fig. 4.1). Under such conditions, aggressive individuals do not monopolise a large share of the food (Bryant and Grant 1995; Grand and Grant 1994; Grant and Guha 1993), growth depensation is low (Ryer and Olla 1995), and hence selection for growth does not select for aggression (Ruzzante and Doyle 1993; Ruzzante 1994).

4.5.2 For populations

At the level of populations, territoriality may limit local density or even population size (Wittenberger 1981). Perhaps the best illustration of these effects have been documented for stream-dwelling salmonids. In juvenile brown trout, intense density-dependent mortality occurs during a critical period that lasts for 1–2 months after the fry emerge from gravel nests (Elliott 1990). A relatively constant number of territorial fish survive this critical period each year, but few non-territorial fish survive.

If space is the factor limiting salmonid populations in streams, then the maximum density of a cohort will decrease as the individuals grow, as predicted by the inverse of an allometric territory size regression (e.g. Fig. 4.6b). Hence, the inverse of the territory size–body size regression may act as a self-thinning line (Lonsdale 1990). Such a line is a good predictor of the maximum density of salmonids in streams and of the occurrence of density-dependent survival, growth and emigration (Grant and Kramer 1990).

4.6 Conclusions

Little is known about the behaviour of deep-water fishes in lakes or oceans. Fortunately, most fish species and biomass are concentrated in shallow-water habitats (Bone and Marshall 1982). Nevertheless, until better observational techniques are developed, the behavioural ecology of fishes will necessarily be the behavioural ecology of shallow-water fishes only.

Whether or not fishes defend territories at all is influenced primarily by intruder pressure, resource density and resource dispersion. As in other taxa (for a review, see Huntingford and Turner 1987), fish defend spawning sites, mates or offspring more often than food, perhaps because reproductive resources are more spatially clumped and defendable than food. However, feeding territories are more common among coral reef fishes than among the freshwater fishes of Canada or the marine fishes of eastern Canada. Coral reefs are productive habitats that may allow individuals to occupy small, permanent, economically defendable home ranges. Many freshwater fishes that do not defend food in the wild do so when it is presented in a spatially clumped, temporally dispersed or spatially predictable manner in the laboratory.

Fishes and other animals often defend contiguous territories. Hence, most optimal territory-size models will not be broadly applicable for many territorial animals in the wild. Although little theoretical work has addressed the optimal size of contiguous territories, experimental work suggests that territorial pressure exerted by neighbours plays a greater role than food abundance in regulating territory size. Furthermore, reef fishes often defend a series of concentric territory boundaries against different species of intruders. Territory radius is positively correlated with the threat the intruder poses to the resources being defended. This phenomenon has not been widely reported in other taxa and deserves investigation.

Comparative studies of space use in fishes are lacking. The few empirical studies to date suggest that fishes occupy relatively small territories or home ranges compared to birds, mammals or lizards.

Acknowledgements

I am particularly grateful to Miles H.A. Keenleyside for introducing me to the study of fish behaviour and providing me with a role model for the worlds of science and everyday life. My research is supported by grants from the Natural Sciences and Engineering Research Council of Canada and the Fonds pour la Formation de Chercheurs et l'Aide à la Recherche of Québec. Michael Bryant, Jean-Guy Godin, Tamara Grand, Lawrence Green, Ernest Keeley, Donald Kramer, two anonymous referees and the Montreal Behavioural Ecology Group made many helpful comments on an earlier version of this chapter.

References

Abbott, J.C. and Dill, L.M. (1985). Patterns of aggressive attack in juvenile steelhead trout (*Salmo gairdneri*). *Can. J. Fish. Aquat. Sci.*, **42**, 1702–1706.

Abrahams, M.V. (1993). The trade-off between foraging and courting in male guppies. *Anim. Behav.*, **45**, 673–681.

Barlow, G.W. (1974a). Hexagonal territories. *Anim. Behav.*, **22**, 876–878.

Barlow, G.W. (1974b). Extraspecific imposition of social grouping among surgeonfishes (Pisces: Acanthuridae). *J. Zool. Lond.*, **174**, 333–340.

Barlow, G.W. (1993). The puzzling paucity of feeding territories among freshwater fishes. *Mar. Behav. Physiol.*, **23**, 155–174.

Begon, M., Harper, J.L. and Townsend, C.R. (1990). *Ecology. Individuals, populations and communities*, 2nd edn. Blackwell Scientific Publ., Oxford.

Bone, Q. and Marshall, N.B. (1982). *Biology of fishes*. Blackie and Son Limited, Glasgow.

Brawley, S.H. and Adey, W.H. (1977). Territorial behavior of threespot damselfish (*Eupomacentrus planifrons*) increases reef algal biomass and productivity. *Environ. Biol. Fish.*, **2**, 45–51.

Breder, C.M. Jr. and Rosen, D.E. (1966). *Modes of reproduction in fishes*. The Natural History Press, Garden City, NJ.

Brett, J.R. (1965). The relation of size to rate of oxygen consumption and sustained swimming speed of sockeye salmon (*Oncorhynchus nerka*). *J. Fish. Res. Board Can.*, **22**, 1491–1501.

Brown, J.A. and Green, J.M. (1976). Territoriality, habitat selection, and prior residency in underyearling *Stichaeus punctatus* (Pisces: Stichaeidae). *Can. J. Zool.*, **54**, 1904–1907.

Brown, J.L. (1964). The evolution of diversity in avian territorial systems. *Wilson Bull.*, **76**, 160–169.

Brown, J.L. and Orians, G.H. (1970). Spacing patterns in mobile animals. *A. Rev. Ecol. Syst.*, **1**, 239–262.

Bryant, M.J. and Grant, J.W.A. (1995). Resource defence, monopolisation and variation of fitness in groups of female Japanese medaka depend on the synchrony of food arrival. *Anim. Behav.*, **49**, 1469–1479.

Buchheim, J.R. and Hixon, M.A. (1992). Competition for shelter holes in the coral-reef fish *Acanthemblemaria spinosa* Metzelaar. *J. Exp. Mar. Biol. Ecol.*, **164**, 45–54.

Chellappa, S. and Huntingford, F.A. (1989). Depletion of energy reserves during reproductive aggression in male three-spined sticklebacks, *Gasterosteus aculeatus* L. *J. Fish Biol.*, **35**, 315–316.

Clayton, D.A. (1987). Why mudskippers build walls. *Behaviour*, **102**, 185–195.

Dill, L.M. (1978). An energy-based model of optimal feeding-territory size. *Theor. Popul. Biol.*, **14**, 396–429.

Dill, L.M., Ydenberg, R.C. and Fraser, A.H.G. (1981). Food abundance and territory size in juvenile coho salmon (*Oncorhynchus kisutch*). *Can. J. Zool.*, **59**, 1801–1809.

Downhower, J.F. and Yost, R. (1976). The significance of male parental care in the mottled sculpin (*Cottus bairdi*). *Amer. Zool.*, **17**, 936.

Doyle, R.W. and Talbot, A.J. (1986). Artificial selection on growth and correlated selection on competitive behaviour in fish. *Can. J. Fish. Aquat. Sci.*, **43**, 1059–1064.

Ebersole, J.P. (1977). The adaptive significance of interspecific territoriality in the reef fish *Eupomacentrus leucostictus*. *Ecology*, **58**, 914–920.

Ebersole, J.P. (1980). Food density and territory size: an alternative model and a test on the reef fish *Eupomacentrus leucostictus*. *Am. Nat.*, **115**, 492–509.

Elliott, J.M. (1984). Growth, size, biomass and production of young migratory trout *Salmo trutta* in a Lake District stream, 1966–83. *J. Anim. Ecol.*, **53**, 979–994.

Elliott, J.M. (1990). Mechanisms responsible for population regulation in young migratory trout, *Salmo trutta*. III. The role of territorial behaviour. *J. Anim. Ecol.*, **59**, 803–818.

Emlen, S.T. and Oring, L.W. (1977). Ecology, sexual selection, and the evolution of mating systems. *Science*, **197**, 215–223.

Enquist, M., Leimar, O., Ljungberg, T., Mallner, Y. and Segerdahl, N. (1990). A test of the sequential assessment game: fighting in the cichlid fish *Nannacara anomala*. *Anim. Behav.*, **40**, 1–14.

Ferguson, M.M. and Noakes, D.L.G. (1982). Genetics of social behaviour in charrs (*Salvelinus* species). *Anim. Behav.*, **30**, 128–134.

Foster, S.A. (1985). Group foraging by a coral reef fish: a mechanism for gaining access to defended resources. *Anim. Behav.*, **33**, 782–792.

Grand, T.C. and Grant, J.W.A. (1994). Spatial predictability of food influences its monopolisation and defence by juvenile convict cichlids. *Anim. Behav.*, **47**, 91–100.

Grant, J.W.A. (1990). Aggressiveness and the foraging behaviour of young-of-the-year brook charr (*Salvelinus fontinalis*). *Can. J. Fish. Aquat. Sci.*, **47**, 915–920.

Grant, J.W.A. (1993). Whether or not to defend? The influence of resource distribution. *Mar. Behav. Physiol.*, **23**, 137–153.

Grant, J.W.A. and Guha, R.T. (1993). Spatial clumping of food increases its monopolisation and defense by convict cichlids, *Cichlasoma nigrofasciatum. Behav. Ecol.*, **4**, 293–296.

Grant, J.W.A. and Kramer, D.L. (1990). Territory size as a predictor of the upper limit to population density of juvenile salmonids in streams. *Can. J. Fish. Aquat. Sci.*, **47**, 1724–1737.

Grant, J.W.A. and Kramer, D.L. (1992). Temporal clumping of food arrival reduces its monopolisation and defence by zebrafish, *Brachydanio rerio. Anim. Behav.*, **44**, 101–110.

Grant, J.W.A., Noakes, D.L.G. and Jonas, K.M. (1989). Spatial distribution of defence and foraging in young-of-the-year brook charr, *Salvelinus fontinalis. J. Anim. Ecol.*, **58**, 773–784.

Grant, J.W.A., Chapman, C.A. and Richardson, K.S. (1992). Defended versus undefended home range size of carnivores, ungulates and primates. *Behav. Ecol. Sociobiol.*, **31**, 149–161.

Grant, J.W.A., Bryant, M.J. and Soos, C.E. (1995). Operational sex ratio, mediated by synchrony of female arrival, alters the variance of male mating success in Japanese medaka. *Anim. Behav.*, **49**, 367–375

Gross, M.R. (1991). Evolution of alternative reproductive strategies: frequency dependent sexual selection in male bluegill sunfish. *Phil. Trans. R. Soc. Lond. B*, **332**, 59–66.

Harestad, A.S. and Bunnell, F.L. (1979). Home range and body weight—a reevaluation. *Ecology*, **60**, 389–402.

Harvey, P.H. and Pagel, M.D. (1991). *The comparative method in evolutionary biology*. Oxford University Press, Oxford.

Hixon, M.A. (1980). Food production and competitor density as the determinants of feeding territory size. *Am. Nat.*, **115**, 510–530.

Hixon, M.A. (1981). An experimental analysis of territoriality in the California reef fish *Embiotoca jacksoni* (Embiotocidae). *Copeia*, **1981**, 653–665.

Howard, H.E. (1920). *Territory in bird life*. John Murray, London.

Huntingford, F.A. and Turner, A.K. (1987). *Animal conflict*. Chapman and Hall, London.

Jones, G.P. (1983). Relationship between density and behaviour in juvenile *Pseudolabrus celiodotus* (Pisces: Labridae). *Anim. Behav.*, **31**, 729–735.

Kawanabe, H. (1969). The significance of social structure in production of the 'ayu', *Plecoglossus altivelis*. In *Symposium on salmon and trout in streams* (ed. T.G. Northcote), pp. 243–251. Institute of Fisheries, University of British Columbia, Vancouver.

Keeley, E.R. and Grant, J.W.A. (1995). Allometric and environmental correlates of territory size in juvenile Atlantic salmon (*Salmo salar*). *Can. J. Fish. Aquat. Sci.*, **52**, 186–196.

Keenleyside, M.H.A. (1978). Parental care behavior in fishes and birds. In *Contrasts in behavior* (eds. E.S. Reese and F.J. Lighter), pp. 3–29. John Wiley and Sons, New York.

Keenleyside, M.H.A. (1979). *Diversity and adaptation in fish behaviour*. Springer-Verlag, Berlin.

Kodric-Brown, A. (1986). Satellites and sneakers: opportunistic male breeding tactics in pupfish (*Cyprinodon pecosensis*). *Behav. Ecol. Sociobiol.*, **19**, 425–432.

Larson, R.J. (1980). Influence of territoriality on adult density in two rockfishes of the genus *Sebastes. Mar. Biol.*, **58**, 123–132.

Lonsdale, W.M. (1990). The self-thinning rule: dead or alive? *Ecology*, **71**, 1373–1388.

Lorenz, K.Z. (1952). *King Solomon's ring*. Crowell, New York.

Lott, D.F. (1991). *Intraspecific variation in the social systems of wild vertebrates*. Cambridge University Press, Cambridge.

Low, R.M. (1971). Interspecific territoriality in a pomacentrid reef fish, *Pomacentrus flavicauda* Whitley. *Ecology*, **52**, 648–654.

McNicol, R.E. and Noakes, D.L.G. (1984). Environmental influences on territoriality of juvenile brook charr, *Salvelinus fontinalis*. *Environ. Biol. Fish.*, **10**, 29–42.

Magnuson, J.J. (1962). An analysis of aggressive behavior, growth, and competition for food and space in medaka (*Oryzias latipes* (Pisces, Cyprinodontidae)). *Can. J. Zool.*, **40**, 313–363.

Magurran, A.E. and Seghers, B.H. (1991). Variation in schooling and aggression amongst guppy (*Poecilia reticulata*) populations in Trinidad. *Behaviour*, **118**, 214–234.

Maher, C.R. and Lott, D.F. (1995). Definitions of territoriality used in the study of variation in vertebrate spacing systems. *Anim. Behav.*, **49**, 1581–1597.

Mahoney, B.M. (1981). An examination of interspecific territoriality in the dusky damselfish, *Eupomacentrus dorsopunicans* Poey. *Bull. Mar. Sci.*, **31**, 141–146.

Metcalfe, N.B., Huntingford, F.A. and Thorpe, J.E. (1986). Seasonal changes in feeding motivation of juvenile Atlantic salmon (*Salmo salar*). *Can. J. Zool.*, **64**, 2439–2446.

Milinski, M. and Parker, G.A. (1991). Competition for resources. In *Behavioural ecology: an evolutionary approach*, 3rd edn. (eds. J.R. Krebs and N.B. Davies), pp. 137–168. Blackwell Scientific Publ., Oxford.

Moran, M.J. and Sale, P.F. (1977). Seasonal variation in territorial response, and other aspects of the ecology of the Australian temperate Pomacentrid fish *Parma microlepis*. *Mar. Biol.*, **39**, 121–128.

Murray, B.C. (1971). The ecological consequences of interspecific territorial behavior in birds. *Ecology*, **52**, 414–423.

Myrberg, A.A. Jr. and Thresher, R.E. (1974). Interspecific aggression and its relevance to the concept of territoriality in reef fishes. *Amer. Zool.*, **14**, 81–96.

Noakes, D.L.G. (1978). Social behaviour as it influences fish production. In *Ecology of freshwater fish production* (ed. by S.D. Gerking), pp. 360–382. Blackwell Scientific Publ., Oxford.

Noble, G.K. (1939). Dominance in the life of birds. *Auk*, **56**, 263–273.

Norman, M.D. and Jones, G.P. (1984). Determinants of territory size in the pomacentrid reef fish, *Parma victoriae*. *Oecologia*, **61**, 60–69.

Nursall, J.R. (1977). Territoriality in redlip blennies (*Ophioblennius atlanticus*—Pisces: Blenniidae). *J. Zool. Lond.*, **182**, 205–223.

Oring, L.W. (1982). Avian mating systems. In *Avian biology*, Vol. 6 (eds. D.S. Farner and J.R. King), pp. 1–92. Academic Press, New York.

Perrone, M. (1978). The economy of brood defence by parental cichlid fishes, *Cichlasoma maculicauda*. *Oikos*, **31**, 137–141.

Puckett, K.J. and Dill, L.M. (1985). The energetics of feeding territoriality in juvenile coho salmon (*Oncorhynchus kisutch*). *Behaviour*, **92**, 97–111.

Ricker, W.E. (1975). Computation and interpretation of biological statistics of fish populations. *Bull. Fish. Res. Board Can.*, **191**, 382 pp.

Roberts, C.M. and Ormond, R.F.G. (1992). Butterflyfish social behaviour, with special reference to the incidence of territoriality: a review. *Environ. Biol. Fish.*, **34**, 79–93.

Robertson, D.R., Sweatman, H.P.A., Fletcher, E.A. and Cleland, M.G. (1976). Schooling as a mechanism for circumventing the territoriality of competitors. *Ecology*, **57**, 1208–1220.

Roff, D.A. (1988). The evolution of migration and some life history parameters in marine fishes. *Environ. Biol. Fish.*, **22**, 133–146.

Rubenstein, D.I. (1981). Population density, resource patterning, and territoriality in the everglades pygmy sunfish. *Anim. Behav.*, **29**, 155–172.

Ruzzante, D.E. (1994). Domestication effects on aggressive and schooling behavior in fish. *Aquaculture*, **120**, 1–24.

Ruzzante, D.E. and Doyle, R.W. (1993). Evolution of social behavior in a resource-rich, structured environment: selection experiments with medaka (*Oryzias latipes*). *Evolution*, **47**, 456–470.

Ryer, C.H. and Olla, B.L. (1991). Agonistic behavior in a schooling fish: form, function and ontogeny. *Environ. Biol. Fish.*, **31**, 355–363.

Ryer, C.H. and Olla, B.H. (1995). The influence of food distribution upon the development of aggressive and competitive behaviour in juvenile chum salmon, *Oncorhynchus keta*. *J. Fish Biol.*, **46**, 264–272.

Sabat, A.M. (1994). Costs and benefits of parental effort in a brood-guarding fish (*Ambloplites rupestris*, Centrarchidae). *Behav. Ecol.*, **5**, 195–201.

Sale, P.F. (1975). Patterns of use of space in a guild of territorial reef fishes. *Mar. Biol.*, **29**, 89–97.

Sale, P.F. (1978). Reef fishes and other vertebrates: a comparison of social structures. In *Contrasts in behavior* (eds. E.S. Reese and F.J. Lighter), pp. 313–346. John Wiley and Sons, New York.

Schoener, T.W. (1968). Sizes of feeding territories among birds. *Ecology*, **49**, 123–141.

Schoener, T.W. (1983). Simple models of optimal feeding-territory size: a reconciliation. *Am. Nat.*, **121**, 608–629.

Scott, W.B. and Crossman, E.J. (1979). Freshwater fishes of Canada. *Bull. Fish. Res. Board Can.*, **184**, 966 pp.

Scott, W.B. and Scott, M.G. (1988). Atlantic fishes of Canada. *Can. Bull. Fish. Aquat. Sci.*, **219**, 731 pp.

Semler, D.E. (1971). Some aspects of adaptation in a polymorphism for breeding colours in the threespined stickleback (*Gasterosteus aculeatus*). *J. Zool., Lond.*, **165**, 291–302.

Slaney, P.A. and Northcote, T.G. (1974). Effects of prey abundance on density and territorial behavior or young rainbow trout (*Salmo gairdneri*) in laboratory stream channels. *J. Fish. Res. Board Can.*, **31**, 1201–1209.

Sokal, R.R. and Rohlf, F.J. (1981). *Biometry*, 2nd edn. W.H. Freeman, San Francisco.

Syarifuddin, S. and Kramer, D.L. (1996). The effect of group size on space use and aggression at a concentrated food source in blue gouramis, *Trichogaster trichopterus* (Pisces: Belontiidae). *Environ. Biol. Fish*, **46**, 289–296.

Symons, P.E.K. (1974). Territorial behavior of juvenile Atlantic salmon reduces predation risk by brook trout. *Can. J. Zool.*, **52**, 677–679.

Taylor, E.B. (1988). Adaptive variation in rheotactic and agonistic behavior in newly emerged fry of chinook salmon, *Oncorhynchus tshawytscha*, from ocean- and stream-type populations. *Can. J. Fish. Aquat. Sci.*, **45**, 237–243.

Thorpe, J.E., Metcalfe, N.B. and Huntingford, F.A. (1992). Behavioural influences on life-history variation in juvenile Atlantic salmon, *Salmo salar*. *Environ. Biol. Fish.*, **33**, 331–340.

Thresher, R.E. (1976). Field analysis of the territoriality of the threespot damselfish, *Eupomacentrus planifrons* (Pomacentridae). *Copeia*, **1976**, 266–276.

Thresher, R.E. (1984). *Reproduction in reef fishes*. T.F.H. Publications, Neptune City, New Jersey.

Tricas, T.C. (1989). Determinants of feeding territory size in the corallivorous butterflyfish, *Chaetodon multicinctus. Anim. Behav.*, **37**, 830–841.

Unger, L.M. (1983). Nest defense by deceit in the fathead minnow, *Pimephales promelas. Behav. Ecol. Sociobiol.*, **13**, 125–130.

Vine, P.J. (1974). Effects of algal grazing and aggressive behaviour of the fishes *Pomacentrus lividus* and *Acanthus sohal* on coral-reef ecology. *Mar. Biol.*, **24**, 131–136.

Ware, D.M. (1982). Power and evolutionary fitness of teleosts. *Can. J. Fish. Aquat. Sci.*, **39**, 3–13.

Warner, R.R. (1980). The coevolution of behavioral and life-history characteristics. In *Sociobiology: beyond nature/nurture?* (eds. G.W. Barlow and J. Silverberg), pp. 151–188. Westview Press, Boulder.

Warner, R.R. and Hoffman, S.G. (1980a). Population density and the economics of territorial defence in a coral reef fish. *Ecology*, **61**, 772–780.

Warner, R.R. and Hoffman, S.G. (1980b). Local population size as a determinant of mating system and sexual composition in two tropical marine fishes (*Thalasomma* spp.). *Evolution*, **34**, 508–518.

Warner, R.R., Robertson, D.R. and Leigh, E.G. Jr. (1975). Sex change and sexual selection. *Science*, **190**, 633–638.

Wilson, E.O. (1975). *Sociobiology*. Harvard University Press, Cambridge.

Wittenberger, J.F. (1981). *Animal social behavior*. Duxbury Press, Boston.

Wyman, R.L. and Hotaling, L. (1988). A test of the model of the economic defendability of a resource and territoriality using young *Etroplus maculatus* and *Pelmatochromis subocellatus kribensis. Environ. Biol. Fish.*, **21**, 69–76.

5 *Foraging tactics*

Paul J.B. Hart

5.1 Introduction

Behavioural mechanisms or tactics of feeding have been largely neglected by behavioural ecologists. Strategic aspects of fish foraging have been studied theoretically and empirically ever since the publication of the theoretical papers of Emlen (1966) and MacArthur and Pianka (1966), which initiated the development of contemporary optimal foraging theory (Stephens and Krebs 1986; Schoener 1987). Optimal foraging theory (OFT) is a collection of mathematical models that each assume that the forager maximises some currency such as the long-term average rate of energy gain (Stephens and Krebs 1986; Fig. 5.1). This currency is in turn assumed to be correlated to Darwinian fitness (see also Hughes, Chapter 6). OFT models specify a strategy that the forager must adopt which will depend on the ambient foraging conditions. The foraging tactics observed will be the behavioural realisation of the strategy. The nature of the tactics, which have also been called modes in the literature (Helfman 1990), is the subject of this chapter. The topic of tactics deserves more attention from both theoreticians and empiricists.

It is worth making some comment on my use of the terms 'strategy' and 'tactics'. The two are often confused in the literature (Wootton 1984; Chapleau *et al.* 1988). For example, O'Brien *et al.* (1990) entitled their article 'Search Strategies of Foraging Animals'. The article describes how animals alter their searching behaviour in response to the properties of the prey. More will be said of this later. I contend that what O'Brien *et al.* are writing about are search tactics that have evolved through natural selection. As Wootton (1984) makes clear, 'strategy' and 'tactic' are two levels in an hierarchy (Fig. 5.1). Within each level, the actions labelled as 'strategy' or 'tactic' are likely to be the most appropriate to achieve the task in hand. In this sense, the term 'foraging strategy' should be used to define the species-specific foraging behaviour patterns which have evolved through natural selection to maximise individual fitness. The 'tactics of foraging' are the behavioural variations within the strategy that allow the animal to vary its behaviour in response to local conditions or to variations in prey type.

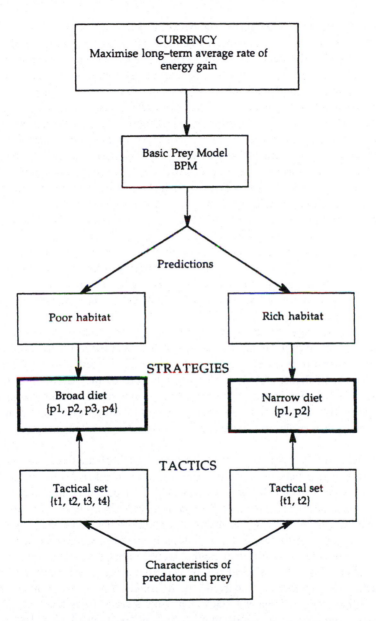

Fig. 5.1 The hierarchical nature of strategies and tactics. The diagram shows the strategic and tactical levels of diet selection behaviour as modelled by the Basic Prey Model. The foraging strategy specifies which prey items should be included in the diet, whilst the foraging tactics determine what behaviours the fish uses to handle the different prey types. The p_i signify the prey types in the diet and t_i signify the tactics used to handle the corresponding type. Any two t_i are not necessarily the same so that one could be 'swallow whole' whilst another could be 'bite and spin'.

It may be argued that the difference between tactics and strategies is semantic and everyone knows what they mean when they talk about one or the other. This is highlighted by Helfman (1990) who writes: 'Many instances of mode (tactic) switching involve the use of different modes for different prey types. Arguably, such switching is determined more by prey type than by mode type and hence can be explained by optimal diet theory ... without reference to mode switching'. The present review will try to show that in Helfman's example dealing with prey choice, the strategic goal will determine which prey types will maximise fitness but the tactics chosen will realise that goal (Fig. 5.1). OFT says little about the tactical behaviours that the forager should use to realise the optimal strategy. Evidence will be presented to show that fish alter their tactics in response to a wide variety of conditions, which presumably allow them to maximise fitness. An analogy might be the choice of tyres in a Formula One car race. The strategy is to win the race by being the fastest and first across the finishing line, but it is the tactic of the number of pit-stops and types of tyre chosen that determines which car makes it first in the context of the tactics employed by other teams and the weather conditions. A fish will choose prey types to maximise fitness, but the tactics of searching, capturing and handling will determine whether the goal is achieved (see also Hughes, Chapter 6).

Tactic and strategy are not independent. The optimal strategy may dictate that a certain subset of prey should be taken, and the calculation of the optimum is made based on the assumptions about the costs of different prey capture tactics. At present, theory does not take enough note of the variation in tactics that are possible. For example, handling time is measured without recognition of the possibility that the fish may change its tactics with prey type to minimise any increase in handling time that might be expected from an increase in prey size.

It is the primary purpose of this chapter to establish that tactical variability in fish foraging behaviour exists and to examine whether the variations have functional significance. The next section of the chapter describes the reasons for the evolution of tactics. Pike, *Esox lucius,* and perch, *Perca fluviatilis,* illustrate how species differ in their tactical flexibility when capturing prey in different ecological conditions, and a section is devoted to recent work that shows how perch are able to use different capture tactics to respond to different feeding environments. Pike are the antithesis of perch showing a one-tactic repertoire to which they are supremely adapted ('A life subdued to its instrument' in the words of the poet Ted Hughes (1982)). The pike's success has much to do with body form and its role in constraining tactics is discussed. I next describe in turn how fish have evolved tactics to effectively search for, capture and then handle prey; these being the main actions during foraging. My final section discusses the role that foraging tactics might play in population and community processes, relates tactics to fitness and attempts to identify topics for future research.

5.2 Tactics to challenge prey defences

The title of this section is designed to convey the idea that to obtain its prey the predator has to run the gauntlet of all the obstacles to success placed in its path by prey defences. For certain species, of which pike is one of the best examples, the predator has specialised in overcoming one aspect of prey defence. Other species, typified by perch, have retained greater flexibility of response. The two species illustrate how the same foraging strategy might be achieved through the employment of different sets of tactics. The pike is a specialist at stationary search and at capturing prey after an ambush. It relies on attacking prey before they have a chance to initiate evasive tactics. The perch has more flexible tactics of capture. It can choose to ambush prey or to chase them in open water. It is also able to enhance its performance in chasing by cooperating with conspecifics.

There are two parts to prey capture. A forager must first perceive and recognise objects in its habitat that are potential prey, and it must then get close enough to be able to capture and ingest them. Perception and recognition of prey are the end point of searching, and they each make heavy demands on sensory and information processing systems. After moving close to a prey, the forager requires an appropriate tactic to make a capture. To enhance its chances of encountering and detecting prey, the predator will often search actively (Pianka 1986).

Fig. 5.2 depicts details of the sequence of behaviours a predator might be expected to show, together with factors potentially influencing them (Endler 1991). There are two types of factor. Factors internal to the fish include its morphology, its capacity to learn and its current physiological state. External factors deriving from the prey are all those features that the prey has evolved to avoid and evade capture (see Smith, Chapter 7 and Godin, Chapter 8). Predation risk and risk of parasitic infection are examples of external biotic factors, whereas temperature and light are examples of abiotic external factors. All these will influence the cycle of events in Fig. 5.2. The figure does not show the capacities the predator possesses to overcome the prey's defences. There is a dynamic interaction between predator and prey through evolutionary time, with each evolving behaviours to outwit the other. In most cases, the interaction is not true coevolution but more an arms race (Endler 1991). For the prey, avoiding the attentions of the predator early in the cycle is preferable (Endler 1991) as it costs less to avoid being seen than, like Jonah, escaping only after having been eaten.

In the context of optimal foraging theory (cf. Introduction), natural selection is assumed to have 'designed' the predator to maximise its fitness over many turns of the predation cycle in Fig. 5.2. OFT has concentrated on limited aspects of the cycle, for example, whether or not to include a particular prey type in the diet, and has had little to say about the behavioural tactics the forager should use to achieve the strategy that maximises fitness. At each stage in the cycle, it is likely that foragers will have evolved one or several tactics that will overcome the prey's defences. Because the fish's foraging environment is likely to vary in

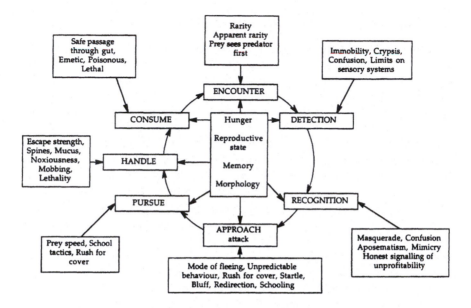

Fig. 5.2   A summary of the predation cycle, which starts when the predator encounters a prey. Items in boxes on the outside of the circle list prey features designed to make it difficult for the predator to catch the prey. Items in the box inside the circle are factors internal to the predator that influence its performance. Derived from Table 6.1 in Endler (1991).

unsystematic ways (Dill 1983), behavioural flexibility is important. This implies that variation in tactics, each tailored to cope with variants of a situation, such as different prey types and different prey escape tactics, will have different fitness payoffs. A central question in OFT should be whether variations in tactics employed at a particular stage in the predation cycle alter the strategic payoff on completion of the cycle. More directly, it can be asked whether different tactics have different fitness payoffs? This question will be returned to at the end of the chapter.

This section has shown that tactics are to be interpreted as the behavioural variations used by a predator to overcome the detection and handling problems posed by the antipredator defences of the prey. In what follows I will concentrate on three aspects of tactical variation. These are the way in which the fish's morphology is correlated with the tactics employed, searching tactics and the way in which fish vary their tactics of capturing prey.

5.3   'A life subdued to its instrument'

In this section I will expand on the example of tactical variability by pike and perch as the work illustrates the subtle way in which tactics vary between and within species. Pike and perch are common predators in temperate lakes and rivers in Europe (Maitland and Campbell, 1992) and in North America with *Perca flavescens* replacing *P. fluviatilis* (Scott and Crossman 1973). Pike eat fish from the start of their lives (Giles *et al.* 1986), whilst only large perch are piscivorous (Eklöv 1995). The foraging behaviour of pike has been studied by Hart and Connellan (1984), Hart and Hamrin (1988, 1990) and Eklöv and Hamrin (1989), whilst Eklöv (1992) and Eklöv and Diehl (1994) have studied pike and perch together.

Although pike are opportunistic predators, they can be selective in their choice of prey under certain conditions (Hart and Hamrin 1988, 1990; Eklöv and Hamrin 1989). Pike prefer rudd, *Scardinius erythrophthalmus*, to perch, but switch to perch when dense vegetation provides cover for rudd (Eklöv and Hamrin 1989). When attacking mixed sizes of rudd, pike prefer smaller fish (Hart and Hamrin 1988) and this may be a result of the faster escape abilities of larger prey (Hart and Hamrin 1990). In these variable conditions pike use mainly ambush as a capture tactic, although they may occasionally switch to chasing (personal observations).

The pike is considered one of the best examples of a species using mainly the sit-and-wait searching tactic (Webb and Skadsen 1980; Webb 1982). Perch can use the same tactic, but they are more flexible and can also search actively and chase prey when sighted. They do this best when together with conspecifics. Whether these two species can or cannot change these tactics have been studied by Eklöv (1992) and Eklöv and Diehl (1994). Eklöv (1992) examined the foraging tactics of perch and pike presented with small perch (age 0+) as prey in enclosures with no refuge. Perch and pike were stocked in each enclosure either alone or with four other conspecifics. Records were then kept of the number of prey consumed, of behaviours (predator and prey locomotion, predator attacks and predator inspection by prey) and of predator growth rate; the latter being an important component of fitness in fishes (Charnov 1993; Schluter 1994). Perch were much more active than pike. When foraging as a shoal, individual perch caught a greater number of prey than when foraging solitarily. Perch lost weight when they were alone and only grew when they were with conspecifics (Fig. 5.3A). In comparison, pike spent little time moving and if they moved they did so slowly, which is typical of sit-and-wait predators (Huey and Pianka 1981; McLaughlin 1989). When there were five pike in the enclosure, the largest was more mobile and faster than the small ones. In contrast to perch, solitary pike caught more prey and grew faster than when they were in groups. The large pike in a group of five grew faster than the other four smaller fish. For pike, being alone meant faster growth.

Under the conditions of Eklöv's (1992) experiment, the perch used social cooperation to improve its performance as a predator. Pike did better alone and they defended their ambush site. Big pike use their size to defend the best hunting sites. In Seibert Lake, Alberta, Canada, Chapman and McKay (1984) followed pike movements using radio tags. Pike covered large distances in a few hours and often ended up in a habitat different from the one from which they started. Once in a new habitat, the fish only make short moves. They chose mostly habitats that

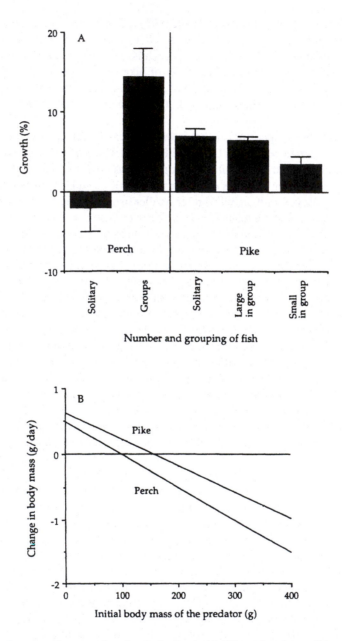

Fig. 5.3 Growth performance of perch and pike faced with different foraging conditions. (A) Perch and pike foraging in enclosures with no refuge and with either one or five conspecifics. (B) Change in body mass for perch (lower line) and pike (upper line) as a function of initial mass. The lines are described by perch: $y = -0.005x + 0.496$, $r^2 = 0.924$, pike: $y = -0.004x + 0.622$, $r^2 = 0.27$ ($n = 4$ in both cases). From Eklöv (1992) and Eklöv and Diehl (1994).

were vegetated and moved offshore on windy days when wave action is likely to have increased water turbidity. Using SCUBA diving, Eklöv (1995) observed pike movements in Degersjön, northern Sweden. He found that pike had home ranges and only moved to areas close by. Differences between the results of Eklöv (1995) and Chapman and McKay (1984) may relate to differences in prey abundance and distribution. The data on pike movement are compatible with the idea that a predator adopting a sit-and-wait tactic will take some care in its choice of a site in which to sit-and-wait. These animals are in effect sampling a patchy environment, but instead of searching for patches they wait for the aggregations of prey ('patches') to swim by. Some sites must have more aggregations passing than others.

The results of Eklöv and Diehl (1994) provide a more complete understanding of the foraging tactics used by pike and perch. Three different sized pike or perch were released together into enclosures that contained 0+ and 1+ perch as prey. In this study, each enclosure contained three small clumps of weed and one end of the area was filled with weed. The dense weed was through a net with a mesh size that was only big enough for prey individuals to move through. In this way, the prey were given an absolute refuge. With prey less available, the perch adopted a sit-and-wait tactic. They tended to aggregate in pairs in the small clumps of weed furthest from the refuge. When prey emerged into open water, the perch waited for them to come near before attacking, chasing and capturing them before they could reach the refuge. In contrast, pike took up station near the refuge and ambushed prey as they emerged. The largest individuals were closest to the refuge. In such a habitat containing an absolute refuge, perch moved much less than when in a weed-free environment. They also caught far fewer prey and were outperformed in this by pike. All sizes of predatory perch grew more slowly in the weeded habitat than did the pike (Fig. 5.3B).

The differences between the pike and perch tactics employed in different conditions of prey availability are summarised in Fig. 5.4. Despite the changes in prey availability, pike are faithful to their sit-and-wait tactic. With prey able to hide in a secure refuge, perch changed from a mobile to a sit-and-wait tactic. The poorer growth in the second experiment indicates that the sit-and-wait tactic had a lower fitness payoff than did the mobile tactic. Later, the existence of flexible search tactics in other fishes will be examined in more detail.

5.4 Body form and foraging tactics

The way in which body form influences the prey capture tactics a predator can employ was examined by Webb (1984). He observed tiger musky (a hybrid between *Esox lucius* male and *E. masquinongy* female), rainbow trout, *Oncorhynchus mykiss* , smallmouth bass, *Micropterus dolomieui,* and rock bass, *Ambloplites rupestris*, attacking fathead minnows, *Pimephales promelas*. The four predators differ significantly in morphology (Fig. 5.5) such that each species has a very different swimming capability (Table 5.1). Hydromechanical principles allowed

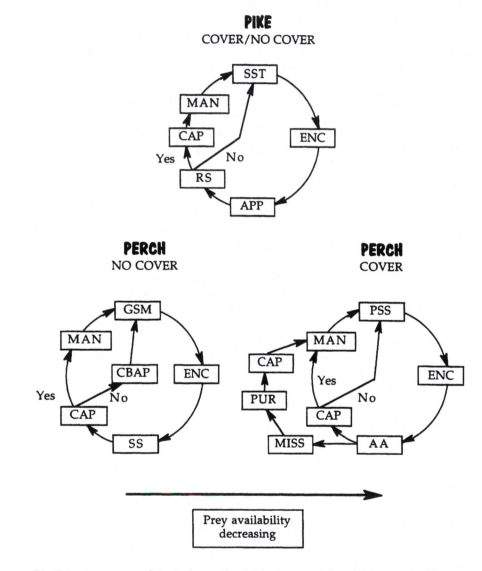

Fig. 5.4 A summary of the tactics employed by pike and perch under varying conditions of prey availability. See text for further details of the experiments forming the basis of the figure. SST = solitary stationary search, ENC = encounter, APP = approach, RS = rapid strike, CAP = capture, MAN = manipulate, GSM = group search (on the) move, SS = split school, CBAP = capture by another perch, PSS = paired stationary search (in cover), AA = ambush attack, PUR = pursuit.

Table 5.1 A comparison of the relative swimming capabilities for four teleost predators when striking at fish prey. An S-start occurs when the fish first bends its body into an S-shape before straightening out and darting forward. In a C-start the body is first bent into a C-shape. From Webb (1984)

| Characteristic | Species | | | |
	Tiger musky	Rainbow trout	Smallmouth bass	Rock bass
Kinematic pattern of fast-starts	S-start & C-start	S-start & C-start	C-start	C-start
Effectiveness of fast-start in reaching specific targets at maximum acceleration	Good	Good	Poor	Poor
Relative performance in body/caudal fin cruising and sprints(chases)	Poor	Good	Good	Poor
Relative manoeuvreability during strikes/chases and braking	Poor	Poor	Good	Good

Webb and Skadsen (1980) to predict the optimum strike tactics, which maximise the probability of an attack leading to a capture, for a predator attacking a moving fish prey. The tactics were to (1) strike at the prey's centre of mass, (2) strike the prey from the side, and (3) strike as rapidly as possible to close with the prey in the least time (see also Webb 1986). Striking in this way allows the predator to aim for the part of the prey that moves least as it accelerates away. As shown in Fig. 5.5 and Table 5.1, the tiger musky and trout have fin positions, body cross-section and body flexibility that renders them the most effective at fast starts and thus allows them to approach the required optimum strike tactics. What each predator did after missing an attacked prey highlights important aspects of their locomotive capabilities. Tiger musky would be moving so fast when it attacked that on missing it overshot the prey to such an extent that it was not able to re-cover quickly enough to give chase. Trout had the ability to turn rapidly and give chase to escaping prey, whereas the two centrarchids did not overshoot the prey and were able to anticipate prey escape moves and use their paired fins to brake, turn and give chase.

The conclusion from Webb's study is that the tiger musky is a lunge/biter that cannot follow up a miss. The trout also has the capacity to be a lunge/biter but spends most of its time chasing. The two bass species are intermediate. All, except the tiger musky, have more options open to them and so are expected to vary their prey catching tactics with prevailing conditions. As we have seen already, the pike retains its single tactic under nearly all conditions.

The structure and function of the jaw mechanism determines the tactics a fish can use to attack, capture and subjugate a prey. Fish have three main methods of capturing prey (Liem 1980); these are inertial suction which is employed by the majority of fishes, ram feeding and manipulation. Many species can use all three

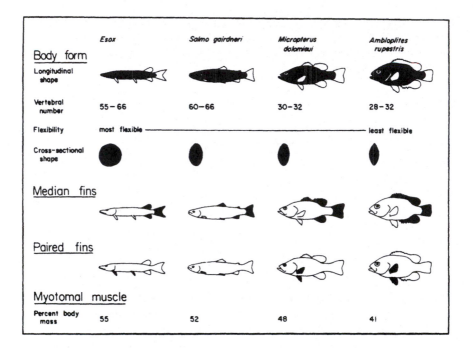

Fig. 5.5   A summary of the functional and morphological features influencing the locomotor performance of four piscivores. Note that *Salmo gairdneri* has been renamed as *Oncorhynchus mykiss*. From Webb (1984).

tactics. In inertial suction, the fish opens its mouth rapidly which creates negative pressure in the buccal cavity (Liem 1980). This causes water, and any objects in it that are close to the fish, to be sucked in. Species such as bream, *Abramis brama*, have specially modified jaws that create a tube–like structure when the mouth is opened. This enhances the suction effect and increases prey capture efficiency. Even a biting predator like the pike relies on the suction effect in the last moments preceding capture. The rapid opening of the mouth as the pike gets close to the prey drags the prey inexorably into the mouth. Many of the different mouth opening tactics observed could be learnt variants of the basic suction feeding mechanism (Wainwright 1986).

Several investigators have examined how the inertial suction mechanism of a species changes in response to different prey properties. Tiger musky vary the pattern of jaw movements in response to how close they are to the prey when capture is initiated (Rand and Lauder 1981). The suction mechanism of four species of centrarchid fish feeding on worms, crickets or fathead minnows was studied by Wainwright and Lauder (1986). They examined the variation in movement of four muscles in response to prey type using electromyography. Of the four species, the bluegill sunfish, *Lepomis macrochirus*, the black crappie, *Pomoxis nigromaculatus*, and the largemouth bass, *Micropterus salmoides*, modulated these four muscles in a pattern that differed across prey types. Rock bass showed only one pattern. The variation between species was attributable to the length of time that the sternohyoideus muscle was active. This muscle, which is in a ventral position, is responsible for pulling on the lower jaw to open the mouth. It has been shown by Liem (1980) and Wainwright and Lauder (1986) to have the greatest effect on volume change of the buccal cavity as the fish strikes at the prey.

The combination of muscle movements can be altered with foraging experience (Wainwright 1986; Galis 1993; Galis *et al.* 1994). Electromyographic recordings were taken from individuals of the pumpkinseed sunfish, *Lepomis gibbosus*, when first offered elusive guppies, *Poecilia reticulata*, as prey and after the fish had had several weeks experience feeding on them. As time passed, the pumpkinseed improved their ability to catch the guppies and did this by changing the pattern of muscle movements. The average duration and amplitude of muscle movement increased, thereby enhancing the suction effect of opening the mouth.

Variation in jaw function presumably has fitness consequences. Liem (1980) argued, on the basis of mechanical models of jaw structure and function, that the particular arrangements of the structures are optimal for prey capture. In other words, the arrangements found are the most efficient from a design point of view. As pointed out by Wainwright and Lauder (1986), no-one has yet shown that this mechanical optimisation is correlated with fitness.

A fascinating experiment by Meyer (1987) with the cichlid *Cichlasoma managuense* has shown that jaw structure is more plastic than might be supposed. Meyer (1987) describes how cichlids that bite at prey have a blunt snout, whereas those that use suction to capture food have a snout that is more pointed. Newly hatched sibling *C. managuense* were reared by Meyer (1987) in two groups. Group I was fed *Artemia* nauplii, which the fish captured by suction, and Group II flaked food and nematodes, which were captured by biting. After eight months on these respective feeding regimes, the fish in Group I had pointed snouts and those in Group II had blunt snouts. The differences were statistically significant. Both groups were then fed on *Artemia* adults for two weeks before the jaws were measured again. Even after such a short time, the snout shapes of fish in Group II had begun to converge on the shape of those in Group I. After a further eight months on the same diet, fish in the two groups had snouts that were indistinguishable in shape. *C. managuense* in nature shows an ontogenetic shift in mouth morphology from blunt (biting) to pointed (sucking), which in turn gives the fish access to different foraging niches. Meyer's experiment shows how this change could be brought about by an interplay between behaviour and jaw morphology. If this phenomenon is common in cichlids, it could have played a significant role in the evolution of the diversity found in cichlids in Lakes Victoria, Malawi and Tanganyika in East Africa (Pitcher and Hart 1995).

The earlier discussion of pike and perch foraging tactics emphasised the dichotomy between a sit-and-wait tactic and active searching. This dichotomy is not absolute (McLaughlin 1989; O'Brien *et al.* 1990) but has been emphasised in the early literature on optimal foraging theory (Schoener 1971). In fishes, it has been frequently assumed that sit-and-wait predators have mostly white muscle needed for rapid acceleration and that active searchers have a higher proportion of red muscle, which is required for sustained swimming. This assumption has been tested by McLaughlin and Kramer (1991), who analysed published data on 84 North Atlantic species, 16 marine species from Australia and 44 North American freshwater species. They argued that if there is a good association between type of muscle and movement tactics then the proportion of red muscle might be expected to show a bimodal distribution. The analysis showed that the freshwater fishes examined had a lower proportion of red muscle than did marine species. The probability of a fish being called 'mobile' as opposed to 'sedentary' increased significantly as the proportion of red muscle in the caudal peduncle increased. Despite this trend, there was no clear separation between the amount of red muscle in the two mobility groups. The data on red muscle did separate into two groups, those fish with no red muscle and those with some. The fishes without red muscle are mostly species that catch stationary prey that would not require the predator to move much. It is unlikely that these species would be able to switch to active foraging in the way some lizards do when prey such as termites become locally abundant (Huey and Pianka 1981). The group with some red muscle contained both sedentary and mobile species. McLaughlin and Kramer (1991) conclude that fish cannot be placed into discrete groups, such as sprinters versus stayers, on the basis of the amount of red muscle they have in their caudal peduncles.

It appears logical to assume that morphology will tightly constrain a fish's foraging tactics. The evidence reviewed in this section undermines this assumption. Although morphological structures set limits to tactics, such structures are not rigid and thus significant behavioural flexibility can be accommodated (Galis *et al.* 1994). The great flexibility in jaw form of *Cichlasoma managuense* is a warning that form is not necessarily fixed for life and can, like behaviour, change in response to short-term changes in the foraging environment.

5.5 Searching for prey

The inclusion of a refuge in the experiment of Eklöv and Diehl (1994) changed the search tactics used by perch to encounter prey. In this section I investigate how general such flexibility is across all foraging behaviours and examine in more detail the dichotomy between sit-and-wait search and cruise search. The function of these two tactics is to enhance the probability that prey will be encountered (Fig. 5.2). The importance of searching in the foraging cycle depends on the relative size and activity of predator and prey (O'Brien *et al.* 1990). For prey that are large relative to the predator, pursuit and handling are more critical to successful capture. When the reverse is true and prey are small relative to the predator, search forms a greater component of the time occupied by the foraging cycle.

5.5.1 Variation in movement

Young-of-the-year brook charr, *Salvelinus fontinalis,* possess two search tactics (Grant and Noakes 1987). These fish live in streams and their invertebrate prey can be found on the bottom or drifting in the current. Some individuals search specifically for benthic prey and this requires a mobile tactic. Others find a sheltered spot and wait for drifting prey to come close before darting out to attempt a capture. Many salmonids employ similar tactics. Grant and Noakes (1987) assumed that individuals were either movers or stayers and that each tactic predisposed fish to foraging in particular areas. They developed a model of encounter rate that predicted how fish employing the two tactics should behave in response to altered foraging conditions. Four predictions were made from the model:

(1) in still water, movers should have a higher foraging rate than stayers,

(2) as the current velocity in the foraging zone increases, the foraging rate of stayers on drifting prey should increase,

(3) in running water, movers should have a higher benthic foraging rate than stayers, and

(4) in running water, there should be no difference in the foraging rate between movers and stayers when feeding on drifting prey.

Observations on wild fish mostly matched these predictions, although movers shifted their foraging effort to the surface in still water (pools), where the bottom was mostly covered with fine silt in which it was harder to find prey.

Grant and Noakes (1987) did not quantify the proportion of movements made by an individual that could be assigned to either the mover or stayer tactics. This was done in a later study by McLaughlin *et al.* (1992). Working with charr in the still pools of streams in Ontario, Canada, they determined the proportion of time recently-emerged fish spent moving, how fast they moved, and the distance over which they pursued prey. Working only in pools reduced the effect of stream flow on behaviour. Individuals differed significantly in all the characteristics of movement measured, and they specialised in one of two patterns of movement, which were differentiated by the proportion of time spent moving. The different variables used to define movement were positively and significantly correlated. Fish that spent more time moving directed a greater proportion of their foraging attempts at the surface and ate more insects than crustacea. The differences in foraging tactics observed by McLaughlin *et al.* (1992) could be explained either by individuals having flexible behaviour and switching from one tactic to the other over time, or by the population containing two stable phenotypes. Their field observations did not allow for a distinction between these two explanations.

A later study by McLaughlin and Grant (1994) showed that recently-emerged charr found to be foraging in faster flowing water were larger and had a smaller body depth and larger caudal fin area than individuals in slow flowing or still water. Occupying different current speeds had dietary consequences, in that fish in faster flowing water had eaten more dipteran larvae than had individuals in slower water. Diptera are found mostly in midwater in streams and this is the zone most exploited by the fast swimming fish. Further features of this system for the same species foraging in still water are presented by McLaughlin *et al.* (1994). These studies indicate that individuals might specialise in either movement or staying, although what is cause and what effect is not clear because individuals that vary in body form might be expected to choose an appropriate habitat so that those performing best in fast flowing water will spend more time in that micro-habitat. Alternatively, observed morphological variation might be an adaptation to the mover or stayer tactics that may reduce intraspecific competition (Milinski 1984; McLaughlin *et al.* 1994). Such slight differences in morphology and behaviour, as shown by the charr, can become exaggerated by natural selection to eventually produce new species. This is well illustrated by the limnetic and benthic morphs of the threespine stickleback, *Gasterosteus aculeatus*, found in six lakes on three islands in British Columbia, Canada (McPhail 1994; Schluter 1994).

5.5.2 A continuum of search tactics

The studies just described reinforce the cruise search and sit-and-wait dichotomy for searching behaviour. A pioneering study by Ware (1978) asked if there is an optimum average speed at which a fish should search. The study was based on the assumption that planktivorous fish move continuously through the water searching always at the limit of their visual field. Observations on many fish species have subsequently revealed that this assumption of continuous movement is possibly false. A threespine stickleback in an aquarium with *Daphnia* as prey will move in short bursts interspersed with brief pauses (personal observations). O'Brien and coworkers have formalised this into a new view of search behaviour which applies to most fish and to a wide range of other taxa (Evans and O'Brien 1988; O'Brien *et al.* 1986, 1989, 1990). Studies of the search paths taken by arctic grayling, *Thymallus arcticus*, (Evans and O'Brien 1988) and white crappie, *Pomoxis annularis*, (O'Brien *et al.* 1986, 1989) looking for planktonic prey are best interpreted as a sequence of short stops during which search is interrupted by longer periods of forward movement. This pattern was called saltatory search by O'Brien *et al.* (1986).

The search cycle for white crappie is shown in Fig. 5.6. The timing of this cycle is altered in response to prey size (Table 5.2). A simulation model was developed incorporating the mechanism proposed to underlie the behavioural variation (O'Brien *et al.* 1989). The model assumed that the currency being maximised was the rate of gain of net energy and that there were three decision variables: run length, run speed and search time. Run lengths of 5.2 cm and 13.5 cm were predicted for fish feeding on small and large prey respectively; observed run lengths were 5.18 ± 0.73 for small prey and 9.8 ± 2.15 for large prey (Table 5.2) . With

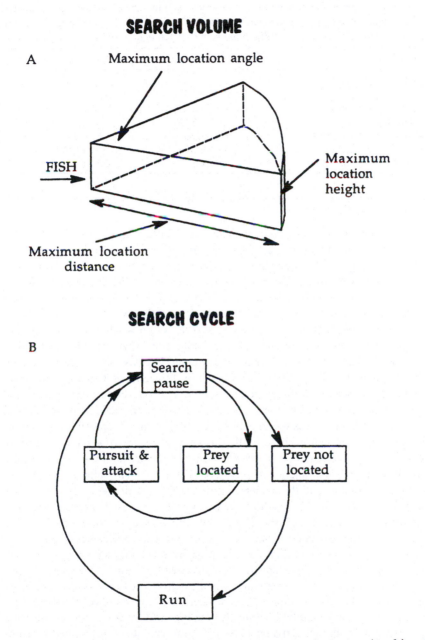

Fig. 5.6 Aspects of white crappie search behaviour. (A) A diagrammatic representation of the volume of water searched by the fish, showing some of the variables whose empirical values are given in Table 5.2. (B) The search cycle of the fish showing how the behaviour that follows search varies with whether a prey is caught or not. Redrawn from O'Brien *et al.* (1989).

Table 5.2 Changes in the timing of the search cycle and the size of the location volume of white crappie catching large and small waterfleas, *Daphnia* spp. The meaning of location height and angle is made clear in Fig. 5.6. From O'Brien *et al.* (1989)

Variable	Small prey	Large prey[1]
Location distance (cm)	8	20
Location height (cm)	4	10
Location angle (degrees)	40	90
Run length (cm)	5.18 ± 0.73	9.8 ± 2.15
Swimming speed (cm s^{-1})	6.3 ± 0.65	14.3 ± 2.03
Unsuccessful search time (s)	1.65 ± 0.56	0.55 ± 0.32
Successful search time (s)	1.36 ± 0.70	0.29 ± 0.20

[1] Small prey were *D. galatea* and *D. ambigua* with sizes from 0.6 to 1.5 mm. Large prey were *D. magna* and *D. pulex* ranging in size from 2 to 4 mm

the large prey, the net energy gain changed only a little over the larger run lengths, so that the optimum was not so clearly determined as it was for the small prey.

The changing search pattern shown by the white crappie was echoed by the arctic grayling (Evans and O'Brien 1988). An extension of the changing pattern of search tactic was developed further by O'Brien *et al.* (1990), who suggested that continuous cruise search and sit-and-wait search are the extremes of a continuum of movement speeds (Fig. 5.7). Saltatory search, which would only exclude continuous cruise in the range shown in Fig. 5.7, allows the searching animal to maximise the efficiency with which space is searched. In cruise search, it is assumed that the animal is looking only for prey at the furthest edge of its perceptual range, which moves ahead with the animal. In saltatory search the forager searches the whole volume in front of it (Fig. 5.6) and then moves on until a new unsearched volume is available. In this way, the overlap of the volume searched is minimised. This process is influenced both by the distance travelled between stops and by the geometry of the searched space. For example, the angle of the searched field in front of the fish is a function of prey size (Table 5.2). The saltatory search hypothesis is attractive and the evidence presented by O'Brien *et al.* (1990) shows that many animals are able to adjust search speed in response to characteristics of prey distribution and abundance. For certain groups such as the pike, charr and many lizard species (Pianka 1986), search behaviour is more stereotyped.

The ability to learn is a final consideration. Bluegill sunfish foraging in open water pause for shorter times during search than when foraging in a weeded habitat (Ehlinger 1989). The longer pause when catching prey in weeds seems to be demanded by the more complex habitat where prey are more difficult to detect and these are encountered less frequently (see Fig. 5.4). With experience, the time spent pausing by the bluegills decreased in both habitats. The fish had presumably learnt through experience how to recognise prey quickly. Such modification in foraging behaviour is likely to apply to a wide range of species. Coupled with the morphological changes described by Meyer (1987), learning could be a powerful force in adapting foragers to short-term spatial and temporal variation in prey properties.

The Basic Prey Model (BPM) of diet choice (Stephens and Krebs 1986; Hughes, Chapter 6) assumes that prey types are encountered sequentially by the forager. This would only be true of an animal employing cruise search. If saltatory search is used, then the forager has the chance to encounter several prey simultaneously. This case is best handled by the theory of Engen and Stenseth (1984) which classifies prey encounters as 'options', with 'option' defined as a set of possible foraging actions characterised by prey properties, their distance from the predator, their energy content and handling time. The predator chooses the option that yields the greatest energy gain per unit time. The BPM assumes that the density of prey is positively correlated with the encounter rate. This might be measured by observing fish in an aquarium and recording the rate at which a particular prey type is reacted to. The actual encounter rate may be much higher if the fish can simultaneously see several prey in the same search volume.

The discussion in the last two sections has shown how search tactics seriously undermine some of the assumptions underlying basic foraging theory. If animals can choose tactics to suit foraging conditions, then the expected diet will be more than just a function of prey abundance and handling times. It is not unreasonable

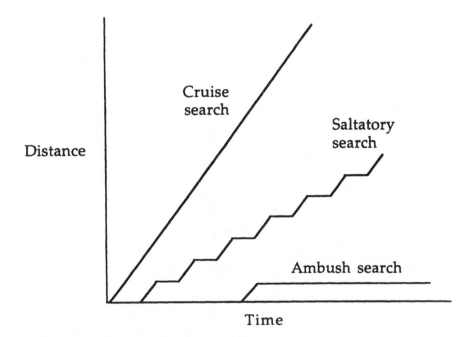

Fig. 5.7 A diagrammatical representation of the distance travelled and its timing for fish predators using cruise search, saltatory search and sit-and-wait. Redrawn from O'Brien *et al.* (1990).

to hypothesise that fish can use a suite of foraging search (and handling, see next section) tactics to cope with the variability of prey or patch types encountered and their spatial distribution (see Fig. 5.1).

5.5.3 Tactics in a neglected habitat—the ocean

All the studies discussed to this point have been of fish living in rivers or lakes. Shallow water subtropical or tropical habitats are the most accessible by using SCUBA equipment for the study of marine fish species. An example is a study of search tactics employed by the blacksmith, *Chromis punctipinnis*, feeding on zooplankton over a reef off Santa Barbara, California (Bray 1981). The blacksmith is diurnally active and spends the night in crevices in the reef. Evidence from marked fish shows that individuals used the same hiding place each night and then travelled out to foraging sites during the day. Adult blacksmith chose productive locations for their foraging sites. The reef is swept by either west flowing or east flowing currents. Plankton samples showed that the water at the upcurrent end of the reef contained a higher density of plankton, regardless of which end of the reef happened to be pointing into the current. Adult blacksmith take up station in the water at the upcurrent end and experiments with caged adults transferred to the downstream end showed that they obtain less prey there. When the current direction switched reef end, the adults switched location too. Juvenile blacksmith did not make this switch; they remained over the shallower portions of the reef where they are close to refuges. This alternative tactic is possibly a response to the energetic cost of a small fish moving upcurrent to take up and maintain station at a reef end and to a higher risk of predation there.

Many deep sea fishes live in a dark and food-poor environment and show extreme adaptations to sit-and-wait foraging, which has been suggested to be the main foraging tactic in the habitat (Fig. 5.8). Many of these fishes have very large mouths, giving them the capacity to consume prey that are as big or bigger than themselves, and lures to tempt remote prey into their jaws (Fig. 5.8). The fish also have low metabolic rates, which allow them to tolerate long periods without sustenance (Marshall 1971; Gage and Tyler 1991). In comparison, there are other abyssal fish that apparently adopt a more mobile search tactic (Mauchline and Gordon 1986; Priede *et al.* 1990; Armstrong *et al.* 1991). The feeding tactics of 33 species of fish caught between 400 and 2900 m in the Rockall Trough to the west of Scotland were inferred from stomach contents by Mauchline and Gordon (1986). Several of these species were feeding on single or multispecies patches of invertebrate prey. The evidence is consistent with the hypothesis that these fish have an active searching tactic. More direct evidence for searching comes from the studies of Priede *et al.* (1990) and Armstrong *et al.* (1991). A closed circuit television camera on a baited rig was dropped at two stations in the North Pacific to depths of 4400 and 5900 m. Some of the bait concealed ingestible transmitters. The cameras picked up two species of abyssal grenadier fish, *Coryphaenoides*

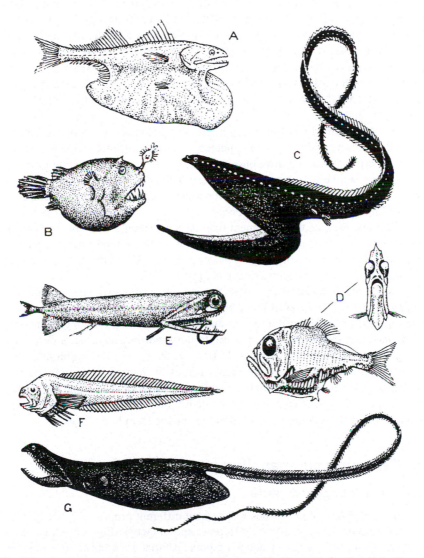

Fig. 5.8 A selection of fish from the deep sea that exemplify characteristic adaptations. Note the specific features such as large mouths, greatly expandable stomachs and lures which adapt the fish to life in an environment where food is scarce. (A) 'Great swallower' *Chiasmodon niger*, ×0.5; (B) *Borophryne apogon*, ×0.5; (C) 'Gulper' *Eurypharynx pelecanoides*, ×1; (D) Hatchet-fish, *Argyropelecus* sp., ×0.5; (E) 'Widemouth' *Malacosteus indicus*, ×0.5; (F) *Paraliparis* sp., ×0.5; (G) 'Gulper', *Saccopharynx ampullaceus*, ×0.25. (From Norman and Greenwood 1975).

yaquinae and *C. armatus*. Unlike most of the species in Fig. 5.8, the grenadiers are large, growing to 0.8 m in length and have a more robust body providing greater muscle power for locomotion (Nelson 1994). Twenty three fish at the first station and 13 at the second ingested the transmitters and were followed for up to 10 h. The data showed that fish were present at a surprisingly high density and moved away from the bait at an average speed of 1 to 15 cm s^{-1}. The second study by Armstrong *et al.* (1991) used the same observation system at 5800 m in the central Pacific. Fish with transmitters moved away from the camera rig at speeds ranging from 1 to 20 cm s^{-1}. For about 13% of total tracking time, fish were more than 15 m off the bottom.

These data, obtained from great depth, provide a glimpse of the search behaviour of these macrourids. The fish appeared to be moving fast and began to arrive at the experimental rig after only an average of 31 min of it landing on the bottom. These active scavengers appear to cruise the depths looking for aggregations of food. The darkness at such depths must mean that the fish use smell or touch to find food. In the study by Priede *et al.* (1990), the amount of bait was less on the first drop of the equipment than on the second. Fish stayed longer at the first site than at the second, which is consistent with the food patch model (the Marginal Value Theorem) of foraging theory (Charnov 1976) that predicts longer patch residence time with decreasing patch profitability.

The rate at which predators search has been the most studied tactical variation in foraging behaviour. The good relationship between tactics and ecology in foraging lizards has led to the belief that the sit-and-wait and widely-foraging tactics are well defined and species specific, leading to the syndrome hypothesis described by McLaughlin (1989). The work of O'Brien *et al.* (1990) has highlighted the possibility that animals use certain search tactics to manipulate search costs in response to changes in prey profitabilities. This possibility shows that the search tactics employed by a species need to be determined by careful observation of behaviour in the wild rather than through deductions from body form.

5.6   Capturing and handling prey

5.6.1   Tactics of prey capture

The capture of different prey types will present different problems for the predator, which might best be solved by different capture tactics (Fig. 5.4). A good illustration of this is tiger musky attacking prey fish that are at different distances away (Webb and Skadsen 1980). A prey that is between 3.8 and 33.9 cm away is mostly likely to be attacked with a tactic in which the fish starts with the body straight. It then bends the body into an S-shape (stage 1), followed by an opposite S-shape (stage 2) before entering stage 3 in which the fish continues to swim with S-shaped propulsive waves passing down the body. Strikes of this nature took from 50–300 ms to complete and were 42% successful. Closer prey are attacked with a different pattern in which the fish first bends its body into an S-shape without forward movement and then progresses with the rest of the attack sequence as described above. These attacks took between 17–100 ms to complete and were 95% successful. It was suggested by Webb and Skadsen (1980) that these two tactics maximise the probability of capture. That the tiger musky's strike tactics achieve their purpose of surprise is illustrated by the fact that fathead minnows responded to tiger musky attacks in only 28% of cases, whereas they responded to 75–90% of attacks by rainbow trout, smallmouth bass and rock bass, which are typical cruising predators (Webb 1982).

Variations in prey capture tactics have been studied in other species. An early study using high speed cinematography investigated the response of largemouth bass to different prey types (Nyberg 1971). Sacramento perch, *Archoplites interruptus*, used two tactics, a slow leisurely approach to capture slow-moving waterfleas, *Daphnia magna,* or a vigorous fast approach to capture the more evasive zooplankter, *Diaphanosoma brachyurum* (Vinyard 1982). The attack tactic for the evasive prey, which was improved by experience, was only 55% successful and used four times as much energy as the tactic used for *Daphnia.*

The foraging tactics of black surfperch, *Embiotica jacksoni,* and striped surfperch, *E lateralis,* on reefs off the southern Californian coast were analysed by Schmitt and Holbrook (1984a, b). Small black surfperch use a visually-mediated 'picking' tactic to collect small prey items individually and the sizes taken are gape limited. Larger and older fish use a less discriminating 'winnowing' tactic in which they take in a mouthful of material and separate in the mouth the food from the detritus. Fish using this tactic took prey sizes in proportion to their occurrence in the habitat. In contrast, the striped surfperch retained the picking tactic into adulthood and became more size selective as they aged. The ontogenetic change in diet shown by the black surfperch was accompanied by a change in habitat use. Young fish fed off small invertebrates located on algae and the surface of a turf-like substratum. As the fish developed the winnowing tactic, they shifted their attention more to the turf which is a rich source of food.

5.6.2 Manipulation and handling tactics

Pike capturing fish prey generally take them at the middle of their flank (Webb and Skadsen 1980). The prey is then manipulated until its head points down the fish's throat (Hart and Connellan 1984). The manipulation takes the form of rapid release of the prey coupled with a swift lateral shake of the head. Threespine stickleback usually catch isopods, *Asellus aquaticus,* by the head (Gill and Hart 1994). They also try to catch the prey with its dorsal surface pointing down. With experience, fish improve their ability to catch prey by the head and upside down. Very large *Asellus* cannot be swallowed easily. Sticklebacks usually spend time spitting a captured prey out of their mouths, reorientating it and catching it again (Gill and Hart 1994). This process can be repeated for several minutes before the prey is finally swallowed. During the process, the prey becomes crumpled and ragged, although it rarely breaks up. A soggy *Asellus* is presumably easier to swallow than a crisp new one. The occurrence of spitting behaviour increases with increasing prey size (Gill and Hart 1994).

The different tactics employed by fish to handle prey are reviewed by Helfman (1990). He mentions that the term 'handling' has often been used to describe any process that occurs between the end of search and after the prey has been swallowed. More precisely, handling is the time between capture and the beginning of the next search (Stephens and Krebs 1986). It includes all behaviours that prepare the prey for swallowing and the process of ingestion itself.

Many fish eat their prey whole, as described for the pike and stickleback. This can only be done for prey that are smaller than the fish's mouth and explains why the maximum size of prey eaten by many species is gape limited. In cases of fish species attacking prey larger than they can swallow whole, various prey manipulation tactics have evolved to overcome this problem. Sharks take bites out of their prey. Many fish twist or rotate their body, whilst holding the prey in their jaws, to break pieces off prey that are too large to swallow. A survey of species that use this rotational tactic is provided by Helfman (1990, Table III). Eels are prime examples of a group that use spinning to break off pieces from large prey. The adoption of this tactic is a function of prey size relative to eel size and of the consistency of the prey (Helfman 1990).

Pursuit and handling times are the main opportunity costs of prey capture built into static foraging models (Stephens and Krebs 1986). The examples in this section show that handling time is largely determined by the tactics a fish uses to deal with particular prey types. Experience with prey types can shorten handling times, so altering their profitability (Hughes 1979, and Chapter 6). In addition, developmental changes in the predator can lead to alterations in prey capture tactics that effectively move the individual to a new niche.

5.7 Conclusions

Three topics are worth considering in the light of the material covered in this chapter:

(1) the influence foraging tactics might play in determining the effect a predator has on prey population dynamics and community structure,

(2) the fitness consequences of different tactics, and

(3) the further development of theory dealing with tactical variation.

The search tactics used by a predator are likely to have significant effects on the predatory mortality imposed on a prey species (Pianka 1986). The extremes of the search tactic continuum, sit-and-wait and widely foraging, will generate different mortality rates on different types of prey and on different size groups within a prey species. The particular mixture of the types of predator search tactic within a community is likely to determine many aspects of community structure, although there is no knowledge yet to verify this conjecture. Future work could usefully determine the frequency of search tactics adopted by the predators in a community and try to measure the prey mortality rate generated by each.

Connected with the pattern of mortality generated by predator search tactics is the phenomenon of ontogenetic niche shifts observed in many fish species (Blaxter 1969; Schmitt and Holbrook 1984a, b; Werner and Gilliam 1984; Werner and Hall 1988; Eklöv and Persson 1995; Kramer *et al.*, Chapter 3; Persson *et al.*, Chapter 12). Bluegill sunfish hatch in the weeded margins of North American lakes and move out into the limnetic zone to feed until they reach about 12 mm in length (Werner and Hall 1988). At this size they move back into the weeds where they remain for several years feeding on benthic invertebrates until they are large enough to be immune to predation from largemouth bass. The bluegills then move back into the limnetic zone. In the marine environment, niche differences throughout a life history are greater still. For example, the plaice, *Pleuronectes platessa*, in the southern North Sea start life in the plankton feeding on *Oikopleura* and copepods, spend their juvenile years in the surf line of the Waddensee feeding on interstitial invertebrates before moving out into the open North Sea to feed on benthic molluscs and annelids (Pitcher and Hart 1982). The different demands of each niche may mean that species showing significant shifts with time retain greater flexibility of foraging tactics. This is certainly true of perch when compared with pike, which uses the same niche from a very young size (Persson and Eklöv 1995). Future research could usefully examine the correlation between ontogenetic niche shifts, niche-specific foraging tactics and fish growth and reproduction.

Size-structured interactions are important in fish population dynamics (Werner and Gilliam 1984; Persson *et al.*, Chapter 12). Such interactions between fish species can be much more complex than a simple predator–prey classification might imply. For example, whilst in the weeded lake margins, bluegill sunfish are sheltering from largemouth bass predation in the limnetic zone but they are also competing with juvenile largemouth bass for the same food resources (Werner and Hall 1988). This interspecific competition can limit the number of largemouth bass that grow large enough to be able to predate on bluegills. This type of complex interaction will be significantly influenced by variability in foraging tactics. If largemouth bass juveniles became more flexible in their tactics, the competitive balance between them and bluegills could be altered, so changing the population dynamics of both species. These types of effect have not yet been examined and there is room for much needed research to investigate the influence of tactical variability on the dynamics of predator–prey interactions.

The previously described tactical changes employed by pike and perch in response to habitat changes illustrate how fitness could change with the different tactics used. Growth rate is an important component of fitness in fishes (Charnov 1993; Schluter 1994) and theoretical studies have shown how growth and diet choice might be linked (Hart 1994). The different tactics employed by perch in habitats with or without weeds had significant effects on their growth (Fig. 5.3). When prey had an absolute refuge and perch were forced to adopt an ambush tactic, they grew more slowly than when the prey had no refuge and the perch could hunt in groups.

A recent experiment by Schluter (1994) examined competition between the benthic and limnetic morphs of the threespine stickleback occurring in lakes in British Columbia, Canada (McPhail 1994). Sticklebacks with similar trophic morphologies competed for the same food types, which caused growth loss in the morphotype least well adapted to the limnetic prey used in the experiment. Growth is taken as a measure of fitness. The negative selective effect of this competition could lead to the observed differences between the limnetic and benthic morphs within 500 years (Schluter 1994). These results also have implications for foraging behaviour. The two morphs have remarkable differences in behaviour (McPhail 1994; personal observation) and some of these differences determine foraging success (Schluter 1993). Consequently, the growth differential caused by competition between similar behavioural phenotypes can be expected to lead to divergence in foraging tactics as well as trophic morphology. As with most of the examples in this chapter, the morphology and behaviour of the predator and prey interact in a feedback system such that changes in one component produces alterations in the others. Changes in trophic morphology could be accompanied by changes in foraging tactics, such as ways of handling awkward prey, that ameliorate the effects of competition.

The material in this chapter emphasises that any new theory on foraging must say something about tactics as well as strategies. Developing tactics to overcome the problems for the predator posed by the prey (Fig. 5.2) must contribute to the fish achieving its strategy of fitness maximisation. Theories dealing with individual foraging behaviour might have to be developed at two levels. Strategic models will set the goals and subsidiary models will define the optimal tactics that will, when strung together in a feeding sequence, achieve the overall strategy. Dynamic modelling (Mangel and Clark 1988) could easily be used to develop such a two-tiered model, because the method divides time into discrete units within which a subset of events can be modelled, each with varying parameters and probabilities.

This review has highlighted the need for models using aspects of game theory (cf. Maynard Smith 1982) to explain the mixture of foraging tactics found in a species population. If the widely foraging and sit-and-wait search tactics are found within a species, which tactic should an individual choose? The best choice will be determined possibly by the number of other individuals using the two tactics, together with the distribution and abundance of prey. The choice of tactic might also depend on the competitive ability of the individual. The sit-and-wait tactic is energetically cheaper than is the widely foraging tactic and this could influence the decision a poor competitor makes. A further factor that the individual should

consider is predation risk. As this is likely to be size dependent in a fish population (Werner and Hall 1988), the tactic chosen may also depend on the fish's size.
Sit-and-wait predators are less exposed to their own predators (Pianka 1986), so
this might characterise smaller individuals. For these reasons it might be expected
that the degree of search mobility in fishes will be a function not only of prey
size, as suggested by O'Brien *et al.* (1990), but also of the size of the predator
itself. There is therefore a need for the development of theory to handle tactical
selection together with detailed observations on the link between size and mobility.
Such studies will also illuminate the size-structured competitive and predator–prey
interactions discussed earlier.

Acknowledgements

A first draft of the manuscript for this chapter was read by Peter Eklöv, Gene
Helfman and Robert McLaughlin. I thank them for the hard work they devoted to
improving my first efforts. Jean-Guy Godin has also been a sympathetic and sup-
portive editor for which I thank him. Any weaknesses in the chapter are of course
my own.

References

Armstrong, J.D., Priede, I.G. and Smith, K.L. Jr. (1991). Temporal change in foraging behaviour of the fish *Coryphaenoides* (*Nematonurus*) *yaquinae* in the central North Pacific.
Mar. Ecol. Prog. Ser., **76**, 195–199.

Blaxter, J.H.S. (1969). Development: eggs and larvae. In *Fish Physiology*, Vol. 3 (eds. W.S.
Hoar and D.J. Randall), pp. 177–252. Academic Press, New York.

Bray, R.N. (1981). Influence of water currents and zooplankton densities on daily foraging
movements of blacksmith, *Chromis punctipinnis*, a planktivorous reef fish. *Fish. Bull.*,
78, 829–841.

Chapleau, F., Johansen, P.H. and Williamson, M. (1988). The distinction between pattern
and process in evolutionary biology: the use and abuse of the term 'strategy'. *Oikos*, **53**,
136–138.

Chapman, C.A. and Mackay, W.C. (1984). Versatility in habitat use by a top aquatic predator,
Esox lucius L. *J. Fish Biol.*, **25**, 109–115.

Charnov, E.L. (1976). Optimal foraging: the marginal value theorem. *Theor. Popul. Biol.*, **9**,
129–136.

Charnov, E.L. (1993). *Life history invariants. Some explorations of symmetry in evolutionary
ecology.* Oxford University Press, Oxford.

Dill, L.M. (1983). Adaptive flexibility in the foraging behaviour of fishes. *Can. J. Fish. Aquat.
Sci.*, **40**, 398–408.

Ehlinger, T.J. (1989). Learning and individual variation in bluegill foraging: habitat-specific
techniques. *Anim. Behav.*, **38**, 643–658.

Eklöv, P. (1992). Group foraging versus solitary foraging efficiency in piscivorous predators:
the perch, *Perca fluviatilis*, and pike, *Esox lucius*, patterns. *Anim. Behav.*, **44**, 313–326.

Eklöv, P. (1995). *Effects of behavioural flexibility and habitat complexity on predator–prey interactions in fish communities.* Doctoral Dissertation, University of Umeå, Umeå, Sweden.

Eklöv, P. and Diehl, S. (1994). Piscivore efficiency and refuging prey: the importance of predator search mode. *Oecologia*, **98**, 344–353.

Eklöv, P. and Hamrin, S.F. (1989). Predatory efficiency and prey selection: interactions between pike, *Esox lucius*, perch, *Perca fluviatilis*, and rudd, *Scardinius erythrophthalmus*. *Oikos*, **56**, 149–156.

Eklöv, P. and Persson, L. (1995). Species–specific antipredator capacities and prey refuges: interactions between piscivorous perch (*Perca fluviatilis*) and juvenile roach (*Rutilus rutilus*). In *Effects of behavioural flexibility and habitat complexity on predator–prey interactions in fish communities* (by P. Eklöv). Doctoral Dissertation, University of Umeå, Umeå, Sweden.

Emlen, J.M. (1966). The role of time and energy in food preference. *Am. Nat.*, **100**, 611–617.

Endler, J.A. (1991). Interactions between predators and prey. In: *Behavioural ecology: an evolutionary approach*, 3rd edn. (eds. J.R. Krebs and N.B. Davies), pp. 169–196. Blackwell Scientific Publ., Oxford.

Engen, S. and Stenseth, N.Chr. (1984). A general version of optimal foraging theory: the effect of simultaneous encounters. *Theor. Popul. Biol.*, **26**, 192–204.

Evans, B.I. and O'Brien, W.J. (1988). A reevaluation of the search cycle of planktivorous arctic grayling, *Thymallus arcticus*. *Can. J. Fish. Aquat. Sci.*, **45**, 187–192.

Gage, J.D. and Tyler, P.A. (1991). *Deep sea biology. A natural history of organisms at the deep–sea floor.* Cambridge University Press, Cambridge.

Galis, F. (1993). Morphological constraints on behaviour through ontogeny: the importance of developmental constraints. In *Behavioural ecology of fishes* (eds. F.A. Huntingford and P. Torricelli), pp. 119–135, Harwood Academic Publ., Reading.

Galis, F., Terlouw, A. and Osse, J.W.M. (1994). The relation between morphology and behaviour during ontogenetic and evolutionary changes. *J. Fish. Biol.*, **45** (*Suppl. A*), 13–26.

Giles, N., Wright, R.M. and Nord, M.E. (1986). Cannibalism in pike fry, *Esox lucius* L.: some experiments with fry densities. *J. Fish Biol.*, **29**, 107–113.

Gill, A.B. and Hart, P.J.B. (1994). Feeding behaviour and prey choice of the threespine stickleback: the interacting effects of prey size, fish size and stomach fullness. *Anim. Behav.*, **47**, 921–932.

Grant, J.W.A. and Noakes, D.L.G. (1987). Movers and stayers: foraging tactics of young–of–the–year brook charr, *Salvelinus fontinalis*. *J. Anim. Ecol.*, **56**, 1001–1013.

Hart, P.J.B. (1994). Theoretical reflections on the growth of three–spined stickleback morphs from island lakes. *J. Fish Biol.*, **45** (*Suppl A*), 27–40.

Hart, P.J.B. and Connellan, B. (1984). Cost of prey capture, growth rate and ration size in pike, *Esox lucius* L., as functions of prey weight. *J. Fish Biol.*, **25**, 279–292.

Hart, P.J.B. and Hamrin, S.F. (1988). Pike as a selective predator. Effects of prey size, availability, cover and pike jaw dimensions. *Oikos*, **51**, 220–226.

Hart, P.J.B. and Hamrin, S.F. (1990). The role of behaviour and morphology in the selection of prey by pike. In *Behavioural mechanisms of food selection* (ed. R.N. Hughes), NATO ASI Series, Vol. G20, pp. 235–254. Springer-Verlag, Berlin.

Helfman, G.S. (1990). Mode selection and mode switching in foraging animals. *Adv. Study Behav.*, **19**, 249–298.

Huey, R.B. and Pianka, E.R. (1981). Ecological consequences of foraging mode. *Ecology*, **62**, 991–999.

Hughes, R. N. (1979). Optimal diets under the energy maximizing premise: the effects of recognition time and learning. *Am. Nat.*, **113**, 209–221.

Hughes, T. (1982). *Selected poems 1957–1981*. Faber and Faber, London.

Liem, K.F. (1980). Acquisition of energy by teleosts: adaptive mechanisms and evolutionary patterns. In *Environmental physiology of fishes* (ed. M.A. Ali), pp. 299–334. Plenum Press, New York.

MacArthur, R.H. and Pianka, E. (1966). On the optimal use of a patchy environment. *Am. Nat.*, **100**, 603–609.

McLaughlin, R.L. (1989). Search modes of birds and lizards: evidence for alternative movement patterns. *Am. Nat.*, **133**, 654–670.

McLaughlin, R.L. and Grant, J.W.A. (1994). Morphological and behavioural differences among recently–emerged brook charr, *Salvelinus fontinalis*, foraging in slow- vs. fast-running water. *Environ. Biol. Fish.*, **39**, 289–300.

McLaughlin, R.L. and Kramer, D.L. (1991). The association between amount of red muscle and mobility in fishes: a statistical evaluation. *Environ. Biol. Fish.*, **30**, 369–378.

McLaughlin, R.L., Grant, J.W.A. and Kramer, D.L. (1992). Individual variation and alternative patterns of foraging movements in recently-emerged brook charr (*Salvelinus fontinalis*). *Behaviour*, **120**, 286–301.

McLaughlin, R.L., Grant, J.W.A. and Kramer, D.L. (1994). Foraging movements in relation to morphology, water–column use, and diet for recently emerged brook trout (*Salvelinus fontinalis*) in still–water pools. *Can. J. Fish. Aquat. Sci.*, **51**, 268–279.

McPhail, J.D. (1994). Speciation and the evolution of reproductive isolation in the sticklebacks (*Gasterosteus*) of south–western British Columbia. In *Evolutionary biology of the threespine stickleback* (eds. M.A. Bell and S.A. Foster), pp. 399–437. Oxford University Press, Oxford.

Maitland, P.S. and Campbell, R.N. (1992). *Freshwater fishes*. Harper Collins, London.

Mangel, M. and Clark, C.W. (1988). *Dynamic modeling in behavioral ecology*. Princeton University Press, Princeton.

Marshall, N.B. (1971). *Explorations in the life of fishes*. Harvard University Press, Cambridge, MA.

Mauchline, J. and Gordon, J.D.M. (1986). Foraging strategies of deep–sea fish. *Mar. Ecol. Prog. Ser.*, **27**, 227–238.

Maynard Smith, J. (1982). *Evolution and the theory of games*. Cambridge University Press, Cambridge.

Meyer, A. (1987). Phenotype plasticity and heterochrony in *Cichlasoma managuense* (Pisces, Cichlidae) and their implications for speciation in cichlid fishes. *Evolution*, **41**, 1357–1369.

Milinski, M. (1984). Competitive resource sharing: an experimental test of a learning rule for ESSs. *Anim Behav.*, **32**, 233–242.

Nelson, J.S. (1994). *Fishes of the world*, 3rd edn. John Wiley and Sons, New York.

Norman, J.R. and Greenwood, P.H. (1975). *A history of fishes*. Ernest Benn, London.

Nyberg, D.W. (1971). Prey capture in largemouth bass. *Amer. Midl. Nat.*, **86**, 128–144.

O'Brien, W.J., Browman, H.I. and Evans, B.I. (1990). Search strategies of foraging animals. *Amer. Sci.*, **78**, 152–160.

O'Brien, W.J., Evans, B.I. and Howick, G. (1986). A new view of the predation cycle of a planktivorous fish, white crappie (*Pomoxis annularis*). *Can. J. Fish. Aquat. Sci.*, **43**, 1894–1899.

O'Brien, W.J., Evans, B.I. and Browman, H.I. (1989). Flexible search tactics and efficient foraging in saltatory searching animals. *Oecologia*, **80**, 100–110.

Persson, L. and Eklöv, P. (1995). Prey refuges affecting interactions between piscivorous perch (*Perca fluviatilis*) and juvenile perch and roach (*Rutilus rutilus*). *Ecology*, **76**, 70–81.

Pianka, E.R. (1986). *Ecology and natural history of desert lizards*. Princeton University Press, Princeton.

Pitcher, T.J. and Hart, P.J.B. (1982). *Fisheries ecology*. Chapman and Hall, London.

Pitcher, T.J. and Hart, P.J.B. (eds.) (1995). *The impact of species changes in African lakes*. Chapman and Hall, London.

Priede, I.G., Smith, K.L. Jr. and Armstrong, J.D. (1990). Foraging behaviour of abyssal grenadier fish: inferences from acoustic tagging and tracking in the North Pacific Ocean. *Deep-Sea Res.*, **37**, 81–101.

Rand, D.M. and Lauder, G.V. (1981). Prey capture in the chain pickerel, *Esox niger*: correlations between feeding and locomotor behaviour. *Can. J. Zool.*, **59**, 1072–1078.

Schluter, D. (1993). Adaptive radiation in sticklebacks: size, shape, and habitat use efficiency. *Ecology*, **74**, 699–709.

Schluter, D. (1994). Experimental evidence that competition promotes divergence in adaptive radiation. *Science*, **266**, 798–801.

Schmitt, R.J. and Holbrook, S.J. (1984a). Gape–limitation, foraging tactics and prey size selectivity of two microcarnivorous species of fish. *Oecologia*, **63**, 6–12.

Schmitt, R.J. and Holbrook, S.J. (1984b). Ontogeny of prey selection by black surfperch *Embiotica jacksoni* (Pisces: Embiotocidae): the roles of fish morphology, foraging behavior, and patch selection. *Mar. Ecol. Prog. Ser.*, **18**, 225–239.

Schoener, T.W. (1971). Theory of feeding strategies. *A. Rev. Ecol. Syst.*, **2**, 369–404.

Schoener, T.W. (1987). A brief history of optimal foraging theory. In *Foraging behavior* (eds. A.C. Kamil, J.R. Krebs and H.R. Pulliam), pp. 5–67. Plenum Press, New York.

Scott, W.B. and Crossman, E.J. (1973). Freshwater fishes of Canada. *Fish. Res. Board Can. Bull.*, **184**, 966 pp.

Stephens, D.W. and Krebs, J.R. (1986). *Foraging theory*. Princeton University Press, Princeton.

Vinyard, G.L. (1982). Variable kinematics of Sacramento perch (*Archoplites interruptus*) capturing evasive and nonevasive prey. *Can. J. Fish. Aquat. Sci.*, **39**, 208–211.

Wainwright, P.C. (1986). Motor correlates of learning behaviour: feeding on novel prey by pumpkinseed sunfish (*Lepomis gibbosus*). *J. exp. Biol.*, **126**, 237–247.

Wainwright, P.C. and Lauder, G.V. (1986). Feeding biology of sunfishes: patterns of variation in the feeding mechanism. *Zool. J. Linn. Soc.*, **88**, 217–228.

Ware, D.M. (1978). Bioenergetics of pelagic fish: theoretical change in swimming speed and ration with body size. *J. Fish. Res. Board. Can.*, **35**, 220–228.

Webb, P.W. (1982). Avoidance responses of fathead minnow to strikes by four teleost predators. *J. comp. Physiol.*, **147A**, 371–378.

Webb, P.W. (1984). Body and fin form and strike tactics of four teleost predators attacking fathead minnow (*Pimephales promelas*) prey. *Can. J. Fish. Aquat. Sci.*, **41**, 157–165.

Webb, P.W. (1986). Locomotion and predator–prey relationships. In *Predator–prey relationships: perspectives and approaches from the study of lower vertebrates* (eds. M. E. Feder and G.V. Lauder), pp. 24–41. University of Chicago Press, Chicago.

Webb, P.W. and Skadsen, J.M. (1980). Strike tactics of *Esox. Can. J. Zool.*, **58**, 1462–1469.

Werner, E.E. and Gilliam, J.F. (1984). The ontogenetic niche and species interactions in size-structured populations. *A. Rev. Γ:ol. Syst.*, **15**, 393–425.

Werner, E.E. and Hall, D.J. (1988). Ontogenetic habitat shifts in bluegill: the foraging rate-predation risk trade-off. *Ecology*, **69**, 1352–1366.

Wootton, R.J. (1984). Introduction: tactics and strategies in fish reproduction. In *Fish reproduction. Strategies and tactics* (eds. G.W. Potts and R.J. Wootton), pp. 1–12. Academic Press, London.

6 Diet selection

Roger N. Hughes

6.1 Introduction

In his classic work on the ecological energetics of freshwater fishes, Ivlev (1961) related feeding behaviour, through diet (prey) choice, to subsequent growth. At that time, a proximate, physiological emphasis was placed on the energy budget, and although this approach generated experimental direction and considerable insight, it eventually ceased to break new ground. Fresh impetus was given by the emerging discipline of evolutionary ecology (Krebs and Davies 1991). Seeking to explain behavioural and life-history phenomena in terms of natural selection, the evolutionary ecologist would view growth as one of many possible surrogate measures of fitness. Thus, by experimentally determining the effect of diet on growth, one can begin to relate behavioural patterns affecting diet selection to fitness. The evolutionary approach continues to have a powerful influence on the questions asked by behavioural ecologists and on the experiments they perform. It is my purpose in this chapter to illustrate the potency of the evolutionary approach as a framework for the study of diet selection in fishes. To do this, I have chosen examples that I know well and that have generality. I have not attempted a comprehensive review of any aspect, but have tried to portray the increasing implication of other areas of ecology by our growing understanding of individual feeding behaviour and the consequent need for an interdisciplinary approach to the subject (Hughes 1993). Thus, I begin with a basic theoretical model of diet selection and consider how its limitations have been overcome by subsequent developments. The subject is in an exciting state of flux, generating many more questions than answers. I draw attention to such questions and suggest some experimental approaches to them.

6.2 Optimal foraging theory

6.2.1 Premises and models

Foraging behaviour may involve 'decisions' (*sensu* Dill 1987) such as where to feed, when to try elsewhere, which food items to accept and which to reject. If foragers do indeed make such decisions, then the process presumably evolved because it promotes fitness. Because food is the source of energy needed for all biological functions of an individual, it is reasonable to assume that feeding rate will be positively correlated with fitness. Although this correlation may be only partial, since other factors such as the need to avoid predators or attend to reproductive matters also are important (Lima and Dill 1990), it is likely to be of central importance. Hence, the development of a theory offering an evolutionary explanation for observed behavioural patterns has been based on the assumption that decisions made while foraging maximise the net rate of energy gain. Using this energy maximisation premise (EMP) together with an economics approach, MacArthur and Pianka (1966) and Emlen (1966) initiated the development of optimal foraging theory (OFT).

Economics models typically concern the maximisation of gain per unit cost. For a foraging animal, the ultimate gain is fitness, represented approximately by energy extracted from ingested food and stored for later allocation to growth or reproduction, or used for more immediate needs. Time is widely assumed to be an important cost, since time spent handling an item may not be spent in pursuit of another, or indeed of any other, time-consuming activity (the principle of 'lost opportunity', Stephens and Krebs 1986). The problem then becomes one of maximising the net energy gained per unit time [max (E/T)]. Of course when commodities other than energy are important contributors to fitness, different objective functions ('currencies' in economics jargon) may become appropriate. For example, when it is necessary to devote time to non-foraging activities such as defending breeding territories, the problem may be inverted to that of minimising the time taken to acquire some necessary amount of energy [min (T/E)] (Schoener 1971). Even the unit of cost may need to be changed. For example, foragers who themselves face a risk of predation could more effectively promote fitness by maximising the energy gained per unit of risk (Gilliam 1990). As theory has expanded to accommodate a greater range of factors suspected of influencing foraging behaviour, there has been a transition from simple, analytical models to more comprehensive, numerical models requiring iterative solution (Stephens and Krebs 1986). Both approaches and their variants, however, continue to be useful.

6.2.2 Basic prey model

Foraging behaviour can be categorised into the exploitation of alternative feeding locations (Kramer *et al.*, Chapter 3; Persson *et al.*, Chapter 12), searching modes (Hart, Chapter 5) and food items (this chapter). OFT has been applied to all three categories, and in each case specific models predict optimal behavioural decisions (i.e. those contributing the most to fitness) by finding solutions that maximise or minimise objective functions of the type described above.

Applied to diet selection, this approach generated the basic prey model (BPM). Developed independently by several authors (listed in Hughes 1980), the BPM

Table 6.1 Some important assumptions of the Basic Prey Model. For other assumptions of this model, see Charnov (1976) and Stephens and Krebs (1986).

---

1. Prey types have characteristic, unchanging profitabilities defined as the yield per unit handling time.

2. Handling time, the period from detection of an item, through capture and ingestion to readiness for another attack, is mutually exclusive of searching time, when the forager is seeking prey.

3. The forager can acquire and retain some expectation of the mean rate of energy gain in the foraging area, against which to compare the profitability of each newly encountered item. This mean rate of energy gain will be a function both of the profitabilities and abundances of the different prey types.

4. The forager uses this comparison to decide which items to accept or reject in order to maximise E/T.

---

makes a large number of simplifying assumptions (Table 6.1) to predict the acceptance or rejection of alternative food types encountered by a forager. Algebraic solution of the model generates three predictions:

(1) foragers should always accept the most profitable prey type,

(2) they should accept the successively less profitable prey types only when E/T would otherwise fall below the mean, and

(3) prey types outside the optimal set should always be rejected; that is, the forager should not show 'partial preferences' by accepting a proportion of inferior items.

Behaviour in accordance with predictions (1) and (2) would cause the diet to expand or contract as abundances of the different prey types change. Behaviour in accordance with prediction (3) would cause prey to be added to or dropped from the diet suddenly rather than gradually.

The first experimental test of the BPM was made by Werner and Hall (1974), who found that bluegill sunfish, *Lepomis macrochirus*, feeding upon waterfleas, *Daphnia* sp., do indeed adjust diet breadth in the predicted way. For individual bluegills used in the experiment, handling time had a characteristic value that was independent of *Daphnia* size, since all sizes could be captured and swallowed easily. Consequently, profitability increased with the size of *Daphnia* and specialisation on the larger items prevailed when these were abundant. Changes in diet breadth, however, were more gradual than specified by prediction (3). Such general concordance with predictions (1) and (2) and discordance with prediction (3) has been the outcome of many experimental studies of foraging behaviour on a wide range of animals (reviewed for example by Pyke 1984; Stephens and Krebs 1986). These results suggest that the BPM captures the reality of certain aspects of diet selection, but that others are misrepresented by some of the assumptions made.

Relaxing assumptions of the BPM

The ubiquity of partial preferences is not surprising (McNamara and Houston 1987a). Violation of almost any of the numerous, simplifying assumptions of the BPM will generate the opposite of prediction (3). Two sources of violation deserve attention here because of their generality and importance in the development of dynamic models, to be considered later. These are hunger state and learning, which separately and collectively, may cause handling times to change within a foraging bout. Selective behaviour may change concomitantly, the forager showing partial preferences when diet selection is averaged over the foraging bout. To understand these relationships, it is necessary to examine features of the predator and prey that determine handling time.

Components of handling time

Some idea of the range of behavioural tasks that a fish must perform when handling a prey item can be gleaned from observations of fifteenspine sticklebacks, *Spinachia spinachia*, feeding on crustaceans. Each prey-handling episode is characterised by a basic sequence of events. First, the stickleback orientates itself towards the prey. There follows a period of fixation when the fish appears to align its vision very precisely, often toward a particular part of the prey's body, and perhaps is assessing factors such as the type or size of the prey and the intervening distance, in order to adopt the appropriate method of attack. During this phase the prey may shift position, whereupon it is tracked by the fish, using movement of the eyes or body. Next, the stickleback stealthily approaches the prey and, once within a critical distance, accelerates into the attack. If the attack is successful, the prey is grasped in the fish's jaws and, if not too awkwardly shaped or too large, swallowed. This basic sequence can be variously modified by the insertion of extra behavioural components needed to cope with prey that are difficult to catch or swallow. Agile prey, such as mysids, require a swift attack aimed to intercept the escape trajectory. Here, the fish's body is primed for high acceleration by bending into an S-shape, and the ensuing attack is directed at an angle of about $32°$ towards the anticipated direction of escape (Kaiser *et al.* 1992). Sometimes the attack is aborted after the S-bend, perhaps if the prey has shifted position slightly. The fish swims backward from the prey, eventually turning away and perhaps starting the whole behavioural sequence over again. Even when carried through, not all attacks will succeed. When the attack fails, the fish will turn away and reinitiate the entire handling sequence if the prey is still within sight.

Other prey such as amphipods, although not agile, may be awkward to handle. Because the fifteenspine stickleback has a relatively narrow buccal cavity, a bulky prey item must be held longitudinally in the mouth before it can be swallowed. This is most likely to be successful if the prey is grasped head-on. When not grasped head-on, the prey must be realigned by holding and turning it, or even by spitting it out and grasping again. Even a correctly aligned item may require dismembering by chewing and spitting if the ratio of prey width to mouth width is too large (Gill and Hart 1994).

All predatory fishes engage in a basic sequence of behavioural components, equivalent to the above, when attacking their prey, although the order and number of components in the sequence may differ ontogenetically and interspecifically (Shaw *et al.* 1994). More importantly in the present context, changes in duration and effectiveness of the basic behavioural components and the insertion or omission of extra components occur routinely for an individual fish within and among foraging bouts. Such changes are the integrated result of fluctuations in hunger level, learning and memory (Croy and Hughes 1990).

Hunger state

Motivation to feed increases with hunger (Colgan 1973), itself defined in terms of stomach fullness, or some physiological state that varies with time since the last meal (Beukema 1968). Three common, but not universal, consequences of this are:

(1) hungrier individuals react to prey at a greater distance (Fig. 6.1a),

(2) hungrier individuals are less selective in what they attack (Ivlev 1961; Kislalioglu and Gibson 1976), and

(3) more satiated individuals take longer to handle prey (Fig. 6.1b) and do so less efficiently (Fig. 6.1c), so lowering prey profitability (Fig. 6.1d).

The first two consequences may reflect the primacy of restoring a positive energy budget as quickly as possible, even at the cost of reduced foraging efficiency. Once this has been achieved, it is energetically optimal to become selective, foregoing opportunities to attack certain items in order to maximise the energy balance in the time available. An alternative explanation for the second consequence is that packing constraints become critical as the stomach fills; this is discussed in further detail later. The third consequence reflects a more leisurely progression through the prey-handling sequence as feeding motivation declines and, to a lesser degree, may also reflect extra time required to evaluate prey when feeding selectively.

Learning and memory

Performance in all categories of handling behaviour, from initial recognition to final ingestion, may improve as a fish gains experience of specific prey (Hughes *et al.* 1992). Improvement begins immediately in a run of encounters with the prey and may reach an asymptote within several foraging bouts. Without reinforcement, the learned skills attenuate. Forgetting occurs more slowly than learning, however, and handling skills fall to minimum levels only after many foraging bouts have elapsed in the absence of the specific prey. Nevertheless, in some cases (Mackney and Hughes 1995), forgetting can have a significant effect over a time scale of

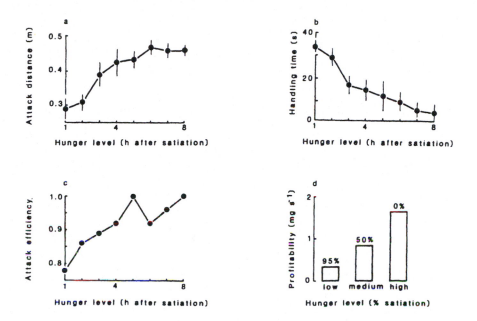

Fig. 6.1 Effects of hunger level on the prey-handling behaviour of fifteenspine sticklebacks, *Spinachia spinachia*, feeding on amphipods, *Gammarus* sp. Different batches of fish were subjected to trials ranging from 1–8 h after feeding to complete satiation. Data are medians, with interquartile ranges where appropriate. After Croy and Hughes (1991b). (a) Attack distance, measured as the distance between fish and prey at orientation. (b) Handling time, measured as the period between orientation and swallowing the prey. (c) Attack efficiency, measured as the proportion of attacks resulting in capture. (d) Profitability, estimated as the dry mass of prey ingested (predicted from regression of prey body mass on length) per unit observed handling time.

days, whence handling efficiency will reflect a balance between learning and forgetting.

The particular aspects of the handling process that are improved by learning will depend on the type of prey. For example, when fifteenspine sticklebacks feed on amphipods, it is important to grasp the prey head-on so that it can be swallowed. The proportion of head-on attacks increases as sticklebacks gain experience with amphipods (Fig. 6.2a). Grasping the amphipod head-on avoids any need to realign the prey, so eliminating extra behavioural components (Fig. 6.2b) and shortening handling time (Fig. 6.2c). Brine shrimp (*Artemia* sp.), on the other hand, are easily swallowed and so never need realigning. In this case, learned handling skills are largely confined to reaction distance and speed of attack (Croy and Hughes 1991a). The net result of learning is to increase the profitability of the specific prey by shortening handling time or by increasing the likelihood of capture and successful ingestion.

A forager will normally encounter mixtures of prey types, and this can reduce learning efficiency. When naive fifteenspine sticklebacks are presented with

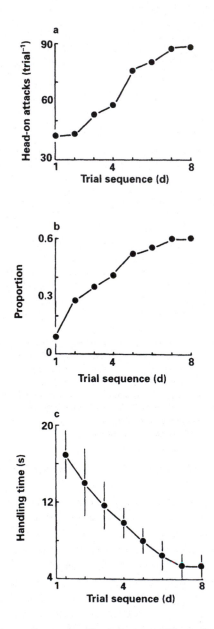

Fig. 6.2　Effects of learning on the prey-handling behaviour of fifteenspine sticklebacks, *Spinachia spinachia*, feeding on amphipods, *Gammarus* sp. Initially naive fish were each subjected to daily feeding trials, using amphipods as prey. Data are medians, with interquartile ranges where appropriate. After Croy and Hughes (1991a). (a) Frequency of attacks that were directed head-on to the prey. (b) Proportion of handling sequences including the additional components Hold, Spit and Chew. (c) Handling time, defined as in Fig. 6.1b.

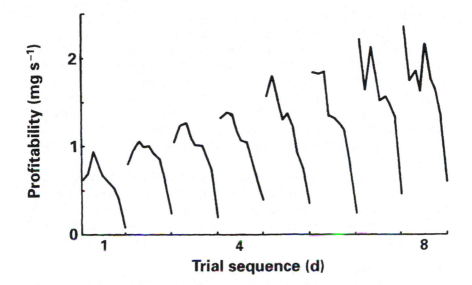

Fig. 6.3 Changes in profitability (dry mass/handling time) of amphipods, *Gammarus* sp., eaten by fifteenspine sticklebacks, *Spinachia spinachia*, during a series of daily trials. Initially, the fish were naive to amphipods. The changes are the combined result of learning and hunger level. After Croy and Hughes (1990).

amphipods and brine shrimp in alternating sequence, the handling time for amphipods, although shortened by learning, remains about 1.4 times longer than the value attained on a pure diet of amphipods (Croy and Hughes 1991a). Of course in nature, encounters with different prey types will not be strictly alternate, but either random or in runs if prey are spatially or temporally segregated. To experimentally determine the effects of patterns of encounter on learning and memory, prey requiring different handling skills might be presented to fishes in alternating runs of constant or randomly varying length.

Combined effects of hunger and learning
The effects of hunger state and learning on prey handling combine additively and interactively. When combining additively, each effect operates independently of the other. For example, the accounts given above for the effects of hunger and learning on the profitability of amphipods eaten by fifteenspine sticklebacks are based on experiments designed to allow only one effect to operate at a time. An experiment allowing both effects to vary simultaneously shows that the previously described patterns are superimposed (Croy and Hughes 1990). Initially, naive fish are run through a series of daily foraging bouts. At the beginning of each foraging bout profitability is at a local minimum, but then begins to rise as learning shortens handling time (Fig. 6.3). Meanwhile, hunger level declines as

the sticklebacks fill their stomachs, which tends to increase handling time in opposition to learning. Satiation exerts increasing dominance over learning, causing profitability to reach a local maximum and then decline towards the end of the foraging bout. By the beginning of the next foraging bout, the previous effect of learning has been reduced somewhat by forgetting, but there is a net gain in learned foraging skills, causing both the minimum and maximum levels of profitability to increase. As learning approaches completion, profitability reaches a stable maximum at the beginning of each successive foraging bout. By this stage, the only remaining change in profitability is the cyclical decrease accompanying satiation.

When combining interactively, hunger state influences the learning process: hungrier fish learn more efficiently than less hungry ones. In the case of fifteenspine sticklebacks feeding on amphipods, previously starved fish learn six times more efficiently than those previously fed to 95% satiation, when presented with single items over a series of daily trials (Croy and Hughes 1991b).

Changing the rank order of profitability

If a lower-ranking prey becomes very abundant and is then encountered much more frequently than the next higher in rank, learning might differentially increase the profitability of the former prey. The profitabilities of the two prey thus would begin to converge and, if initially not too dissimilar, even become transposed in rank (Hughes 1979). When fifteenspine sticklebacks are presented with changing relative abundances of brine shrimp and amphipods, profitabilities change concurrently as a result of learning. In this case, however, the profitabilities converge but hardly become transposed, being initially too different (Fig. 6.4).

Variable profitabilities and the BPM

It is clearly unrealistic to assume that profitabilities will necessarily remain constant throughout a foraging bout. Variations caused by changing hunger level and learning will make the unambiguous ranking of profitabilities impossible, unless there are wide, intrinsic differences between prey. Partial preferences therefore are to be expected when diet selection is averaged across a foraging bout. The BPM, however, may still capture the essence of selective behaviour for short periods within a foraging bout.

6.2.3 Implications of time scale

The BPM assumes that encounter rates and profitabilities of the prey remain constant for the duration of a foraging bout, which therefore is the characteristic time scale of the model. As shown above, however, processes influencing diet selection range from the behavioural to the physiological. At one extreme, encounter rates may change within seconds or minutes. Hunger level perhaps changes within minutes and, at the other extreme, assimilation of energy from the food

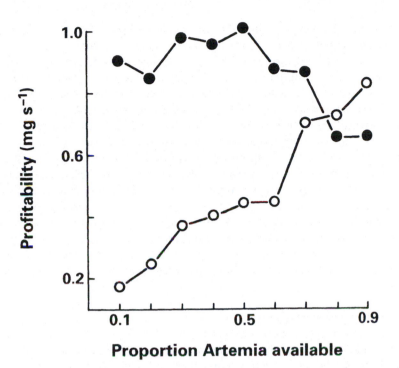

Proportion Artemia available

Fig. 6.4 Changes in profitability (dry mass/handling time) of brine shrimp, *Artemia* sp. (open circles), and amphipods, *Gammarus* sp. (closed circles), across trials, when fifteenspine stickleback, *Spinachia spinachia*, were presented with sequential changes in relative abundances of the two prey. After Hughes and Croy (1993).

may require hours. Time scale therefore deserves careful attention, as discussed below.

Diet selection

It has been stressed that the apparent occurrence of partial preferences may result from using the foraging bout as the time unit for predicting diet selection. The BPM assumes that the forager is able to measure the average rate of energy gain in the feeding locality and use this as a basis for rejecting or accepting lower-ranking prey. The average should predict the current probabilities of encounter for different prey. When beginning to search for prey, the forager only has information that it may have remembered from previous bouts. Prey abundances, however, may have changed in the meantime. Consequently, the forager will need to update continually its estimation of the average rate of energy intake. How useful, therefore, is past information? Theoretical considerations (McNamara and Houston 1987b) show that the faster the circumstances change the shorter should be the memory. The memory window could operate in either of two ways. It could give equal weighting to all information within its limit, beyond which information is completely forgotten, or it could give exponentially decreasing weighting to information receding into the past. The latter is the more general model and has been used to predict the exploitation of alternative food sources by goldfish, *Carassius auratus* (Lester 1984), and threespine sticklebacks, *Gasterosteus aculeatus* (Milinski and Regelmann 1985). The constant of exponential decay should be adjusted to the rate at which the environment changes. Fishes encountering small aggregations of specific prey as they search the habitat, particularly if the prey themselves are highly mobile, should place great emphasis on the most recent experiences. An extreme form of decremental emphasis placed on past information would be to choose whichever prey the forager last had success in capturing (Bergelson 1985). Such a 'rule-of-thumb' would not yield exactly the optimal diet, but it might achieve a close approximation by tracking changes in the abundances of prey during a foraging bout. No critical data are yet available; the theory needs to be tested by observing sequential encounters under controlled conditions.

Profitability and energy maximisation
The generally adopted, operational definition of handling time excludes the time taken to digest the food, because many foragers continue to feed while digestion is in progress. The rate of energy return from each item ingested, however, will partly depend on digestion time, and so profitabilities calculated using a strictly behavioural definition of handling time will differ from those based also on physiology.

Gut-emptying time has been measured, as an approximation for digestion time, for fifteenspine sticklebacks feeding on different prey (Kaiser *et al.* 1992). Amphipods pass through the gut 1.7 times faster than mysids of similar individual mass, perhaps because the slender mysids can be packed more tightly in the gut, so taking longer to digest. Based on handling behaviour alone, the ratio of profitabilities for amphipods and mysids is 0.01/1.00. When digestion time is included, the ratio becomes 1.00/0.58. The rank order is reversed, depending on whether profitabilities are measured on a behavioural or a physiological time scale.

On which time scale is diet selection based? Fifteenspine sticklebacks strongly prefer mysids to amphipods, clearly demonstrating that, in their case, diet selection is based on the behavioural rather than the physiological time scale. Other cases may be different, depending on whether or not digestive kinetics limit the rate at which successive items can be processed. Fishes such as pike (*Esox* spp.), which consume relatively large and highly nutritious prey items, will not feed again until digestion of the last item has reached some critical stage. This is because so much energy is directed to digestion that the forager is metabolically incapable of sustaining attacks on prey (Armstrong *et al.* 1992). Measurements of handling time therefore must include the quiescent phase during digestion and so may amount to hours or days, depending on the relative size of the prey (Hart and Hamrin 1990). If, in such cases, diet selection is indeed based on the rate of energy intake, then it must involve the physiological time scale. More work at this interface between behavioural and physiological ecology needs to be done.

Digestive constraints on gut evacuation rate need to be studied much more thoroughly in the context of foraging behaviour. Both intra- and interspecific comparisons will be instructive. Intraspecific comparisons would reveal effects of digestibility, caused by skeletal material or packing configuration, on gut evacuation rate. Moreover, ontogenetic shifts in digestive constraints are likely to be widespread, as exemplified by the shift from planktivory to molluscivory in pumpkinseed sunfish, *Lepomis gibbosus*, (Osenberg *et al.* 1994) and from carnivory to herbivory in the Chilean clingfish, *Sicyases sanguineus* (Cancino and Castilla 1988). We should also examine the shorter time scale, looking for physiological adjustments, perhaps in suites of enzymes, or morphological adjustments in trophic musculature or gut length, when fish are reared on contrasted diets (e.g. soft-bodied versus hard-shelled prey). The effects of any such modifications on dietary specialisation and subsequent ability to switch diets should be investigated. Interspecific comparisons should reveal phylogenetically-determined morphological components of constraint. Interacting with ontogenetic constraints, phylogenetic digestive constraints are central to the integration of foraging theory and community ecology (Osenberg *et al.* 1994).

6.2.4 The importance of relative prey abundance

The BPM predicts that the set of prey types to be included in the optimal diet depends only on the abundance of higher ranking prey. Even if lower ranking prey are extremely abundant, this in itself should not influence diet selection. The prediction rests on three assumptions:

(1) discrimination costs are insignificant,

(2) prey items are encountered one at a time, and

(3) encounter rates with different prey are mutually independent.

There are circumstances when any of these assumptions may be invalid, causing diet selection to become sensitive to the relative abundances of prey.

Evaluation and discrimination costs

The decision to reject a prey item could occur at any time between orientation and ingestion. Rejection before the item is grasped will often be based solely on visual information. If the period between orientation and grasping comprises only a small fraction of total handling time, visual evaluation may be an insignificant cost. This inference, nevertheless, should be tested by using high-speed videography to measure the time between orientation and rejection. If, on the other hand, further evaluation involving the use of tactile or gustatory stimuli occurs after grasping, evaluation time is likely to become a significant cost. Threespine sticklebacks feeding on the isopod, *Asellus aquaticus*, may reject an item after orientation, but usually grasp the item in the mouth before deciding to accept or reject it. Indeed, larger *Asellus* may be grasped and spat out several times before the decision is made (Hart and Gill 1992). Surprisingly, the time taken to make the decision is approximately constant (about 3 s), irrespective of fish or prey size (Gill and Hart 1994). Incorporation of evaluation (recognition) time into the BPM makes the optimal diet sensitive to relative prey abundance (Hughes 1979).

If a fish sees more than one prey item at once, encounters will be simultaneous (contrary to assumption by the BPM) and the fish is then faced with a problem of discrimination. As more prey items appear within the visual field, the nervous system will have to cope with more information, perhaps at a cost of poorer discrimination or, as discussed below, of decreased attention paid to other matters such as vigilance for predators.

Impaired foraging efficiency caused by 'information overload' (*sensu* Milinski 1990) has been demonstrated by presenting threespine sticklebacks with pairs of *Daphnia* moving at right angles about a central point of intersection (Ohguchi 1981). The sticklebacks attacked the prey less often when they were of similar colour than when different. Moreover, the sticklebacks could not discriminate between a red *Daphnia* and a similar-sized piece of carrot moving simultaneously at right angles to each other, although discrimination did occur when the two were moved in the same plane (horizontally). When the items moved at right angles, the sticklebacks were apparently confused by too much information and could no longer discriminate between them. Information-processing constraints such as this may be expected to increase as the variety and population density of alternative prey increase and as their relative abundances become more even.

Perch, *Perca fluviatilis*, forage less efficiently on *Daphnia* and phantom midge, *Chaoborus* sp., when these are presented simultaneously rather than separately (Persson 1985). The two prey types have different evasive mechanisms. *Daphnia* relies on inconspicuousness, effective at the perception stage of attack, whereas *Chaoborus* relies on fleeing, involving the pursuit stage of attack. Combined information from these contrasted prey may result in sensory confusion or impair the learning of specific handling skills. Further experiments are needed to assess the relative importance of information overload and impaired learning when foragers encounter complex situations. The ability of fishes to cope with multiple sources of information, presented simultaneously or in sequence, should be a priority for investigation, because complex tasks are likely to typify the field situation. How do categories of information, such as prey recognition, prey handling, navigation through the habitat, and recognition of predators interact? Do they mutually interfere with learning and memory? Can they be prioritised in any of their effects? Such questions could be investigated by presenting fish with different behavioural tasks in various sequences under controlled conditions.

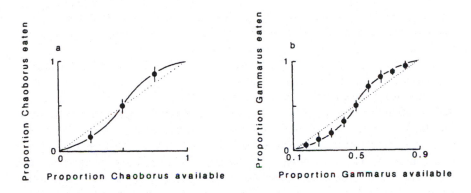

Fig. 6.5 Frequency-dependent selection, shown by sigmoid curves of the proportion of an alternative prey eaten by fifteenspine stickleback, *Spinachia spinachia*, plotted as a function of the proportion available. Data are medians with interquartile ranges. Dotted lines indicate expectations from non-selective feeding. After Hughes and Croy (1993). (a) Midge larvae, *Chaoborus* sp., presented in combination with brine shrimp, *Artemia* sp. (b) Amphipods, *Gammarus* sp., presented in combination with brine shrimp.

Frequency-dependent selection

Many types of predator tend to feed disproportionately on any acceptable prey whose relative abundance is high (Hughes and Croy 1993). If those prey subsequently become scarce, the predator will, after a lag, switch preference to alternatives that have meanwhile become common and feed disproportionately on them in turn. The selection exercised by the predator therefore depends on the relative abundances, or frequencies, of the prey. The BPM itself predicts such frequency-dependent selection if it is modified to allow learning to transpose profitabilities (Hughes 1979). No evidence for this, or any other explanation based on OFT, has yet been demonstrated in fishes. There are, however, many other possible explanations for frequency-dependent selection (Hughes and Croy 1993). One of them, with the longest pedigree, is that the forager develops a 'searching image' enabling it to pick out a specific prey more effectively from the inevitable background of irrelevant stimuli (Tinbergen 1960). The searching image, if such a thing exists, is acquired and maintained through repeated experience and hence depends on the prey being relatively abundant. Despite the large number of investigations (mostly with animals other than fishes) that have been made on frequency-dependent selection, there has been no direct evidence of searching images (Guilford and Dawkins 1987; Hughes and Croy 1993). Rather, the searching image has been invoked merely because data have ruled out other specific explanations, as for example with fifteenspine sticklebacks foraging on brine shrimp and midge larvae (Fig. 6.5a). Without direct evidence in its favour, however, the searching image cannot be accepted as factual. The searching-image concept has proved to be remarkably intangible and other ways of investigating frequency-dependent selection might prove to be more profitable, at least in the short term. Particularly worthy of consideration are 'rules-of-thumb' simple behavioural responses that would have approximately the same effects on diet selection as a more complex, theoretical mechanism.

Another possible mechanism for frequency-dependent selection is the learned improvement of attack efficiency. This does not necessarily shorten handling time, in which case intrinsic profitability will remain unaltered. It does, however, increase the probability of successfully capturing the prey and so the expected profitability per attack (intrinsic profitability × probability of capture) will increase. Whether or not the forager perceives an increase in profitability per attack and uses this information for selective feeding, the increase in attack efficiency will automatically generate disproportionate feeding on the specific prey (Juanes 1994). Both mechanisms might operate synergistically and require distinguishing by careful experimentation. Changes in attack efficiency accompanying frequency-dependent selection have been recorded in fifteenspine sticklebacks foraging on amphipods and brine shrimp (Fig. 6.5b), but yet again without further experimentation the mechanisms most responsible for frequency-dependent selection cannot be unambiguously identified.

6.3   Mechanical constraints on diet selection

6.3.1   Size and morphological constraints

Zooplanktivores tend to have relatively small mouths, piscivores have relatively large and wide gapes, and bottom feeders sometimes have relatively delicate, extensible jaws forming a tube to suck invertebrates out of the sediment. Such morphological adaptations allow foragers to handle particular kinds of prey with great efficiency, but render them less capable of handling other kinds of prey. This well known trade-off between specialisation and generalisation limits the set of prey types that can be handled by any particular forager. Morphological constraints on diet breadth are obvious among taxa at or above the species level, but may also occur intraspecifically during ontogeny and among populations from contrasting habitats. Such intraspecific differences may need to be taken into account when interpreting foraging behaviour, particularly when using laboratory observations to predict diet selection in the field.

For any given body form, body size itself influences the set of prey types that can be handled. Small foragers generally can handle only small prey, or those with feeble defensive or evasive capabilities. Larger foragers can handle larger, stronger or swifter prey, in addition to somewhat smaller items. The set of potential prey types therefore tends to increase with the size of the forager. It is uncertain whether or not handling efficiency for smaller items simultaneously decreases, although the available evidence suggests that it does not (Persson and Greenberg 1990; Juanes 1994).

Ontogenetic changes in body size and morphology

Because of the relationship between handling capability and forager size described above, the prey size at maximum profitability will increase as a fish grows from larva to adult. Larger fish are therefore not only capable of handling a greater range of prey types, but may also be expected to prefer larger prey. In Lake Victoria, Tanzania, the cichlid *Haplochromis piceatus* feeds predominantly on cladocerans and copepods, but also on other organisms, notably midge larvae and pupae (Galis 1990). The average size of prey eaten in the field increases with fish size. Larger fish also tend to eat more prey that are relatively difficult to catch. There is thus a tendency for fish, as they grow, to change preference from cladocerans and copepods to insects. Laboratory experiments show that this dietary shift corresponds to a change in relative profitability of prey, with the net rate of energy intake on a diet of *Chaoborus* larvae becoming increasingly greater than on a diet of *Daphnia* as fish size increases (Galis 1990).

The change in diet described above originates from size constraints on handling ability. There is another change, by contrast, that is attributable to a morphological constraint. During ontogeny, one of the pharyngeal jaw muscles, the retractor dorsalis, assumes an increasingly more vertical position, thereby imparting a greater capacity for crushing during mastication. This enables larger fish to efficiently handle *Chaoborus* pupae, a prey type denied to smaller fish with lesser crushing ability (Galis 1990). Changes in diet accompanying ontogenetic changes in trophic morphology can have important ecological repercussions (Osenberg *et al.* 1994). Small pumpkinseed sunfish feed on soft-bodied littoral invertebrates, whereas larger individuals with stronger jaws and teeth specialise on snails. Small bluegill sunfish feed on the same prey as the small pumpkinseeds, but larger individuals forage among the zooplankton. Although older and larger fish have separate food niches, the juveniles are forced through a competitive bottleneck, which may also affect their foraging performance and food resources available to them during subsequent ontogenetic stages (Osenberg *et al.* 1994).

Population differences in morphology

When populations of the same species live in isolated habitats with different ecological characteristics, natural selection may cause morphological divergence. If this involves the trophic apparatus, then characteristic differences in foraging behaviour may be expected (Skúlason and Smith 1995).

A study of threespine stickleback populations in British Columbia, Canada has revealed the existence of different morphotypes in small and large lakes (Lavin and McPhail 1986); these morphotypes may even have been differentiated to the species level (McPhail 1992; Day *et al.* 1994). In small, eutrophic lakes dominated by benthic habitats, sticklebacks have relatively longer jaws with associated wider gape, which facilitate the handling of bulky, benthic prey. In large, oligotrophic lakes dominated by open-water habitat, sticklebacks have relatively longer and more closely-spaced gill rakers, which increases handling efficiency for small,

zooplanktonic prey. Rearing experiments verify the genetic control of these morphological differences, and behavioural experiments show a corresponding difference in foraging success, with sticklebacks from large lakes performing better
than similar-sized individuals from small lakes when presented with zooplanktonic
prey. Moreover, rearing on diets of pelagic or benthic prey can modify the jaw
structure and foraging behaviour of the sticklebacks. The pelagic (limnetic) form
is more labile than the benthic form in this respect, perhaps being genetically
programmed for greater phenotypic plasticity in correspondence with their shift
from pelagic to benthic habitats during the breeding season (Day *et al.* 1994). Similar differences in habitat, morphology and foraging efficiency have been found in
Scottish populations of the threespine stickleback (Ibrahim and Huntingford 1988).
Such trophic polymorphisms are of considerable interest with regard to sympatric speciation, since they may occur even within populations when these inhabit
lakes with extensive, structurally contrasted habitats (Wimberger 1994; Skúlason
and Smith 1995).

6.3.2 Packing constraints

As satiation approaches, there may be insufficient space left in a fish's stomach to
accommodate larger items, which therefore may be rejected whatever their profitability. In an experiment testing predictions of the BPM (Hart and Gill 1992),
threespine sticklebacks with empty stomachs readily accepted 8-mm *Asellus*. As
their stomachs filled, they became increasingly reluctant to accept these larger
items, but continued to readily accept 5-mm prey. The data, however, are con-
founded by the possibility that the sticklebacks were responding indirectly to
profitability, since 8-mm items were less profitable on average than 5-mm ones.
Another experiment in which sticklebacks were presented with a wider range of
prey sizes (Hart and Ison 1991) showed that the fish most readily accepted 3-mm
prey, even though these were less profitable than 5-mm items. Moreover, 3-mm
items were not in the optimal set predicted by the BPM. These considerations seem
to support the packing-constraint hypothesis (Hart and Gill 1992). These results
should be interpreted with caution, however, since the handling time, and hence
profitability, of 3-mm prey was much less variable than that of larger items. Possibly foragers perceive the profitability of an item as an expected value, equivalent to the statistical mean or median for that type of prey, discounted by some
measure of its variance. More experiments addressing this problem are needed.

The assumption, made by the BPM, that prey types have unique predictable
profitabilities clearly is an approximation to reality. In many cases it may be a very
poor approximation. For example, if the vulnerability to capture, ease of handling
or yield of assimilable energy of prey are not strongly correlated with stimuli that
lead the fish to attack them, then prey quality will vary considerably. By offering
fish with distinguishable artificial prey of constant mean profitability but with different variances in profitability, it should be possible to test whether or not the

forager is influenced by variance per se. This is a problem in 'risk-sensitive for-aging' (Caraco *et al.* 1980) and may be expected to implicate the fish's energetic state. Fish with precarious energy budgets, particularly small individuals with low energy stores and high specific metabolic rates, would be expected to be particu-larly sensitive to the variance in quality of any prey type. Conversely, larger in-dividuals might be expected to be less sensitive. Measurements of the variance in the profitability of natural prey types need to be made in order to assess the gen-erality of this problem.

6.4 Indicators of prey profitability

The energy yield of a prey item is proportional to its mass and digestibility. Handling time may be a complex, partial function of size relative to the forager, perhaps remaining constant over a range of smaller sizes, then accelerating towards infinity as the maximum capacity of the forager's feeding apparatus is reached. It may also partly depend on body shape, presence or absence of appendages and defensive mechanisms, speed and agility, and on the environmental context (e.g. in the water column, among weed, within sediment, attached to stones) in which the prey item occurs.

Since profitability may be the result of so many variables, how can it be assessed quickly enough to influence the rapid decisions that an 'optimal forager' would have to make during a foraging bout? One possibility is that, through experience or genetic programming, foragers come to respond to a small set of key stimuli in-dicating the probable profitability of each item encountered. Variance associated with the correlation between key stimuli and profitability would mean that some errors of judgement might be made, but if the average accuracy were high enough, response to a simple set of stimuli would result in a close approximation to the-oretically optimal diet selection.

If prey are small relative to the size of the forager, handling time will generally be unimportant and size of the prey should reliably indicate profitability, provided that there are no marked differences in defensive or evasive capabilities among prey. Accordingly, threespine sticklebacks exhibit size-selective predation when foraging on cladocerans, preferring larger items and achieving close to the the-oretically optimal diet (Gibson 1980). In this case, size is a sufficient predictor of profitability.

Variables additional to size become important when a greater diversity of prey is involved. In such a situation, laboratory trials showed that threespine stickle-backs responded preferentially and in rank order to red colour, fast movement, straight shape and larger size, except if all prey were pale, when movement be-came more important than colour (Ibrahim and Huntingford 1989). Given choices of paired prey in the laboratory, the sticklebacks showed consistent preferences within pairs; these preferences sometimes corresponded to the more profitable prey, but in other cases they did not. (The experiments were designed to examine

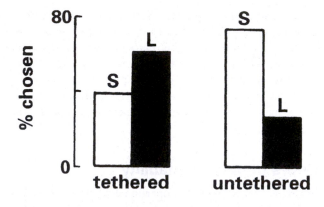

Fig. 6.6   Prey-size selection by blennies, *Lipophrys pholis*, feeding on shore crabs, *Carcinus maenas*. Left panel: tethered prey presented in pairs, each comprised of a smaller and a larger individual ($n = 542$). Right panel: untethered prey presented in mixed groups of smaller and larger individuals ($n = 90$). After ap Rheinallt (1982).

the role of visual cues in diet selection, not to test the BPM, which requires encounters to be sequential rather than simultaneous). Combinations of prey presented to the sticklebacks, however, did not necessarily represent those likely to be encountered in the field, where certain types of prey tend to be segregated among different habitats, even on small spatial scales. In field experiments using cages that enclosed different types of habitat (Ibrahim and Huntingford 1989), sticklebacks showed the same preferences for prey stimuli as in the laboratory, but they preferentially selected zooplankton over benthic prey in the field, corresponding to their general rank order of profitability. Evidently, the use of simple sets of stimuli for optimal diet selection will be reliable only in a realistic context, incorporating the essential features of natural habitat structure and prey distribution. For this reason alone, laboratory observations on diet selection should be interpreted cautiously, paying heed to possible limitations imposed by experimental simplicity.

Moreover, differences in experimental protocol, by affecting the manifestation of specific stimuli, may alter the outcome of selective feeding. When blennies, *Lipophrys pholis*, were presented with a tethered pair of smaller and larger shore crabs, *Carcinus maenas*, they usually chose the larger (Fig. 6.6) even though it had a lower profitability. Conversely, when presented with larger numbers of untethered crabs, blennies tended to prefer the smaller ones (Fig. 6.6). Conducted on its own, the second experiment might have led to the conclusion that blennies behaved in a way concordant with OFT, by selecting the more profitable prey type. The two experiments, however, showed that selective feeding was based on a hierarchical response to stimuli. Movement was the strongest stimulus, followed by size. Because movement was absent when prey were tethered, the blennies responded to the next strongest stimulus and chose the larger prey. When prey were untethered, movement became the dominant stimulus and smaller prey were attacked more frequently because they were more active than larger prey. The reason for this difference in activity was that larger crabs monopolised refuges round the edges of the aquarium, aggressively repelling smaller crabs that were then forced to keep searching for places to hide. Changes in diet selection according to environmental context clearly deserve further investigation.

6.5 Competition among foragers

Competition during a foraging bout could affect diet selection in two ways: it could accelerate the depletion of prey, thereby reducing encounter rates and so discouraging selective feeding (= exploitative or scramble competition), and it could alter the behaviour of foragers independently of prey availability (= interference or contest competition). Accelerated prey depletion does not necessarily involve interaction among competitors and, other things being equal, all competitors will be affected in a similar way. Interactions among competitors, in contrast, may cause differential changes in foraging behaviour that depend on competitive rank.

Prey depletion in a competitive situation could be perceived by the forager simply as a reduction in encounter rates with prey. Here, the BPM would predict a broadening of the diet in proportion to the scale of depletion of the more profitable items. On the other hand, depletion could be anticipated by the forager, having noticed the presence of competitors. This might cause the forager to adopt an opportunistic strategy on a 'first-come-first-served' basis, whence the diet should broaden independently of encounter rates with prey. For example, isolated juvenile coho salmon, *Oncorhynchus kisutch*, showed an increased readiness to attack prey, and a corresponding broadening of the diet, when they could see their own reflection in a mirror (Dill and Fraser 1984).

If foragers differ considerably in competitive ability, the optimal feeding strategy might then depend on the individual's rank. Competition may be of the exploitative type, not involving aggressive dominance, or of the interference type involving a dominance hierarchy. Exploitative competition in threespine sticklebacks causes weaker competitors to accept prey that are intrinsically less profitable (Milinski 1982). This maximises the inferior competitor's rate of energy intake by avoiding specialisation on prey likely to be pre-empted by superior competitors. Feeding opportunities for superior competitors are unlikely to be significantly diminished by the behaviour of other competitors, and so superior individuals may be expected to retain preference for the intrinsically most profitable prey, as shown for threespine stickleback (Milinski 1982). Although competitively weaker sticklebacks accept a proportion of intrinsically less profitable prey, they do so in a selective manner, consistent with the effective profitability of the prey in competitive circumstances (Milinski 1982). Aggressive (interference) competition, on the other hand, may force subdominants to feed opportunistically, and therefore unselectively, between threats from dominants (Croy and Hughes 1991c). Possible differences in the ways that exploitative and interference competition influence foraging behaviour have not yet adequately been studied. In contrast to exploitative competition, which may be expected to have relatively simple consequences through the depletion of local resources, interference competition is likely to have powerful, potentially complex effects on foragers that form a dominance hierarchy. Particular attention should be paid to individual foraging efficiency and in the way this might change as removal experiments allow individuals to move upwards in the dominance hierarchy.

6.6 Risk of predation while foraging

The risk of falling prey to another forager while foraging oneself may influence many aspects of behaviour, including timing and location of feeding, searching mode, prey-handling method, vigilance and social behaviour (Lima and Dill 1990). Through these mechanisms, risk of predation while foraging may directly or indirectly influence diet selection, causing a forager to broaden its diet or even to prefer safer but less profitable prey (Godin 1990; Sih 1993). Only direct influences are considered in the present section; the indirect influence through choice of feeding location is considered in Section 6.7.

The BPM itself can be adapted to predict optimal diet selection in the face of predation risk by adding several further assumptions (Gilliam 1990). These are:

(1) there is an energy-budget constraint requiring the forager to achieve a certain minimum rate of energy gain per bout,

(2) the individual is no longer at risk to predation once it stops foraging, and

(3) fitness is promoted both by increasing the rate of energy intake and increasing survivorship.

If selecting the energetically most profitable diet places the forager at greater risk to predation, this last assumption changes the objective function (see Section 6.2.1) from one of maximising the energy gain per unit foraging time to one of maximising the energy gain per unit risk; that is, from [max (E/T)] to [max $(E/T)*(T/M) = \max (E/M)$], where M is the average number of individuals killed by predators while foraging. The optimal diet, therefore, is envisaged to be a compromise between maximising E and minimising M. The model can be used to predict the dietary consequences when prey-handling becomes more hazardous than searching, or vice versa. In the first case, the forager should have narrower diets than predicted by the BPM and broader in the second. Gilliam (1990) warns that it would be an insufficient test of the model simply to compare diets in the presence and absence of predation risk. This is because the forager might adjust its searching mode and prey-handling methods on perceiving the threat of predation, whereas the present model inherits from the BPM the assumptions of constant encounter rates and profitabilities. An appropriate test would be to measure any changes in searching and handling behaviour and then compare the relative success of the BPM and the present model in predicting the observed diets, noting particularly whether the BPM makes systematic errors in prediction.

Previous experiments have not been designed in ways suitable for Gilliam's model to be rigorously tested, but they do show that perceived risk of predation can influence diet selection in a manner consistent with the model. For example, after being subjected to a simulated threat of predation, both juvenile coho salmon (Dill and Fraser 1984) and juvenile Atlantic salmon, *Salmo salar*, (Metcalfe *et al.* 1987) become more reluctant to move away from their refuge to intercept prey items drifting past in the current. This results in more discrimination errors and tends to make the fish take closer, safer prey items in preference to more distant but more profitable ones. The data, therefore, provide evidence of behaviour optimising the balance between rate of energy gain and survivorship.

In more specific agreement with Gilliam's model, guppies, *Poecilia reticulata*, forage selectively on smaller sizes of waterfleas, *Daphnia magna*, when a predator is in sight, but when it is not they feed on larger prey size as predicted by the BPM for the specified conditions (Godin 1990). Contrary to the assumptions of the latter model, encounter rates, handling times, profitability, reaction distance and attack efficiency were not constant, but changed according to the presence or absence of the simulated threat of predation. More such experiments are needed to examine the detailed changes in behaviour when a forager perceives predation risk and causally relate this to any changes in diet selection.

6.7 Location of food sources

Because spatial and temporal patterns of prey abundance tend to be non-random and because different prey types are often segregated by habitat, choice of foraging location may strongly influence diet and opportunities for selective feeding. One result of this may be the differential use of habitats by foragers, a topic that is treated fully by Kramer *et al.* (Chapter 3) and Persson *et al.* (Chapter 12). The present discussion will examine ways in which fishes maximise their ability to locate sources of food.

To discover sources of food, fishes must explore their habitats. This may involve frequent, even almost continual, movement from place to place in those species that search out their prey, or it may involve occasional shifts of location in ambushers (see Hart, Chapter 5). Each case is a form of sampling. Location of food will be made more efficient by any process that reduces the necessary amount of sampling. This has been studied in fishes that search for their food, for which two mechanisms of minimising sampling effort have been identified.

The simpler of these two mechanisms applies to shoaling species and has been termed 'forage-area copying'. Here, members of the shoal seeing other members actively feeding will join to feed in the same area. Each member thereby benefits from the combined searching capability of the entire shoal. Forage-area copying has been demonstrated in goldfish, by using a transparent barrier to separate two shoals. One shoal was presented with a good and a poor food patch and the other with two equally productive food patches. The second shoal showed significant preference for the patch next to the 'good' patch on the opposite side of the barrier, where the first shoal could be seen actively foraging (Pitcher and House 1987). The other mechanism is topographical learning, which enables a forager to relocate productive sources of food by navigation and homing, using visual information about the habitat. Goldfish, for example, will learn to associate local visual cues, such clumps of weed, with the proximity of food (Warburton 1990). Once experienced in this way, the fish more quickly locate the food on entering the foraging area. Location efficiency is enhanced further by cognisance of 'global' cues, which can be used to navigate before the local cues are in sight. Global cues could include any general topographical features of the habitat.

6.8 Dynamic programming

The BPM is static: it assumes that all attributes of the forager, its prey and its immediate environment remain constant during the foraging bout. From the above discussions, it is clear that many of these attributes do not remain constant, but change in ways that influence diet selection. A dynamic model is needed to accommodate such changes. One technique that has proved to be useful for modelling a wide range of behavioural problems involving changing attributes is stochastic dynamic programming (SDP) (Mangel and Clark 1988).

SDP first specifies a measure of fitness that is to be maximised. In the case of diet selection, fitness could be measured as the probability of being alive (maximum = 1.0) or as stomach fullness (maximum = total stomach capacity) at the end of the foraging bout. If a time period encompassing more than one foraging bout were to be modelled, then stomach fullness might be replaced by stored energy (maximum = total storage capacity). A fitness function is then specified in the form of a difference equation, relating fitness to the state of the forager.

To identify optimal behavioural actions, the state of the forager at the end of the foraging bout is assumed to be at a level maximising fitness (e.g. survivorship = 1.0, or stomach is full). Knowing this terminal value for the left-hand side of the difference equation, the right-hand side can be solved, giving the state of the forager one time unit into the past, together with the optimal behavioural action taken at that time. Now that the state of the forager one time unit into the past is known, this can be substituted into the left-hand side of the difference equation and the state at two time units into the past computed. Iterating further into the past gives a set of values for the state variable and associated optimal behavioural actions. Generally, the optimal policy becomes consistent when sufficient iterations into the past have been made and this stable result is taken to represent the optimal behavioural action sought.

Hart and Gill (1993) used SDP to predict the behaviour of threespine sticklebacks feeding on smaller and larger isopods (see Section 6.3.2). In their model, fitness is represented as survivorship, with a value of 1.0 taken as the terminal value and used to begin the iteration. The state variable is stomach fullness, with a maximum value set by volumetric constraint and a minimum, non-zero value below which the fish dies of starvation (an unrealistic assumption within a single foraging bout, but one which allows other constraints to be explored). The decision to accept larger prey is constrained by the remaining stomach capacity; if this is too small, then larger prey must be rejected. If during one time unit no prey item is encountered, the stomach empties by a specified amount. If a prey item is eaten, the stomach fills by an amount equal to the difference between the yield from the prey and the depletion during the pre-ingestive handling phase. If a prey item is rejected, the stomach empties by an amount proportional to the handling time prior to rejection. Stomach fullness therefore is a dynamic variable, responding to the stochastic process of prey encounter (hence stochastic dynamic programming). The model predicts increased selectivity towards smaller prey as the stomach fills and gives a much better fit to the observed diets than the BPM.

Godin (1990) used SDP to predict the optimal behaviour of guppies facing the choice of different sized prey, *Daphnia magna*, and the perceived risk of predation (a cichlid placed behind a transparent screen). In his model, fitness is represented as the net energy gained from foraging and equals the energetic value of the stomach content. Maximum stomach capacity is taken to be the fitness value at the end of the foraging bout and the starting value for iteration. Energetic content of the stomach is also the state variable, responding, as in the previous example, to the stochastic process of prey encounter. There is assumed to be no risk of starvation and therefore no lower limit above zero for stomach fullness. The state of the forager, however, is weighted by survivorship, which takes values less than 1.0 when modelling various levels of perceived risk to predation. The model predicts that when free from predation risk guppies should feed unselectively until approaching satiation, when volumetric constraint dictates that first the largest and then progressively smaller items should be ignored. When threatened by predation, guppies should begin the dietary exclusion of progressively smaller items at a much lower level of stomach fullness, ceasing to feed altogether when close to complete satiation. Again, the SDP model gives a closer fit to observed diets than the BPM.

6.9 Conclusions

The increasing use of dynamic models of foraging behaviour in place of the original static models spells the increasing willingness of behavioural ecologists to address the natural complexities that they have always recognised to be important. Initial progress required the stark simplicity of the static models, and for certain specific applications this still may be advantageous: it should not be a question of fashion, the old models should be retained in our theoretical repertoire. Nevertheless, an increasingly multidisciplinary approach to foraging behaviour is inevitable and the conceptual distinctness of OFT will fade (Hughes 1993).

Among the developments that are likely to be important are the following. First, there will be increasing attention paid to interactions at the population and community levels. Some models have predicted that optimal foraging behaviour would have no significant impact on prey-population structure (Pitcher and Turner 1990), but others have predicted strong interactions between trophic levels (Gilliam 1990; Osenberg *et al.* 1994; Kramer *et al.*, Chapter 3; Persson *et al.*, Chapter 12).

Second, learning will be seen to play an important role in all aspects of foraging behaviour. Of particular interest will be the rate of learning and the length of the memory window in relation to the type of information being learned and to the life history of the fish. Does memory depend on whether the information is about habitat features, characteristics of prey or the identity of predators? Do learning and memory characteristics change during ontogeny and do they differ between shorter-lived and longer-lived species (Hughes *et al.* 1992)?

Third, there will be a focus of interest on genetically determined differences in behavioural traits among individuals. Such traits are susceptible to selection, whether this be natural or a product of human activities. There is mounting evidence suggesting that fish farming generates selection pressures resulting in phenotypes that have lowered fitness under natural circumstances. If allowed by accident or design to interbreed with them, farmed fish could threaten natural stocks. For example, on the fish farm, there is little or no risk from natural predators but there is intense competition for food. In the case of rainbow trout, *Oncorhynchus mykiss*, larger size increases competitive ability. This regime selects for strong feeding motivation, increasing food intake and body growth. The enhanced feeding motivation, however, makes the individuals less responsive to the sight of predators and increases their risk of mortality (Milinski 1984; Godin and Smith 1988). Farming therefore selects 'high gain-high risk' phenotypes that may be disadvantageous in natural habitats (Johnsson 1993).

Such demonstration of genetically based variation in foraging behaviour vindicates the evolutionary approach described in this chapter. The genetics of foraging behaviour promise to be a rich area for development. Using molecular techniques it should be possible to identify genetically based variation in foraging behaviour at all scales, from intrapopulation to geographical (Shaw *et al.* 1994).

Acknowledgements

I am most grateful to Jean-Guy Godin, Paul Hart, Felicity Huntingford, Michel Kaiser, Manfred Milinski and an anonymous referee for suggesting ways of improving the manuscript.

References

ap Rheinallt, T. (1982). *The foraging behaviour of some marine predators*. Ph.D. thesis, University of Wales, Bangor.

Armstrong, J.D, Priede, I.G. and Lucas, M.C. (1992). The link between respiratory capacity and changing metabolic demands during growth of northern pike, *Esox lucius* L. *J. Fish Biol.*, **41** (*Suppl.* B), 65–75.

Beukema, J.J. (1968). Predation by the three-spined stickleback (*Gasterosteus aculeatus*): the influence of hunger and experience. *Behaviour*, **31**, 1–126.

Bergelson, J.M. (1985). A mechanistic interpretation of prey selection by *Anax junius* larvae (Odonata: Aeschnidae). *Ecology*, **66**, 1699–1705.

Cancino, J.M. and Castilla, J.C. (1988). Emersion behaviour and foraging ecology of the common Chilean clingfish *Sicyases sanguineus* (Pisces, Gobiesocidae). *J. Nat. Hist.*, **22**, 249–261.

Caraco, T., Martindale, S. and Whittam, T.S. (1980). An empirical demonstration of risk-sensitive foraging preferences. *Anim. Behav.*, **28**, 820–830.

Charnov, E.L. (1976). Optimal foraging: attack strategy of a mantid. *Am. Nat.*, **110**, 141–151.

Colgan, P.W. (1973). Motivational analysis of fish feeding. *Behaviour*, **45**, 38–66.

Croy, M.I. and Hughes, R.N. (1990). The combined effects of learning and hunger in the feeding behaviour of the fifteenspine stickleback (*Spinachia spinachia* L.). on the acquisition of learned foraging skills by the fifteenspine stickleback, *Spinachia spinachia* L. *Anim. Behav.*, **41**, 161–170.

Croy, M.I. and Hughes, R.N. (1991a). The role of learning and memory in the feeding behaviour of the fifteenspine stickleback, *Spinachia spinachia* L. *Anim. Behav.*, **41**, 149–159.

Croy, M.I. and Hughes, R.N. (1991b). The influence of hunger on feeding behaviour and on the acquisition of learned foraging skills in the fifteenspine stickleback, *Spinachia spinachia* L. *Anim. Behav.*, **41**, 161–170.

Croy, M.I. and Hughes, R.N. (1991c). Effects of food supply, hunger, danger and competition on choice of foraging location by the fifteenspine stickleback, *Spinachia spinachia* L. *Anim. Behav.*, **42**, 131–139.

Day, T., Pritchard, J. and Schluter, D. (1994). Ecology and genetics of phenotypic plasticity: a comparison of two sticklebacks. *Evolution*, **48**, 1723–1734.

Dill, L.M. (1987). Animal decision making and its ecological consequences: the future of aquatic ecology and behaviour. *Can. J. Zool.*, **65**, 803–811.

Dill, L.M. and Fraser A.H.G. (1984). Risk of predation and the feeding behaviour of juvenile coho salmon (*Oncorhynchus kisutch*). *Behav. Ecol. Sociobiol.*, **16**, 65–71.

Emlen, J.M. (1966). The role of time and energy in food preference. *Am. Nat.*, **100**, 611–617.

Galis, F. (1990). Ecological and morphological aspects of changes in food uptake through the ontogeny of *Haplochromis piceatus*. In *Behavioural mechanisms of food selection* (ed. R.N. Hughes), NATO ASI Series Vol. G20, pp. 281–302. Springer-Verlag, Berlin.

Gibson, R.M. (1980). Optimal prey-size selection by three-spined sticklebacks (*Gasterosteus aculeatus*): a test of the apparent-size hypothesis. *Z. Tierpsychol.*, **52**, 291–307.

Gill, A.B. and Hart, P.J.B. (1994). Feeding behaviour and prey choice of the threespine stickleback: the interacting effects of prey size, fish size and stomach fullness. *Anim. Behav.*, **47**, 921–932.

Gilliam, J.F. (1990). Hunting by the hunted: optimal prey selection by foragers under predation hazard. In *Behavioural mechanisms of food selection* (ed. R.N. Hughes), NATO ASI Series Vol. G20, pp. 797–819. Springer-Verlag, Berlin.

Godin, J.-G.J. (1990). Diet selection under the risk of predation. In *Behavioural mechanisms of food selection* (ed. R.N. Hughes), NATO ASI Series Vol. G20, pp. 739–770. Springer-Verlag, Berlin.

Godin, J.-G.J. and Smith S.A. (1988). A fitness cost of foraging in the guppy. *Nature, Lond.*, **333**, 69–71.

Guilford, T and Dawkins, M.S. (1987). Search image not proven: a reappraisal of recent evidence. *Anim. Behav.*, **35**, 1838–1845.

Hart, P.J.B. and Gill, A.B. (1992). Constraints on prey size selection by the three-spined stickleback: energy requirements and the capacity and fullness of the gut. *J. Fish Biol.*, **40**, 141–314.

Hart, P.J.B. and Gill, A.B. (1993). Choosing prey size: a comparison of static and dynamic foraging models for predicting prey choice by fish. *Mar. Behav. Physiol.*, **23**, 91–104.

Hart, P.J.B. and Hamrin, S.F. (1990). The role of behaviour and morphology in the selection of prey by pike. In *Behavioural mechanisms of food selection* (ed. R.N. Hughes), NATO ASI Series Vol. G20, pp. 235–254. Springer-Verlag, Berlin.

Hart, P.J.B. and Ison, S. (1991). The influence of prey size and abundance, and individual phenotype, on prey choice by three-spined stickleback, *Gasterosteus aculeatus* L. *J. Fish Biol.*, **38**, 359–372.

Hughes, R.N. (1979). Optimal diets under the energy maximisation premise: the effects of recognition time and learning. *Am. Nat.*, **113**, 209–221.

Hughes, R.N. (1980). Optimal foraging theory in the marine context. *Oceanogr. Mar. Biol. Ann. Rev.*, **18**, 423–481.

Hughes, R.N. (ed.) (1993). *Diet selection: an interdisciplinary approach to foraging behaviour.* Blackwell Scientific Publ., Oxford.

Hughes, R.N. and Croy, M.I. (1993). An experimental analysis of frequency-dependent predation (switching) in the 15-spined stickleback, *Spinachia spinachia. J. Anim. Ecol.*, **62**, 341-352.

Hughes, R.N., Kaiser, M.J., Mackney, P.A. and Warburton, K. (1992). Optimizing foraging behaviour through learning. *J. Fish Biol.*, **41** (*Suppl.* B), 77–91.

Ibrahim, A.A. and Huntingford, F.A. (1988). Foraging efficiency in relation to within-species variation in morphology in three-spined sticklebacks, *Gasterosteus aculeatus. J. Fish Biol.*, **33**, 823–824.

Ibrahim, A.A. and Huntingford, F.A. (1989). Laboratory and field studies on diet choice in three-spined sticklebacks, *Gasterosteus aculeatus* L., in relation to profitability and visual features of prey. *J. Fish Biol.*, **34**, 245–257.

Ivlev, V.S. (1961). *Experimental ecology of the feeding of fishes.* Yale University Press, New Haven.

Johnsson, J.I. (1993). Big and brave: size selection affects foraging under risk of predation in juvenile rainbow trout, *Oncorhynchus mykiss. Anim. Behav.*, **45**, 1219–1225.

Juanes, F. (1994). What determines prey size selectivity in piscivorous fishes? In *Theory and application in fish feeding ecology* (eds. D.J. Stouder, K.L. Fresh and R.J. Feller), pp. 79–100. University of South Carolina Press, Columbia, South Carolina.

Kaiser, M.J., Westhead, A.P., Hughes, R.N. and Gibson, R.N. (1992). Are digestive characteristics important contributors to the profitability of prey? *Oecologia*, **90**, 61–69.

Kislalioglu, M. and Gibson, R.N. (1976). Some factors governing prey selection by the fifteen-spined stickleback, *Spinachia spinachia. J. Exp. Mar. Biol. Ecol.*, **25**, 159–169.

Krebs, J.R. and Davies, N.B. (eds.) (1991). *Behavioural ecology: an evolutionary approach*, 3rd edn. Blackwell Scientific Publ., Oxford.

Lavin, P.A. and McPhail, J.D. (1986). Adaptive divergence of trophic phenotype among freshwater populations of the threespine stickleback (*Gasterosteus aculeatus*). *Can. J. Fish. Aquat. Sci.*, **43**, 2455–2463.

Lester, N.P. (1984). The feed: feed decision: how goldfish solve the patch depletion problem. *Behaviour*, **89**, 175–199.

Lima, S.L. and Dill, L.M. (1990). Behavioral decisions made under the risk of predation: a review and prospectus. *Can. J. Zool.*, **68**, 619–640.

MacArthur, R.H. and Pianka, E.R. (1966). On optimal use of a patchy environment. *Am. Nat.*, **100**, 603–609.

Mackney, P.A. and Hughes, R.N. (1995). Foraging behaviour and memory window in sticklebacks. *Behaviour*, **132**, 1241–1253.

McNamara, J.M. and Houston, A.I. (1987a). Partial preferences and foraging. *Anim. Behav.*, **35**, 1084–1099.

McNamara, J.M. and Houston, A.I. (1987b). Memory and the efficient use of information. *J. theor. Biol.*, **125**, 385–396.

McPhail, J.D. (1992). Ecology and evolution of sympatric sticklebacks (*Gasterosteus*): evidence for a species-pair in Paxton Lake, Texada Island, British Columbia. *Can. J. Zool.*, **70**, 361-369.

Mangel, M. and Clark, C.W. (1988). *Dynamic modeling in behavioral ecology*. Princeton University Press, Princeton.

Metcalfe, N.B., Huntingford, F.A. and Thorpe, J.E. (1987). The influence of predation risk on the feeding motivation and foraging strategy of juvenile Atlantic salmon. *Anim. Behav.*, **35**, 901–911.

Milinski, M. (1982). Optimal foraging: the influence of intraspecific competition on diet selection. *Behav. Ecol. Sociobiol.*, **11**, 109–115.

Milinski, M. (1984). A predator's costs of overcoming the confusion-effect of swarming prey. *Anim. Behav.*, **32**, 1157–1162.

Milinski, M. (1990). Information overload and food selection. In *Behavioural mechanisms of food selection* (ed. R.N. Hughes), NATO ASI Series Vol. G20, pp. 721–737. Springer-Verlag, Berlin.

Milinski, M. and Regelmann, K. (1985). Fading short-term memory for patch quality in sticklebacks. *Anim. Behav.*, **33**, 678–680.

Ohguchi, O. (1981). Prey density and selection against oddity by three-spined sticklebacks. *Adv. Ethol.*, **23**, 1–79.

Osenberg, C.W., Olson, M.H. and Mittelbach, G.G. (1994). Stage structure in fishes: resource productivity and competition gradients. In *Theory and application in fish feeding ecology* (eds. D.J. Stouder, K.L. Fresh and R.J. Feller), pp. 151–170. University of South Carolina Press, Columbia, South Carolina.

Persson, L. (1985). Optimal foraging: the difficulty of exploiting different feeding strategies simultaneously. *Oecologia*, **67**, 338–341.

Persson, L. and Greenberg, L.A. (1990). Interspecific and intraspecific size class competition affecting resource use and growth of perch, *Perca fluviatilis*. *Oikos*, **59**, 97–106.

Pitcher, T.J. and House, A.C. (1987). Foraging rules for group feeders: area copying depends upon food density in shoaling goldfish. *Ethology*, **76**, 161–167.

Pitcher, T.J. and Turner, G.F. (1990). The role of the optimal diet predator in multispecies fishery assessment. In *Behavioural mechanisms of food selection* (ed. R.N. Hughes), NATO ASI Series Vol. G20, pp. 847–862. Springer-Verlag, Berlin.

Pyke, G.H. (1984). Optimal foraging theory: a critical review. *A. Rev. Ecol. Syst.*, **15**, 523–575.

Schoener, T.W. (1971). Theory of feeding strategies. *A. Rev. Ecol. Syst.*, **2**, 369–404.

Shaw, P.W., Carvalho, G.R., Magurran, A.E. and Seghers, B.H. (1994). Factors affecting the distribution of genetic variability in the guppy, *Poecilia reticulata*. *J. Fish Biol.*, **45**, 875–888.

Sih, A. (1993). Effects of ecological interactions on forager diets: competition, predation risk, parasitism and prey behaviour. In *Diet selection: an interdisciplinary approach to foraging behaviour* (ed. R.N. Hughes), pp. 182–211. Blackwell Scientific Publ., Oxford.

Skúlason, S. and Smith, T.B. (1995). Resource polymorphisms in vertebrates. *Trends Ecol. Evol.*, **10**, 366–370.

Stephens, D.W. and Krebs, J.R. (1986). *Foraging theory*. Princeton University Press, Princeton.

Tinbergen, L. (1960). The natural control of insects in pinewoods. I. Factors influencing the intensity of predation by songbirds. *Arch. Neerland. Zool.*, **13**, 265–343.

Warburton, K. (1990). The use of local landmarks by foraging goldfish. *Anim. Behav.*, **40**, 500–505.

Werner, E.E. and Hall, D.J. (1974). Optimal foraging and the size selection of prey by the bluegill sunfish (*Lepomis macrochirus*). *Ecology*, **55**, 1042–1052.

Wimberger, P.H. (1994). Trophic polymorphisms, plasticity and speciation in vertebrates. In *Theory and application is fish feeding ecology* (eds. D.J. Stouder, K.L. Fresh and R.J. Feller), pp. 19–43. University of South Carolina Press, Columbia, South Carolina.

7 *Avoiding and deterring predators*

R. Jan F. Smith

7.1 Introduction

The threat of predation has profound and pervasive effects on prey fishes. If we begin with structure, the body shape, size, the degree and type of body armour, and the size and arrangement of spines are all subject to selection by predators. Colours that may attract mates can also make an animal conspicuous to predators, leading to seasonal changes in appearance or to colours that may be less than optimal for mate attraction. Control of seasonal colour changes is probably a physiological response to predation, as is the production of toxins and venoms. Behavioural responses to predation are very diverse and include avoiding dangerous areas, hiding at certain times of day and recognising predators and fleeing or hiding from them, as well as evasive tactics at the moment of attack.

Despite the direct and potentially fatal nature of predatory attack, fishes cannot devote all their time and energy to defence. They must also feed and mate, fundamental requirements that are often in conflict with safety from predators (e.g. Lima and Dill 1990). It is better to skip a meal than to be eaten, but if a fish skips too many meals it will never get a chance to breed. This means that the antipredator benefits of each defensive mechanism must be balanced against its effects on reproductive success and requirements for that success, including foraging (see Hart, Chapter 5; Hughes, Chapter 6), migration (see Dodson, Chapter 2) and the defence of spawning sites or young (see Grant, Chapter 4; Sargent, Chapter 11).

The predators will be different in different habitats and at different stages in a prey individual's life cycle. Fishes must therefore have the flexibility to respond to new predators and to shift the balance between defence and other needs as they grow and migrate. This means that learning may play a large role in antipredator behaviour, but there may also be unlearned ontogenetic shifts in behaviour, physiology or morphology.

The task of this chapter and the one by Godin (Chapter 8) on defence against predators is to examine our current knowledge of how fish cope with predation. On the basis of that examination, we will attempt a synthesis of that knowledge and suggest directions for future research. This is a subject that is ripe for theoretical analysis and experimentation. Too often traits have simply been assumed to be beneficial without testing the benefits or considering the costs, particularly the costs of lost opportunities.

Lima and Dill (1990) have recently formalised the individual behavioural decisions and outcomes that could occur in an encounter between predator and prey Fig. 7.1). Each behavioural or morphological feature that I discuss in this chapter has some effect on one or more of the decision points in such an encounter. Because this chapter will deal with the avoidance and deterrence of predation, while Godin (Chapter 8) deals with escape from predators, I will emphasise the early stages of an encounter and mechanisms that allow prey to avoid encounters completely (increasing $1 - p - q$ in Fig. 7.1). However, it is impossible to draw an absolute line between the various stages of the predation sequence. For example, effective defences such as toxins, spines or good escape ability may not only increase the prey's chances of escape but also increase the probability that predators will ignore prey (i_1 and i_2 in Fig. 7.1), particulary if defences are combined with aposematic (warning) or pursuit deterrence signals.

There are relatively few direct tests of the effectiveness of defensive mechanisms. Benefits are often assumed, perhaps correctly, but without specific tests and the costs of presumed defensive adaptations are virtually never addressed. Yet to understand the evolution of defensive mechanisms we must have some idea of both the real benefits and costs. I will use selected examples to illustrate the types of study that can be done. Experimental and theoretical work in this area has tended to concentrate on a few situations, predator inspection for example, while many other interesting problems, such as costs and benefits of crypsis or mimicry, remain virtually unstudied.

7.2 Predator avoidance

One might initially think that prey should emphasise defences that work in the early stages of a potential encounter and thus minimise the chance of injury or death. However, 'early defences' have significant costs. Avoiding habitats occupied by predators, for example, may mean forgoing the use of good foraging areas, while crypsis restricts movement and may interfere with mating or territorial displays. These costs have to be balanced against the use of defences that act early in the sequence.

7.2.1 Habitat selection as a mechanism of predator avoidance

Prey often avoid the places where predators forage (reducing p and q in Fig. 7.1; e.g. Cerri and Fraser 1983; Werner *et al.* 1983; Savino and Stein 1989; Huntingford and Wright 1992; see also Kramer *et al.*, Chapter 3; Godin, Chapter 8; Persson *et al.*, Chapter 12). The game-theory analysis of Hugie and Dill (1994) suggests prey should also avoid habitats where risk of capture is high, because of lack of cover, etc., regardless of predator distribution. The costs of habitat avoidance include loss of foraging area and breeding sites. Werner *et al.* (1983), for example,

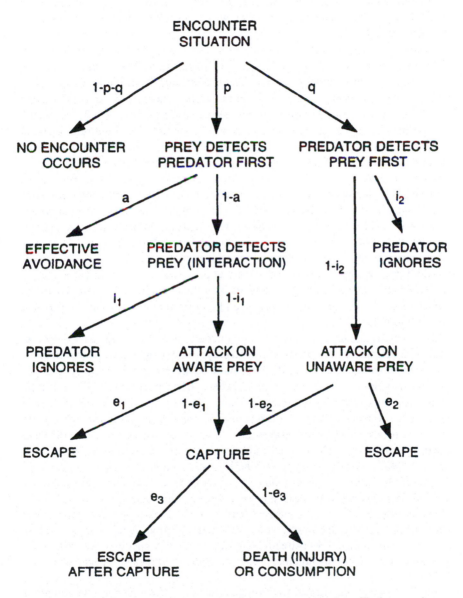

Fig. 7.1 The possible outcomes of an encounter between prey and predator. The letters and numbers beside the arrows represent the probability of an encounter following a particular pathway (see text). From Lima and Dill (1990).

found that the presence of largemouth bass, *Micropterus salmoides*, caused small bluegill sunfish, *Lepomis macrochirus*, to spend more time in dense cover, with fewer plankton, leading to slower growth. The addition of pike, *Esox lucius*, to a small lake led to immediate and dramatic movement of small cyprinids and brook stickleback, *Culaea inconstans*, from midwater to shallow shoreline habitats (He and Kitchell 1990).

In some experimental studies, predators were in contact with the prey and even consumed some (Cerri and Fraser 1983; Savino and Stein 1989). Stimuli available to the prey would include visual and chemical stimuli from the predator, alarm pheromones from prey, and others. Other researchers have confined live predators so that they could not catch the prey (e.g. Gotceitas and Colgan 1990), but visual and perhaps chemical stimuli were present. Others have used entirely artificial 'predator' stimuli such as toy hammers (Huntingford and Wright 1989) or model bird heads (Giles and Huntingford 1984; Huntingford and Wright 1992).

Avoidance of high predation areas is often short term. Juvenile bluegill sunfish returned to high risk areas less than 24 hours after bass were removed (Gotceitas and Colgan 1990). Juvenile Atlantic salmon, *Salmo salar*, avoided open water when a fibreglass model of a predatory brown trout, *Salmo trutta*, was pulled through their tank, but recovered within two hours (Metcalfe *et al.* 1987). Because experiments are often designed to test if predators exclude prey from high-quality foraging sites, it is not surprising that prey return quickly to these sites of high food abundance. In nature, predators may thus benefit by hunting where the prey's food is most abundant, as predicted by the game-theory analysis of Hugie and Dill (1994).

Prey may avoid areas without actually encountering the predator by detecting chemicals released by predators or injured prey. For example, European minnows, *Phoxinus phoxinus*, (von Frisch 1941) and fathead minnows, *Pimephales promelas*, (Mathis and Smith 1992) avoid regions where alarm pheromone (Schreckstoff) is present. This alarm pheromone is released only by mechanical injury (Smith 1992) and is thus an accurate indicator of predation. A similar alarm pheromone occurs in brook stickleback and they also avoid areas where they have detected alarm pheromones, including fathead minnow alarm pheromone (Chivers and Smith 1994a; Wisenden *et al.* 1994, 1995b). Fathead minnows avoid areas where they detect Iowa darter, *Etheostoma exile*, skin extract if they come from a population that is sympatric with darters, but not if they come from a darter-free region (Chivers *et al.*, 1995c). This suggests that minnows learn heterospecific alarm pheromones and use that information in habitat selection. Adult Pacific salmon, *Oncorhynchus* spp., delay upstream migration when they encounter water that has washed L-serine from mammalian predators, such as humans or bears (Idler *et al.* 1956).

Predator detection and learned avoidance of dangerous patches (Huntingford and Wright 1989) would contribute to local knowledge of the risk in the prey's home range. A resident should form a personal map of the safe and dangerous regions in its home range, including good refuges. The tidepool goby *Bathygobius soporator*, for example, learns the topography of its home range at high tide and then escapes predators by jumping from pool to pool at low tide (Aronson 1971). The blackeye goby, *Coryphopterus nicholsi*, can find a refuge hole more quickly if it has had preliminary experience in a region (Markel 1994). The value of such local knowledge should lead to strong selection for homing by displaced fish.

Alarm signals may play a role in learning to recognise dangerous areas. Fathead minnows learn to recognise the odour of a particular stream site when it is paired once with alarm substance or with a fright reaction by a conspecific, while they continue to ignore water from elsewhere in the same stream (Chivers and Smith 1995a, b). Alarm pheromones in the faeces of predators may also indicate high risk habitats (Brown *et al.* 1995a) and pike, at least, counter by defecating away from their foraging areas when they are eating minnows (Brown *et al.* 1995b).

When a fish avoids an area in response to predation risk it inevitably is selecting an area of lower risk. The risk may be lower because there are fewer predators or because predators are less effective due to more or better hiding places. A special case of selecting safe habitat occurs in fishes such as the anemone fishes, *Amphiprion* sp. (Mariscal 1970), or the man-of-war fish, *Nomeus* sp. (Tinker 1978), that selectively associate with dangerous invertebrates. The pilot fishes (e.g. *Naucrates* sp.) and remoras (Family Echeneididae) that associate with large pelagic vertebrates (Tinker 1978) may also gain protection from predators. Presumably one trade-off for these fishes is that they are restricted to the location and habitat of their protective species.

7.2.2 Crypsis

Some fish avoid predators by resembling the background, eucrypsis or general resemblance (Endler 1986), thus reducing the value of q (Fig. 7.1). The flatfishes, for example, are masters at changing their colour pattern to match their background. Fish may also resemble some specific inedible object such as a leaf (e.g. leaf-fish, *Monocirrhus polyacanthus*; Fig. 7.2) or sea weed (e.g. juvenile rockmover wrasse, *Novaculichthys taeniourus*; Randall and Randall 1960), thus increasing i_2 (Fig. 7.1). Crypsis could also increase escape success e_1 by making it harder for a predator to relocate escaping prey. Seventeen genera of fishes have been reported to resemble plants (Breder 1946). Endler (1986) termed this 'masquerade'. In eucrypsis the predator does not realise the prey is present, whereas in masquerade the searcher perceives the prey but mistakes it for something inedible.

Crypsis is not just visual. Predators can also sense chemicals, sound, water currents and electricity associated with prey. The mucous envelopes produced at night by some parrotfishes may reduce chemical conspicuousness (Winn and Bardach 1959). A concealed fish can ambush its prey, deceive its own predators,

Fig. 7.2 A 'leaf-fish', *Monociorrhus polyacanthus*. The shape, down to the 'stem' on the lower lip, and the colour and swimming behaviour resemble a drifting leaf. This masquerade can deceive both predators and prey, but may restrict the leaf fish to habitats that have suitable leaf models. Drawing by Mark Wilson.

or both (Breder 1946; Randall and Randall 1960). Predatory crypsis should be tuned to the senses of the predator's prey, and defensive crypsis in prey tuned to their predators' senses.

Any movement, whether foraging, migration or mating, will break crypsis against a static background. Moving backgrounds or masquerading as moving objects will impose special requirements. Leaf-mimicking fish and juvenile rockmover wrasse, for example, exhibit drifting patterns of movement that resemble their models (Randall and Randall 1960). This will limit foraging since a direct approach to food is incompatible with the cryptic movement pattern. A cryptic or masquerading fish must select habitats that match its method of concealment. These requirements become particularly stringent for masquerade and may impose severe restrictions on the habitats available to such fishes.

Crypsis conflicts with any behaviour that requires conspicuousness. Territorial and mating displays, for example, must be obvious to be effective. The restriction of bright breeding colours to distinct seasons through hormonal control is probably driven by predation. Male guppies, *Poecilia reticulata*, from high predation areas are less brightly coloured than males from safer areas (Endler 1986), and females from high predation areas prefer duller males than females from low risk areas (Breden and Stoner 1987). Male coloration in wild populations of guppies in Trinidad generally matches the preference of females in the same populations, probably due to female preference determining male success (Houde and Endler 1990). Endler (1995) has recently reviewed the coevolution of multiple interrelated traits in guppies, with particular emphasis on the interactions between predation, environmental gradients and guppy coloration and behaviour.

The interactions between crypsis and visual display lead to fine scale adaptations to the visual characteristics of habitat and the sensory physiology of conspecifics and predators (Endler 1992, 1995). For example, the ambient light spectra vary from one guppy habitat to another and those spectral (colour) differences will influence which colour patterns are conspicuous and which ones are cryptic. A male pattern with patches of grey, blue, yellow-green and red gives good contrast of brightness and colour in white light, such as occurs in large openings in the forest or under cloudy conditions (Endler 1992, 1995). When the same pattern is viewed in the yellow-green light of forest shade, the yellow-green patches would appear bright but the blue and red patches would be relatively dark. Displaying animals can manipulate the conspicuousness of their colours by selecting the habitat in which they perform the displays, as well as by changing colour (Endler 1992).

In cases where predators differ from prey in their sensory abilities, the prey may be able to use relatively 'private' communication channels. Endler (1992) suggests that one particular guppy predator, the pike cichlid, *Crenicichla alta*, may be less sensitive to blue than guppies, so selection should favour blue patches in male guppies for sexual displays in areas of pike cichlid predation. Similarly, Endler (1992) found that prawns are less sensitive to orange and thus orange display patches should be favoured in areas where prawn predation is high.

Endler (1992) argues that the sensory physiology of animals can bias the direction of evolution by favouring certain stimulus characteristics over others. This 'sensory drive' will mean that behavioural and morphological traits associated with crypsis or conspicuousness will tend to evolve together, influenced by the physical properties of the sensory medium (e.g. the spectra of available light) and the sensory abilities of prey and predators.

Radabaugh (1989) tested predator avoidance responses of males of three darter species as they changed from cryptic non-breeding colours to breeding colours. Non-breeding fantail darters, *Etheostoma flabellare*, greenside darters, *E. blennoides*, and orangethroat darters, *E. spectabile*, all 'froze' when a model of a predatory bass was moved overhead. After entering breeding coloration the relatively dull male fantail darters and the greenside darters, whose bright green breeding colours match the green algae they select as breeding habitat, continued to freeze as their preferred antipredator tactic. In contrast, the conspicuously coloured orangethroat darters switched to fleeing rather than freezing when in breeding colours. Thus, antipredator behaviour changed as the degree of crypsis changed with breeding season and species (see also Godin, Chapter 8).

Patterns that appear conspicuous at close range may act as camouflage at longer range. The parr marks and coloured spots on juvenile salmonids may disrupt body outlines at intermediate distances and merge into background textures when viewed from greater distances (Donnelly and Dill 1984). Countershading, dark upper surfaces and light undersurfaces, will counteract the bright top illumination from downwelling light and the shadow beneath an animal, thus greatly decreasing visual detection range.

Crypsis in a pelagic environment

Countershading and matching the general background are also common in pelagic environments, as are mirror-like body surfaces. Ventral bioluminescence occurs in many mesopelagic marine fishes. The light from ventral photophores is regulated to match the downwelling light from the surface and thus obscures the fish's silhouette when viewed from below (McFall-Ngai 1990).

Transparency is very common among pelagic larvae and juveniles and also occurs in adults of several species (McFall-Ngai 1990). However, transparent fish that eat conspicuous food can be revealed to their own predators by their visible stomach contents (McFall-Ngai 1990). These fishes should avoid conspicuous food even if it is nutritious. Some transparent fish have silvery pigment that conceals the gut contents and possibly other conspicuous internal organs (McFall-Ngai 1990). Transparency in larvae may be incompatible with high ultra-violet exposure in surface waters during the summer, resulting in interactions between transparency, habitat use and spawning season (Hunter *et al.* 1981). Eyes need pigment shielding and are incompatible with complete transparency. A transparent cladoceran crustacean, *Bosmina longirostris*, produces large-eyed and small-eyed morphs and fish prey more on large-eyed individuals, resulting in small eye size in populations sympatric with fish (Zaret and Kerfoot 1975). Similar effects likely occur in transparent fishes. Some marine fish larvae hatch without eye pigments, then pigmentation occurs over the next few days (Fuiman and Magurran 1994); this probably precludes image formation and may reflect a conflict between vision and crypsis. Transparency is actively maintained by physiological processes (McFall-Ngai 1990) and may fail due to parasites or stress.

7.2.3 Behavioural avoidance of detection by predators

Freezing, a common component of crypsis, removes the visual stimulus of relative movement against a stationary background, thereby reducing q (Fig. 7.1). (Moving in synchrony with a moving background is discussed under crypsis.) Immobility also reduces the 'detection space' for chemicals, noise, water currents and electrical activity. As with crypsis, immobility can be defensive or predatory and often plays a dual role, hiding from predators and lurking for prey, with the potential for conflict between the two roles. An obvious cost of immobility is that it interferes with actions requiring movement, as discussed under crypsis.

Guppies sink and freeze when an aerial predator model swoops over their tank (Goodey and Liley 1985). Johnny darters, *Etheostoma nigrum*, reduce activity for up to 24 hours after a smallmouth bass, *Micropterus dolomieui*, has been in their tank (Rahel and Stein 1988). Many otherwise active fishes, including darters, gobies and cyprinids, freeze in response to alarm pheromones from injured conspecifics (Smith 1992). Freezing by individuals before they are attacked probably reduces the chance that they will be detected by predators (see Godin, Chapter 8, for post-attack immobility).

Role of timing in predator avoidance

Timing is a critical aspect of predator avoidance. Prey can seek refuge when predators are active, thereby reducing q (Fig. 7.1). The elaborate diel timing of movements by species that occur on reefs (Helfman 1993) is almost certainly driven by the need to forage or mate while minimising predation. Diurnal and nocturnal fishes each make a round trip between their refuges and foraging areas twice a day, often passing one another on the way and sharing refuge sites on a shift-work basis. There may be additional daily trips by breeding reef fishes to spawning sites. Some predators, with good crepuscular vision, are active in the twilight as nocturnal and diurnal specialists commute with eyes that are not fully adapted to the ambient illumination. These predators may cause a 'quiet period' at twilight on tropical reefs when neither nocturnal nor diurnal prey fish dare to move (e.g. Munz and McFarland 1973). Freshwater European minnows do not detect an approaching pike model in twilight illumination as soon as they do in daylight levels (Pitcher and Turner 1986).

There is remarkably little direct evidence that timing of prey activity can be altered by predation. In one exception, Helfman (1986) pulled models of the predatory lizardfish, *Synodus intermedus*, toward juvenile grunts (Haemulidae) as they passed a particular point on their regular daily migration route. The simulated attacks delayed grunt migration. Good defences may permit longer periods of activity. Gladstone and Westoby (1988) concluded that toxic eggs and flesh allowed the sharpnose puffer, *Canthigaster valenti*, to utilise breeding times that were not available to palatable fish.

Predator recognition

Many fish can recognise predators, thus avoiding false alarms while still responding appropriately to predators (e.g. Coates 1980a,b; Karplus and Algom 1981). Since failing to identify a predator may lead to death, while failing to identify a non-predator may only mean a loss of foraging opportunity (the death vs. dinner dilemma), fish should initially fear all large unfamiliar animals. They could then learn which species are dangerous and continue to avoid those while habituating to species that do not attack (Csányi 1985). For example, humbug damselfish, *Dascyllus aruanus*, respond defensively to strange objects such as empty plastic

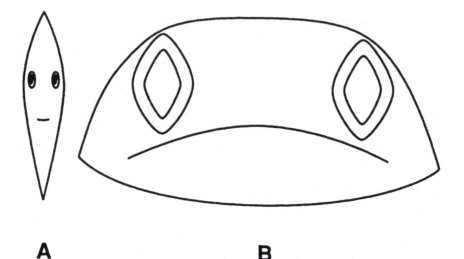

A   **B**

Fig. 7.3   Ideal non-predatory (A) and predatory (B) fish faces constructed by Karplus and Algom (1981) after comparing features of numerous predatory and non-predatory fishes. From Karplus and Algom (1981).

bags and divers but gradually habituate, while retaining fleeing and inspecting responses to predators and ignoring familiar non-predators (Coates 1980a,b).

Karplus and Algom (1981) analysed 20 facial features of 105 species of fishes from 35 families and found that piscivores and non-piscivores could be distinguished on the basis of the shape and size of the mouth and placement of the eyes. They constructed 'ideal' predatory and non-predatory model faces (Fig. 7.3) and showed experimentally that the pomacentrid, *Chromis caeruleus*, responded appropriately to them (Karplus *et al.* 1982).

Some fishes do not seem to fear unfamiliar predators, even ones that are large and have typical predatory characteristics. For example, wild-caught fathead minnows from a stream with no pike do not show a defensive reaction to the sight or smell of pike (Mathis *et al.* 1993), while fathead minnows from nearby populations that co-occur with pike show fright reactions to both stimuli. There can even be sex differences in predator recognition, linked to differences in vulnerability. Breeding male fathead minnows defending spawning sites and developing embryos among boulders along the shore of a pond were found to be more vulnerable to predation by garter snakes, *Thamnophis radix* and *T. sirtalis*, than were female minnows that spent more time in midwater (Matity *et al.* 1994). When they were tested for responses to chemical stimuli from snakes, the males responded defensively but the females did not. Minnows from a nearby lake with fewer snakes were unresponsive, regardless of their sex.

Most studies of predation have been conducted on wild-caught fishes from regions where predators are present (e.g. Coates 1980a, b; Karplus *et al.* 1982) and do not tell us how the prey acquired recognition ability. This can be determined using prey that are unfamiliar with specific predators, either from natural predator-free populations or artificial rearing. Predator-free populations have not recently experienced selection for defences against the predator. Fish from a predator-sympatric population that are reared without experience with predators have a genotype selected by predation, but lack individual experience.

Wild-caught fish from pike-free populations of threespine sticklebacks, *Gasterosteus aculeatus*, (Giles and Huntingford 1984; Huntingford *et al.* 1994), brook stickleback (Mathis and Smith 1993d; Gelowitz *et al.* 1993), European minnows (e.g. Magurran and Pitcher 1987) and fathead minnows (e.g. Mathis *et al.* 1993) have all been compared with pike-sympatric populations for their recognition and avoidance of pike. Similarly, populations of guppies that differ in exposure to two natural predators, Hart's rivulus, *Rivulus hartii*, and prawn, *Macrobrachium* spp., have been examined (e.g. Magurran and Seghers 1990). The general result is that fish from populations with predators are better able to recognise and respond appropriately to those predators.

Differences in antipredation behaviour can be specific to the predator. For example, guppies that co-occur with prawns but not predatory fish are wary of prawns but not fish and, conversely, guppies that are experienced with piscivorous fish but not prawns avoid the 'attack cone' of fish but not prawns (Magurran and Seghers 1990). Predators do not always induce caution. Guppies and Hart's rivulus from high predation areas stay more tenaciously on a feeding patch in the presence of predators compared with guppies from low predation areas (Fraser and Gilliam 1987). Because of such population differences in behaviour, studies on fish of uncertain origin should be interpreted with caution.

Genetic differences in predator-related learning

European minnows from pike-free and pike-sympatric populations differ genetically in schooling and evasion tactics. Minnows from the pike-sympatric population form more cohesive schools, inspect more often and have more synchronised evasion tactics, even when reared without experience with pike (Magurran 1990). As well, fish from the pike-sympatric population learned to respond to the predator more readily than individuals from the pike-free populations. Similarly, threespine stickleback from a high predation region, but raised in captivity without experience with predators, learn more readily to avoid a dangerous food patch than do stickleback from a low predation region (Huntingford and Wright 1992). There is an interesting twist in this story. The male chases and recovers its young as they try to leave the nest. The young can learn antipredator responses during these paternal chases and, in populations subjected to fish predation, the chases of the male are more vigorous and the young learn predator avoidance more effectively (Huntingford *et al.* 1994).

Defensive responses triggered by stimuli associated with predation can teach prey to recognise predators. For example, European minnows from a pike-free population, raised without experience with pike, do not respond defensively to pike odour (Magurran 1989). However, when pike odour is presented at the same time as minnow alarm pheromone the minnows learn to respond to pike odour. The minnows also learn to respond defensively if chemicals from a non-predatory fish, *Tilapia mariae*, are paired with alarm pheromone but the response is weaker, suggesting that learning is 'constrained' to favour learned recognition of natural predators (Magurran 1989). Similarly, fathead minnows from pike-free regions do not respond defensively to pike odour or the sight of pike, in contrast to wild-caught minnows from a pike-sympatric population (Mathis *et al.* 1993). When the smell or sight of pike is paired with alarm pheromone the minnows learn to respond to the pike stimuli (Chivers and Smith 1994a, b). A parallel situation occurs in brook stickleback. Pike-naive stickleback learn to recognise pike odour when it is paired, once, with stickleback skin extract (Chivers *et al.* 1995b). There is also a possible constraint on learning in the fathead system. The minnows learned to respond to the sight of pike and non-predatory goldfish, *Carassius auratus*, with equal facility. However, they retained a stronger response to pike than to goldfish two months later (Chivers and Smith 1994b).

Predator labelling

If alarm pheromones can teach a fish to fear predators, can prey pheromone leaking from a predatory fish 'label' that predator so that naive prey can recognise it as a predator? Fathead minnows from a pike-free creek do not initially respond defensively to chemical stimuli from a pike if it has been eating fish, such as swordtails, *Xiphophorus helleri*, that lack minnow alarm pheromone (Mathis and Smith 1993a). However, if the pike has recently eaten female or non-breeding male fathead minnows that possess alarm substance cells, then the naive minnows do respond to the pike's scent. Breeding male minnows, which lack alarm substance cells, do not cause this predator labelling effect (Mathis and Smith 1993b). The minnows learn to respond to the smell of unlabelled pike from only one exposure to labelled pike (Mathis and Smith 1993a). Similarly, pike-naive brook stickleback respond to conspecific alarm pheromone in the diet of pike (Gelowitz *et al.* 1993).

The fright response of a conspecific can be an effective stimulus. Zebra danios, *Danio rerio*, that have been conditioned with alarm pheromone to respond to a formerly neutral odourant, morpholine, can teach this response to naive conspecifics (Suboski *et al.* 1990). Similarly, fathead minnows can learn the odour of a risky site (Chivers and Smith 1995a), or a predator (Mathis *et al.* 1996), when paired with an experienced minnow that has previously encountered the stimulus odour combined with alarm pheromone.

Predator inspection

When fishes encounter a strange object or even a known predator they often swim toward it, pause, then return to their former location (Dugatkin and Godin 1992b; Pitcher 1992). They may do this alone, in pairs or in larger groups. This behaviour was initially called an 'approach' (George 1960), but 'predator inspection' has come to be the general label for this behaviour (Pitcher 1992). The term may be unfortunate since it implies a function that may not be the only or even predominant benefit favouring prey approaching predators. Dugatkin and Godin (1992a, b) used the more general term 'approaching'.

Costs to the inspector could include death, lost opportunity costs and the metabolic cost of locomotion (Dugatkin and Godin 1992a, b). Benefits could include increased information about the potential predator, informing others (including kin) of the predator's presence and status, deterring the predator, or advertising quality to others (Dugatkin and Godin 1992a, b; Pitcher 1992). Numerous studies indicate that prey can use vision alone to approach predators that are in separate aquaria (e.g. Licht 1989; Murphy and Pitcher 1991). We usually do not know if approaches are modified because of the absence of non-visual stimuli. Since some prey fish can identify predators by smell (e.g. Chivers and Smith 1993; Mathis and Smith 1993a, b), sniffing the predator may be an important part of inspection, especially if prey pheromones can label the predator as dangerous.

Approachers habituate more rapidly to non-predators than to real predators or realistic models (Csányi 1985; Magurran and Girling 1986; Murphy and Pitcher 1991). Guppies can distinguish by visual inspection between hungry and satiated predators (Licht 1989), and paradise fish, *Macropodus opercularis*, are able to learn and remember the characteristics of a non-predatory species on the basis of one encounter (Csányi 1985). Information can be transmitted from fish to fish once one individual is alarmed (e.g. Verheijen 1956; Godin 1986; Magurran and Higham 1988; Smith and Smith 1989; Ryer and Olla 1991).

Pitcher (1992) argued that approaching may be a signal, perhaps alarm or pursuit deterrence, as well as an information gathering mechanism, in which case approaches by pairs and multiples could be seen as cooperative signals. Magurran (1990) has provided evidence that pike are less likely to attack when they receive inspection visits by minnows (see also Dugatkin and Godin 1992b). Similarly, the blue acara cichlid, *Aequidens pulcher*, is less likely to attack guppies that are inspecting than ones that are not inspecting (Godin and Davis 1995). Inspection often occurs in the same context as other potential alarm signals (e.g. Smith and Smith 1989). Approaching the predator could act as an alarm signal, showing the predator's location to other prey and possibly signalling to a predator that it, specifically, has been detected. The benefits of such alarm signalling to the sender are not entirely clear (Smith 1992). Approaching predators may serve several functions including inspection, pursuit deterrence and alarm signalling.

7.3 Predator deterrence

To deter predators, thus increasing i_1 and i_2 (Fig. 7.1), prey must be able to decrease predator fitness. They could increase handling time or reduce predator success, with heavy armour, spines, dangerous fighting capability or effective escape, for example. Other approaches are venomous spines, toxic secretions, toxic flesh, and strong electrical discharge. Several mechanisms can be combined, as in the boxfishes, which are heavily armoured, and secrete toxic chemicals when disturbed (Thomson 1963).

Here, I will deal only with the role of such defences in deterring predators before they attack and attempt capture (see also Godin, Chapter 8). If prey are not readily useable by a predator, through toxins, etc., then it will benefit a predator to select other prey. This shift away from the protected species could be mediated by natural selection favouring avoidance behaviour, or by individual predators learning to avoid certain prey types.

7.3.1 Morphological deterrents

Whole taxonomic groups are characterised by their sharp spines (e.g. the spiny-rayed fishes, Superorder Acanthopterygii) or bony armour (e.g. boxfishes, Family Ostraciidae). These mechanisms likely reduce predator success but this has seldom been specifically tested, perhaps because it seemed so obvious.

Early studies indicated that the spines on threespine stickleback persuade pike to attack spineless fishes first when a choice is available (Hoogland *et al.* 1957). This is supported by studies indicating that spines are more strongly developed in populations subjected to fish predation than in populations subjected only to predation by birds and invertebrates (Reimchen 1994). It remains unknown whether predators differ in their probability of attacking prey types known, by the predator, to differ in vulnerability due to morphological defences. However, the immediate response of threespine stickleback when they see a predatory fish is to fixate it visually and erect their spines (Huntingford *et al.* 1994). This could serve as a predator deterrent signal from the stickleback to the predator indicating that the predator is detected and advertising the spine size of the stickleback.

Many fishes have venom glands on their spines, toxic glandular secretions, toxic flesh or eggs (Halstead 1988). Gladstone and Westoby (1988) compared the sharpnose puffer, which possesses toxic flesh and eggs, with other reef fishes. They concluded that toxicity frees this species from some of the constraints imposed by predation on more edible species. The puffers spawn over a longer season than many other species, perform 'unhurried' courtship and spawn throughout the day, leaving their toxic, demersal eggs unattended. Spawning was unrelated to tides. In contrast, many other species in the same area are forced by predation to follow distinct diel, tidal and seasonal cycles of reproduction and to protect their eggs.

Several groups of fishes have independently evolved electric organs that can deter predators. These strongly electric fish include electric rays, *Torpedo* spp, the electric 'eel', *Electrophorus electricus*, electric catfish, *Malapterurus electricus*, and stargazers, *Astroscopus* spp. (Bennett 1971). There is little or no study of the defensive effects of these strong discharges or the specific stimuli that elicit them. As with crypsis, one must distinguish between aggressive and defensive selection pressures since electric organs can stun prey as well as deter predators (e.g. Bray and Hixon 1978). The particular position of defensive electrical discharges in the predation sequence is not known in nature. The risk of shock could, at the very least, deter knowledgeable predators before they attacked.

In at least one fish, detecting predators induces changes in morphology that make it less vulnerable to predation. In the absence of predators, crucian carp, *Carassius carassius*, assume a slim, hydrodynamically efficient, body shape (Brönmark and Miner 1992). When predatory northern pike are present, the carp develop a deeper, less efficient, body shape that is presumably harder to swallow. Apparently the carp's response is induced by chemical cues from predators that have eaten a piscivorous diet rather than the predator alone (Brönmark and Pettersson 1994).

Inducible defences are favoured when predation varies on a time scale that allows slow morphological responses. Crucian carp, for example, often live in small lakes that may become deoxygenated over winter, wiping out pike while the carp survive by anaerobic respiration. This leaves the carp in a pike-free environment where the slim morph will have an advantage, until the next colonisation by pike. Inducible morphological defences occur in several invertebrates (Harvell 1990).

This phenomenon includes both pre-attack and post-attack components. A body that is hard to swallow certainly operates at the post-attack stage (see Godin, Chapter 8), but it may also deter predators from attacking if they are capable of judging their gape limit against prey. The system also depends on prey detecting predation far enough in advance that relatively slow growth processes can change their vulnerability. Therefore, there must be pre-attack detection mechanisms that trigger the morphological changes.

7.3.2 Behavioural deterrents

Aposematism and mimicry

Dangerous or unpalatable prey may deter predators by aposematic or 'warning' signals, such as bright colours or other readily detected stimuli. Aposematic signals make intuitive sense, but it has been difficult to pin down the evolutionary processes involved (e.g. Guilford 1990). From the prey's viewpoint, there are three possible selective benefits:

(1) if the prey survives, it may benefit by reducing future attacks,

(2) even if it dies, its kin, if they are nearby or resemble the sender, may benefit and enhance the sender's inclusive fitness, and

(3) there may be 'synergistic' benefits to non-kin that share common conspicuous signals (Guilford 1990).

Such synergism could lead to Batesian or Müllerian mimicry. There are many

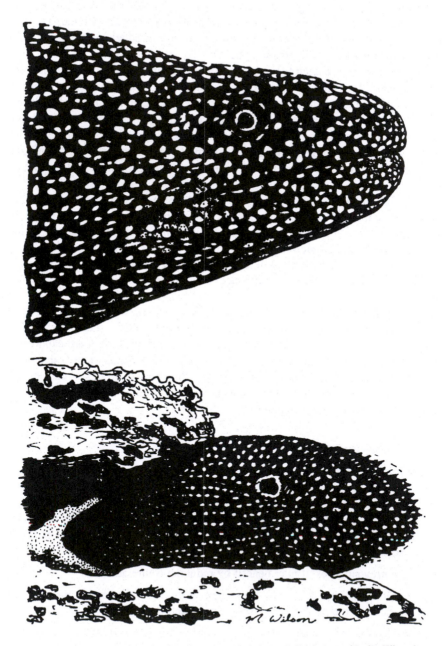

Fig. 7.4 A moray eel, *Gymnothorax meleagris*, and its mimic, *Calloplesiops altivelis*. When threatened by a predator, *Calloplesiops* flees to a crevice but leaves its tail exposed and spreads its dorsal fin to reveal a conspicuous eyespot. Drawing by Mark Wilson based on a photograph by Tom McHugh.

toxic, venomous or distasteful fishes and many brightly coloured species, but little direct evidence that aposematic signals, colours or other types have evolved. This likely reflects lack of study rather than a real dearth of aposematic displays in fishes.

Mimicry provides indirect evidence for aposematic signalling in fishes. For example, the plesiopid fish, *Calloplesiops altivelis*, (Fig. 7.4) has colour patterns and a pigment ocellus on the posterior portion of the dorsal fin that make the caudal portion of the fish resemble the head of the moray eel, *Gymnothorax meleagris* (McCosker 1977). When approached by predator, *Calloplesiops* flees to a refuge but leaves its tail exposed with the fins expanded revealing the ocellus. The moray has skin toxins and is fiercely aggressive toward potential predators. Since the moray normally leaves its head exposed when approached by predators, its conspicuous colour and behaviour may be aposematic and the non-toxic *Calloplesiops* a Batesian mimic (McCosker 1977).

Other mimicry systems occur in fishes. For example, the blenny, *Meiacanthus atrodorsalis*, with a venomous defensive bite is mimicked by two non-venomous blennies, *Ecsinius bicolor* and *Runula laudanus* (Losey 1972). Again, this implies that the colour or behaviour of the model, *M. atrodorsalis*, is a deterrent to predators. Some eels (e.g. *Ophichthys colubrinus*) may mimic dangerous sea snakes (Breder 1946).

Pursuit deterrence and pursuit invitation displays

If prey detect a predator in time, they will be able to escape and the predator's pursuit will be wasted (Hasson 1989). If the prey communicates its detection to the predator, the predator avoids a wasted pursuit and the prey avoids the costs of fleeing. Of course, prey could cheat by signalling when they were, in fact, vulnerable. This should lead to selection for 'honest' signals that are reliable indicators of escape ability. The stotting display of Thomson's gazelles (Caro 1986a, b; FitzGibbon and Fanshawe 1988) is apparently an honest signal of escape ability that deters predator pursuit. There are few clear cut cases of pursuit deterrence signalling in fishes but there are some good candidates, such as predator inspection (Dugatkin and Godin 1992b, Pitcher 1992; Godin and Davis 1995) and goby bobbing (Smith and Smith 1989).

Pursuit invitation might benefit prey if, by encouraging pursuit when the predator would not win, it shortened the total time spent dealing with a predator (Smythe 1977). The prey, having eluded pursuit, could go on with their other activities. Pursuit invitation seems unlikely to persist in a sustained relationship between vertebrate predators and prey. More likely, it would lead to pursuit deterrence as the predator learned that the 'invitation' indicates the prey is uncatchable. Another type of pursuit invitation occurs when parents invite pursuit to distract predators from young. Parental male threespine stickleback lure egg-raiding shoals of stickleback away from their nest by simulating either feeding or nest fanning (Ridgway and McPhail 1988; Whoriskey 1991).

Alarm signalling

It is often assumed that alarm signals function primarily by warning conspecifics of danger. But, how could this 'altruism' evolve through natural selection? Why should an animal incur costs, such as making itself conspicuous, in order to save others? And yet, many fishes perform displays or release chemicals when threatened or attacked and conspecifics respond to these signals with appropriate defensive behaviour (Smith 1992). For example, alarm pheromone (Schreckstoff), released by epidermal injury, is characteristic of the Superorder Ostariophysi (minnows, catfish, etc.) which includes thousands of species. Other alarm pheromone systems occur among the percid darters, gobies, cottids and at least one species of stickleback (Smith 1992; Mathis and Smith 1993d). Iowa darters possess a two stage chemical warning system. Individuals that see a predator model but are not injured release a 'disturbance pheromone' that elicits an alert posture in receivers (Wisenden *et al.*, 1995b) while injury to the epidermis releases an alarm pheromone that induces freezing in receivers (Smith 1979).

Longspine squirrelfish, *Holocentrus rufus*, utter a distinctive staccato call when they detect a predator and a separate distress call when captured (Winn *et al.* 1964). Some gobies bob conspicuously when they see or smell a predator (Smith and Smith 1989), and other gobies warn their symbiotic burrowing shrimp with fin flicks when a predator is near (Karplus 1987). European minnows 'skitter' (suddenly change in position in a shoal) when they detect predators (Magurran 1990).

These potential alarm signals fit the theme of this chapter in two ways. First, some receivers avoid predation because they detect the alarm signal. Fathead minnows survive longer in the presence of pike if they are 'warned' with alarm pheromone (Mathis and Smith 1993c). Fish respond to alarm signals with the typical defensive behaviour patterns of their species such as schooling, freezing and assuming cryptic coloration. As well, alarm signals assist prey in learning to recognise predators, and even label predators by altering the predator's chemical signature. Predator recognition learned through associating an alarm signal with predator characteristics may be transmitted culturally to fish that have never been exposed to the associated stimuli (Suboski *et al.* 1990; Mathis *et al.* 1996). These, and possibly other, mechanisms can lead to rapid changes in the antipredator responses of a population. When 10 small pike were added to a pond containing 20 000 pike-naive fathead minnows, the minnows, which had been unresponsive to pike odour, showed a strong response to pike odour after only two weeks (Chivers and Smith 1995c).

Benefits to the sender are less clear, but if some warned prey are their kin senders may gain inclusive fitness (Smith 1992). If senders survive after giving warning, as alarm calling squirrelfish and bobbing gobies often will, then they may benefit if the warning reduces predator success and discourages hunting that prey type in that region (Trivers 1971) or if alarm signals save group members or territorial neighbours (Smith 1986). Alarm signals may also deter pursuit by informing the predator that it has been detected and that success is unlikely.

Distress signals

Distress signals are given when capture is imminent or already under way (see Godin, Chapter 8). At least 24 families of fishes make distinctive sounds when captured or prodded (Myrberg 1981). The function of such distress signals is not clear. Högstedt (1983), discussing birds, suggests that such calls may summon mobbers or provide escape opportunities by attracting secondary predators that would interfere with the first predator. Injury released alarm pheromones in fishes are also released at the time of actual capture and could attract secondary predators. Following this line of reasoning, Mathis *et al.* (1995) tested the response of northern pike to artificial and natural minnow alarm pheromone. The pike were preferentially attracted to skin extract containing alarm substance over extract from skin that lacked alarm substance cells. As well, predatory diving beetles were trapped more frequently in traps baited with extracts that contained alarm substance. Predators could learn to avoid prey that give distress signals and prey could learn to recognise predators that elicited distress signals from their conspecifics.

Mobbing

One way to deter a predator is to attack it. There are a number of examples of fishes mobbing predators, threatening and biting (Dugatkin and Godin 1992b). Helfman (1989), for example, found that damselfish, *Stegastes planifrons*, adjusted their avoidance or mobbing to the degree of threat posed by a trumpet fish, *Aulostomus maculatus*. As well as signalling to the predator or directly injuring it, mobbing may serve to 'educate' other prey in a region about the characteristics of the predator. In birds, mobbing is very effective in transmitting predator recognition among conspecifics (Curio *et al.* 1978).

Shoaling and schooling

Shoals and schools, cohesive social groups of fish that often move together with impressive coordination, provide several levels of antipredator defence (Godin 1986, Chapter 8; Magurran 1990; Pitcher and Parrish 1993). They fit in this chapter because the effectiveness of shoals and schools in reducing predation may deter predators from attacking, and information gained from shoal mates may permit prey to recognise and avoid predators in the future. A 'shoal' is a group of fish that remain together through social attraction (Pitcher and Parrish 1993). 'School' is now used to refer to synchronised and polarised groups, making it a subcategory within the broader term shoaling.

Fish often increase school cohesion in response to predators (e.g. Magurran and Pitcher 1987) and alarm pheromones (e.g. Mathis and Smith 1993c). This indicates that, in general, schooling is an effective defence. The presence of shoal mates may increase the probability of detecting predators, dilute the chance of capture for an individual, and confuse predators, and information from inspectors about predators may benefit other members of the shoal (Godin 1986; Magurran 1990; Pitcher 1992; Pitcher and Parrish 1993). Alarm signals, combined with the multiplied sensory capability of the shoal, may permit prey to detect predators at greater distances and to learn to recognise predatory species if some shoal mates are experienced with the predator (Suboski *et al*. 1990; Smith 1992; Mathis and Smith 1993b).

Two recent lines of evidence reinforce the idea that shoals may facilitate learned responses to danger. First, fathead minnows recognise former members of the same natural shoal after months of separation (Brown and Smith 1994) and they perform antipredator behaviour more effectively when in the company of familiar shoal mates (Chivers *et al*. 1995a). Second, there is direct experimental evidence that fathead minnows learn to recognise predators (Mathis *et al*. 1996) and dangerous locations (Chivers and Smith 1995b) when they witness the response of experienced conspecifics to chemical stimuli from predators or dangerous sites. Taken together these findings suggest that some shoals may be long-term associations of familiar, cooperating individuals, benefiting from shared knowledge of the hazards in their environment.

Positions within a shoal may carry different costs and benefits (Krause 1993c). For example, Krause (1993a) found that juvenile roach, *Rutilus rutilus*, feeding at the front of a shoal had higher feeding rates than conspecifics at the rear of the shoal. When he selectively frightened one fish in a shoal by exposing it to alarm pheromone after the rest of the shoal had been habituated to the alarm pheromone, the frightened minnow moved to a more central position in the shoal (Krause 1993b). Thus, individuals may gain pre-attack benefits by adjusting their position in a shoal when they detect a predator.

7.4 Conclusions

It is difficult to grasp the magnitude of the effects of predation on fishes. Major phylogenetic lineages are defined by their antipredator defences, the puffers by their inflation mechanism, the sticklebacks by their stickles and the spiny rayed fishes by their sharp spines, for example. Perhaps surprisingly, some of the most abundant groups of fishes (e.g. cyprinids, salmonids and clupeids) lack obvious morphological defences against large predators. Their evolutionary success, however, suggests that they can avoid predation effectively. Their defences are apparently largely behavioural, such as schooling, recognising predators, and avoiding times and areas where predators are active. It seems easy to understand how a porcupine fish that has toxic skin secretions (Halstead 1988) and can inflate a spike-covered body can deal with predation, although we know virtually nothing about costs and specific benefits in such a defence system. But can we understand the defence system of a herring or a minnow with no spines or puffing ability and palatable flesh? Yet minnows and herring are, on the basis of abundance, more successful.

Research on predator defence has the potential to reveal profound insights into factors that affect thousands of species in over vast ranges of habitat. The Schreckstoff alarm pheromone system, for example, is probably present in about

70% of the world's freshwater fish species. Schooling specialists such as the herrings may be among the most abundant vertebrates in the world. It behoves us to try to understand how they have reached this level of success in the face of the many predators that find them edible.

There are many potential opportunities for the study of predator avoidance and deterrence in fishes. Such study should concentrate on attempting to measure associated costs and benefits of the defences, using experimental or comparative approaches, and on the use of theoretical models to guide research. Too often benefits are assumed without testing, when careful analysis may reveal that the situation is more complex, and interesting, than initially expected.

The costs of defensive mechanisms are seldom addressed. This is particularly true for morphological and chemical defences. The presence of inducible defences in plants, invertebrates (Harvell 1990) and now fishes (Brönmark and Miner 1992) suggests that there is a cost to retaining chemical and morphological defences that is offset by benefits when predators are present, but leads to the loss of the defence in the absence of predators.

Knowledge of the biology of fishes can provide novel opportunities for testing hypotheses in behavioural ecology. For example, minnows with or without alarm pheromone can be produced by hormonal manipulation and used in experimental comparisons (e.g. Mathis and Smith 1993b). Similarly, puffer fish that are reared without symbiotic bacteria lack tetrodotoxin and can become toxified by administering bacteria (Matsui *et al.* 1990). One could presumably use toxic and non-toxic puffers to examine the physiological and metabolic costs of toxicity and to test the responses of predators.

Practical applications may also emerge from the study of fish defences. For example, since predator avoidance, including avoidance of dangerous habitats, is apparently a widespread characteristic of fishes there is the potential to manipulate fish in order to protect them from hazards such as dams or water intakes or direct them to certain areas by changing cover characteristics or predator regimes. Understanding the role of learning in predator defence may be critical in successful propagation of fish for release into the wild. Training programs pairing alarm stimuli with predator stimuli could substantially improve survival or prey fishes (Suboski and Templeton 1989).

Acknowledgements

Preparation of this chapter was supported in part by the University of Saskatchewan and the Natural Sciences and Engineering Research Council of Canada. I thank S.L. Lima, L.M. Dill, I. Karplus and D. Algom for permission to use their illustrations. Alicia Mathis, Doug Chivers, Grant Brown, Brian Wisenden and Jean Smith provided useful comments on the chapter and contributed to numerous interesting discussions of the subject material. The editor and referees made suggestions that greatly improved the chapter.

References

Aronson, L.R. (1971). Further studies on orientation and jumping behavior in the gobiid fish, *Bathygobius soporator. Ann. N.Y. Acad. Sci.*, **188**, 378–407.

Bennett, M.V.L. (1971). Electric organs. In *Fish physiology*, Vol. 5. (eds. W.S. Hoar and D.J. Randall), pp. 347–492. Academic Press, New York.

Bray, R.N. and Hixon, M.A. (1978). The night shocker: predatory behavior of the Pacific electric ray (*Torpedo californica*). *Science*, **200**, 333–334.

Breden, F. and Stoner, G. (1987). Male predation risk determines female preference in the Trinidad guppy. *Nature, Lond.*, **329**, 831–833.

Breder, C.M. (1946). An analysis of the deceptive resemblances of fishes to plant parts, with critical remarks on protective mimicry and adaptation. *Bull. Bingham. Oceanogr. Collect.*, **10**, 1–49.

Brönmark, C. and Miner, J.G. (1992). Predator-induced phenotypical change in body morphology in crucian carp. *Science*, **258**, 1348–1350.

Brönmark, C. and Pettersson, L.B. (1994). Chemical cues from piscivores induce a change in morphology in crucian carp. *Oikos*, **70**, 396–402.

Brown, G.E. and Smith, R.J.F. (1994). Fathead minnows use chemical cues to discriminate natural shoalmates from unfamiliar conspecifics. *J. Chem. Ecol.*, **20**, 3051–3061.

Brown, G.E., Chivers, D.P. and Smith, R.J.F. (1995a). Fathead minnows avoid conspecific and heterospecific alarm pheromones in the faeces of northern pike. *J. Fish Biol.*, **47**, 387–393.

Brown, G.E., Chivers, D.P. and Smith, R.J.F. (1995b). Localized defecation by pike: a response to labelling by cyprinid alarm pheromones? *Behav. Ecol. Sociobiol.*, **36**, 105–110.

Caro, T.M. (1986a). The functions of stotting, a review of the hypotheses. *Anim. Behav.*, **34**, 649–662.

Caro, T.M. (1986b). The functions of stotting in Thomson's gazelles: some tests of the predictions. *Anim. Behav.*, **34**, 663–684.

Cerri, R.D. and Fraser, D.F. (1983). Predation and risk in foraging minnows: balancing conflicting demands. *Am. Nat.*, **121**, 552–561.

Chivers, D.P. and Smith, R.J.F. (1993). The role of olfaction in chemosensory-based predator recognition in the fathead minnow, *Pimephales promelas. J. Chem. Ecol.*, **19**, 623–633.

Chivers, D.P. and Smith, R.J.F. (1994a). The role of experience and chemical alarm signalling in predator recognition by fathead minnows (*Pimephales promelas*, Rafinesque). *J. Fish Biol.*, **44**, 273–285.

Chivers, D.P. and Smith, R.J.F. (1994b). Fathead minnows (*Pimephales promelas*) acquire predator recognition when alarm substance is associated with the sight of unfamiliar fish. *Anim. Behav.*, **48**, 597–605.

Chivers, D.P. and Smith, R.J.F. (1995a). Fathead minnows (*Pimephales promelas*) learn to recognise chemical stimuli from high risk habitats by the presence of alarm substance. *Behav. Ecol.*, **6**, 155–158.

Chivers, D.P. and Smith, R.J.F. (1995b). Chemical recognition of risky habitats is culturally transmitted among fathead minnows, *Pimephales promelas* (Osteichthyes, Cyprinidae). *Ethology*, **99**, 286–296.

Chivers, D.P. and Smith, R.J.F. (1995c). Free-living fathead minnows (*Pimephales promelas*) rapidly learn to recognise pike (*Esox lucius*) as predators. *J. Fish Biol.*, **46**, 949–954.

Chivers, D.P., Brown, G.E. and Smith, R.J.F. (1995a). Familiarity and shoal cohesion in fathead minnows (*Pimephales promelas*): implications for antipredator behaviour. *Can. J. Zool.*, **73**, 955–960.

Chivers, D.P., Brown, G.E. and Smith, R.J.F. (1995b). Acquired recognition of chemical stimuli from pike, *Esox lucius*, by brook sticklebacks, *Culaea inconstans*, (Osteichthyes, Gasterosteidae). *Ethology*, **99**, 234–242.

Chivers, D.P., Wisenden, B.D. and Smith, R.J.F. (1995c). The role of experience in the response of fathead minnows (*Pimephales promelas*) to skin extract of Iowa darters (*Etheostoma exile*). *Behaviour*, **132**, 665–674.

Coates, D. (1980a). The discrimination and reactions towards predatory and non-predatory species of fish by humbug damselfish, *Dascyllus aruanus* (Pisces, Pomacentridae). *Z. Tierpsychol.*, **52**, 347–354.

Coates D. (1980b). Anti-predator defense via interspecific communication in humbug damselfish, *Dascyllus aruanus* (Pisces, Pomacentridae). *Z. Tierpsychol.*, **52**, 355–364.

Csányi, V. (1985). Ethological analysis of predator avoidance by the paradise fish (*Macropodus opercularis* L.) 1. Recognition and learning of predators. *Behaviour*, **92**, 227–240.

Curio, E., Ernst, U. and Vieth, W. (1978). The adaptive significance of avian mobbing. II. Cultural transmission of enemy recognition in blackbirds: effectiveness and some constraints. *Z. Tierpsychol.*, **48**, 184–202.

Donnelly, W.A. and Dill, L.M. (1984). Evidence for crypsis in coho salmon, *Oncorhynchus kisutch* (Walbaum) parr: substrate colour preference and achromatic reflectance. *J. Fish Biol.*, **25**, 183–195.

Dugatkin, L.A. and Godin, J.-G.J. (1992a). Predator inspection, shoaling and foraging under predation hazard in the Trinidadian guppy, *Poecilia reticulata*. *Environ. Biol. Fish.*, **34**, 265–276.

Dugatkin, L.A. and Godin, J.-G.J. (1992b). Prey approaching predators: a cost–benefit perspective. *Ann. Zool. Fennici*, **29**, 233–252.

Endler, J.A. (1986). Defense against predators. In *Predator-prey relationships* (eds. M.E. Feder and G.V. Lauder), pp. 109–134. University of Chicago Press, Chicago.

Endler, J.A. (1992). Signals, signal conditions, and the direction of evolution. *Am. Nat.*, **139**, S125-S153.

Endler, J.A. (1995). Multiple-trait coevolution and environmental gradients in guppies. *Trends Ecol. Evol.*, **10**, 22–29.

FitzGibbon, C.D. and Fanshawe, J.H. (1988). Stotting in Thomson's gazelles: an honest signal of condition. *Behav. Ecol. Sociobiol.*, **23**, 69–74.

Fraser, D.F. and Gilliam, J.F. (1987). Feeding under predation hazard: response of the guppy and Hart's rivulus from sites with contrasting predation risk. *Behav. Ecol. Sociobiol.*, **21**, 203–209.

Frisch, K. von (1941). Über einen Schreckstoff der Fischhaut und seine biologische Bedeutung. *Z. vergl. Physiol.*, **29**, 46–145.

Fuiman, L.A. and Magurran, A.E. (1994). Development of predator defences in fishes. *Rev. Fish Biol. Fish.*, **4**, 145–183.

Gelowitz, C.M., Mathis, A. and Smith, R.J.F. (1993). Chemosensory recognition of northern pike (*Esox lucius*) by brook stickleback (*Culaea inconstans*): population differences and the influence of predator diet. *Behaviour*, **127**, 105–118.

George, C.J.W. (1960). *Behavioral interaction of the pickerel (Esox niger Lesueur and Esox americanus Lesueur) and the mosquitofish (Gambusia patruelis Baird and Girard)*. Ph.D thesis, Harvard University, Cambridge, MA.

Giles, N. and Huntingford, F.A. (1984). Predation risk and inter-population variation in anti-predator behaviour in the three-spined stickleback, *Gasterosteus aculeatus* L. *Anim. Behav.*, **32**, 264–275.

Gladstone, W. and Westoby, M. (1988). Growth and reproduction in *Canthigaster valentini* (Pisces, Tetraodontidae): a comparison of a toxic reef fish with other reef fishes. *Environ. Biol. Fish.*, **21**, 207–221.

Godin, J.-G.J. (1986). Antipredator function of shoaling in teleost fishes: a selective review. *Naturaliste can.*, **113**, 241–250.

Godin, J.-G.J. and Davis, S.A. (1995). Who dares, benefits: predator approach behaviour in the guppy (*Poecilia reticulata*) deters predator pursuit. *Proc. R. Soc. Lond. B*, **259**, 193–200.

Goodey, W. and Liley, N.R. (1985). Grouping fails to influence the escape behaviour of the guppy (*Poecilia reticulata*). *Anim. Behav.*, **33**, 1032–1033.

Gotceitas, V. and Colgan, P. (1990). The effects of prey availability and predation risk on habitat selection by juvenile bluegill sunfish. *Copeia*, **1990**, 409–417.

Guilford, T. (1990). Evolutionary pathways to aposematism. *Acta Oecologia*, **11**, 835–841.

Halstead, B.W. (1988). *Poisonous and venomous marine animals of the world*. Darwin Press, Princeton.

Harvell, C.D. (1990). The ecology and evolution of inducible defenses. *Q. Rev. Biol.*, **65**, 323–329.

Hasson, O. (1989). The effect of uncertainty on the relationship between the frequency of warning signals and prey density. *Theor. Popul. Biol.*, **36**, 241–250.

He, X. and Kitchell, J.F. (1990). Direct and indirect effects of predation on a fish community: a whole-lake experiment. *Trans. Am. Fish. Soc.*, **119**, 825–835.

Helfman, G.S. (1986). Behavioral responses of prey fishes during predator–prey interactions. In *Predator-prey relationships* (eds. M.E. Feder and G.V. Lauder), pp. 135–156. University of Chicago Press, Chicago.

Helfman, G.S. (1989). Threat-sensitive predator avoidance in damselfish-trumpetfish interactions. *Behav. Ecol. Sociobiol.*, **24**, 47–58.

Helfman, G.S. (1993). Fish behaviour by day, night and twilight. In *Behaviour of teleost fishes*, 2nd edn. (ed. T.J. Pitcher), pp. 479–512. Chapman and Hall, London.

Högstedt, G. (1983). Adaptation unto death: function of fear screams. *Am. Nat.*, **121**, 562–570.

Hoogland, R.D., Morris, D. and Tinbergen, N. (1957). The spines of sticklebacks (*Gasterosteus* and *Pygosteus*) as a means of defence against predators (*Perca* and *Esox*). *Behaviour*, **10**, 205–237.

Houde, A.E. and Endler, J.A. (1990). Correlated evolution of female mating preferences and male color pattern in the guppy *Poecilia reticulata*. *Science*, **248**, 1405–1408.

Hugie, D.M. and Dill, L.M. (1994). Fish and game: a game theoretic approach to habitat selection by predators and prey. *J. Fish Biol.*, **45** (Suppl. A), 151–169.

Hunter, J.R., Kaupp, S.E. and Taylor, J.H. (1981). Effects of solar and artificial ultraviolet-B radiation on larval northern anchovy, *Engraulis mordax*. *Photochem. Photobiol.*, **34**, 477–486.

Huntingford, F.A. and Wright, P.J. (1989). How sticklebacks learn to avoid dangerous feeding patches. *Behav. Processes*, **19**, 181–189.

Huntingford, F.A. and Wright, P.J. (1992). Inherited population differences in avoidance conditioning in 3-spined sticklebacks, *Gasterosteus aculeatus. Behaviour*, **122**, 264–273.

Huntingford, F.A., Wright, P.J. and Tierney, J.F. (1994). Adaptive variation in antipredator behaviour in threespine stickleback. In *The evolutionary biology of the threespine stickleback* (eds. M.A. Bell and S.A. Foster), pp. 277–296. Oxford University Press, Oxford.

Idler, D.R., Fagerlund, U. and Mayoh, H. (1956). Olfactory perception in migrating salmon. I. L-serine, a salmon repellent in mammalian skin. cJ. *Gen. Physiol.*, **39**, 889–892.

Karplus, I. (1987). The association between gobiid fishes and burrowing alpheid shrimps. *Oceanogr. Mar. Biol. Ann. Rev.*, **25**, 507–562.

Karplus, I. and Algom, D. (1981). Visual cues for predator face recognition by reef fishes. *Z. Tierpsychol.*, **55**, 343–364.

Karplus, I., Goren, M. and Algom, D. (1982). A preliminary experimental analysis of predator face recognition by *Chromis caerulus* (Pisces, Pomacentridae). *Z. Tierpsychol.*, **58**, 53–65.

Krause, J., (1993a). The relationship between foraging and shoal position in a mixed shoal of roach (*Rutilus rutilus*) and chub (*Leuciscus cephalus*)—a field study. *Oecologia*, **93**, 356–359.

Krause, J. (1993b). The effect of 'Schreckstoff' on the shoaling behaviour of a minnow: a test of Hamilton's selfish herd theory. *Anim. Behav.*, **45**, 1019–1024.

Krause, J. (1993c). Positioning behaviour in fish shoals: a cost–benefit analysis. *J. Fish Biol.*, **43** (*Suppl.* A), 309–314.

Licht, T. (1989). Discriminating between hungry and satiated predators: the response of guppies (*Poecilia reticulata*) from high and low predation sites. *Ethology*, **82**, 238–243.

Lima, S.L. and Dill, L.M. (1990). Behavioral decisions made under the risk of predation: a review and prospectus. *Can. J. Zool.*, **68**, 619–640.

Losey, G.S. (1972). Predation protection in the poison-fang blenny, *Meiacanthus atrodorsalis*, and its mimics, *Escenius bicolor* and *Runula laudadus* (Blenniidae). *Pac. Sci.*, **26**, 129–139.

McCosker, J.E. (1977). Fright posture of the plesiopid fish *Calloplesiops altivelis*: an example of Batesian mimicry. *Science*, **197**, 400–401.

McFall-Ngai, M.J. (1990). Crypsis in the pelagic environment. *Amer. Zool.*, **30**, 175–188.

Magurran, A.E. (1989). Acquired recognition of predator odour in the European minnow (*Phoxinus phoxinus*). *Ethology*, **82**, 216–223.

Magurran, A.E. (1990). The adaptive significance of schooling as an anti-predator defence in fish. *Ann. Zool. Fennici*, **27**, 51–66.

Magurran, A.E. and Girling, S.L. (1986). Predator model recognition and response to habituation in shoaling minnows. *Anim. Behav.*, **34**, 510–518.

Magurran, A.E. and Higham, A. (1988). Information transfer across fish shoals under predator threat. *Ethology*, **78**, 153–158.

Magurran, A.E. and Pitcher, T.J. (1987). Provenance, shoal size and the sociobiology of predator-evasion behaviour in minnow shoals. *Proc. R. Soc. Lond. B*, **229**, 439–465.

Magurran, A.E. and Seghers, B.H. (1990). Population differences in predator recognition and attack cone avoidance in the guppy *Poecilia reticulata. Anim. Behav.*, **40**, 443–452.

Mariscal, R.N. (1970). The nature of the symbiosis between Indo-Pacific anemone fishes and sea anemones. *Mar. Biol.*, **6**, 58–65.

Markel, R.W. (1994). An adaptive value of spatial learning and memory in the blackeye goby, *Coryphopterus nicholsi. Anim. Behav.*, **47**, 1462–1464.

Mathis, A. and Smith, R.J.F. (1992). Avoidance of areas marked with a chemical alarm substance by fathead minnows (*Pimephales promelas*) in a natural habitat. *Can. J. Zool.*, **70**, 1473–1476.

Mathis, A. and Smith, R.J.F. (1993a). Fathead minnows (*Pimephales promelas*) learn to recognise pike (*Esox lucius*) as predators on the basis of chemical stimuli from minnows in the pike's diet. *Anim. Behav.*, **46**, 645–656.

Mathis, A. and Smith, R.J.F. (1993b). Chemical labelling of northern pike (*Esox lucius*) by the alarm pheromone of fathead minnows (*Pimephales promelas*). *J. Chem. Ecol.*, **19**, 1967–1979.

Mathis, A. and Smith, R.J.F. (1993c). Chemical alarm signals increase the survival time of fathead minnows (*Pimephales promelas*) during encounters with northern pike (*Esox lucius*). *Behav. Ecol.*, **4**, 260–265.

Mathis, A. and Smith, R.J.F. (1993d). Intraspecific and cross-superorder responses to chemical alarm signals by brook stickleback. *Ecology*, **74**, 2395–2404.

Mathis, A., Chivers, D.P. and Smith, R.J.F. (1993). Population differences in responses of fathead minnows (*Pimephales promelas*) to visual and chemical stimuli from predators. *Ethology*, **93**, 31–40.

Mathis, A., Chivers, D.P. and Smith, R.J.F. (1995). Chemical alarm signals: predator-deterrents or predator attractants? *Am. Nat.*, **145**, 994–1005.

Mathis, A., Chivers, D.P. and Smith, R.J.F. (1996). Cultural transmission of predator recognition in fishes: intraspecific and interspecific learning. *Anim. Behav.*, **51**, 185–201.

Matity, J.G, Chivers, D.P. and Smith, R.J.F. (1994). Population and sex differences in antipredator responses of breeding fathead minnows (*Pimephales promelas*) to chemical stimuli from garter snakes (*Thamnophis radix* and *T. sirtalis*). *J. Chem. Ecol.*, **20**, 2111–2121.

Matsui, T., Taketsugu, S., Sato, H., Yamamori, K., Kodama, K., Ishii, A., Hirose, H. and Shimizu, C. (1990). Toxification of cultured puffer fish by the administration of tetrodotoxin producing bacteria. *Nippon Suisan Gakkaishi*, **56**, 705.

Metcalfe, N.B., Huntingford, F.A. and Thorpe, J.E. (1987). The influence of predation risk on the feeding motivation and foraging strategy of juvenile Atlantic salmon. *Anim. Behav.*, **35**, 901–911.

Munz, F.W. and McFarland, W.N. (1973). The significance of spectral position in the rhodopsins of tropical marine fishes. *Vision Res.*, **13**, 1829–1874.

Murphy, K.E. and Pitcher T.J. (1991). Individual behavioural strategies associated with predator inspection in minnow shoals. *Ethology*, **88**, 307–319.

Myrberg, A.A. (1981). Sound communication and interception in fishes. In *Hearing and sound communication in fishes* (eds. W. Tavolga, A.N. Popper and R.R. Fay), pp. 395–452. Springer-Verlag, New York.

Pitcher, T.J. (1992). Who dares, wins—the function and evolution of predator inspection behaviour in shoaling fish. *Neth. J. Zool.*, **42**, 371–391.

Pitcher, T.J. and Parrish, J.K. (1993). Functions of shoaling behaviour in teleosts. In: *Behaviour of teleost fishes*, 2nd edn. (ed. T.J. Pitcher), pp. 364–439. Chapman and Hall, London.

Pitcher, T.J. and Turner, J.R. (1986). Danger at dawn: experimental support for the twilight hypothesis in shoaling minnows. *J. Fish Biol.*, **29**, 59–70.

Radabaugh, D.C. (1989). Seasonal colour changes and shifting antipredator tactics in darters. *J. Fish Biol.*, **34**, 679–685.

Rahel, F.J. and Stein, R.A. (1988). Complex predator–prey interactions and predator intimidation among crayfish, piscivorous fish, and small benthic fish. *Oecologia*, **75**, 94–98.

Randall, J.E. and Randall, H.A. (1960). Examples of mimicry and protective resemblance in tropical marine fishes. *Bull. Mar. Sci. Gulf Caribb.*, **10**, 444–480.

Reimchen, T.E. (1994). Predators and morphological evolution in threespine stickleback. In *The evolutionary biology of the threespine stickleback* (eds. M.A. Bell and S.A. Foster), pp. 240–276. Oxford University Press, Oxford.

Ridgway, M.S. and McPhail, J.D. (1988). Raiding shoal size and a distraction display in male sticklebacks (*Gasterosteus*). *Can. J. Zool.*, **66**, 201–205.

Ryer, C.H. and Olla, B.L. (1991). Information transfer and the facilitation and inhibition of feeding in a schooling fish. *Environ. Biol. Fish.*, **30**, 317–323.

Savino, J.F. and Stein, R.A. (1989). Behavior of fish predators and their prey: habitat choice between open water and dense vegetation. *Environ. Biol. Fish.*, **24**, 287–293.

Smith, R.J.F. (1979). Alarm reaction of Iowa and johnny darters (*Etheostoma*, Percidae, Pisces) to chemicals from injured conspecifics. *Can. J. Zool.*, **57**, 1278–1282.

Smith, R.J.F. (1986). Evolution of alarm signals: role of benefits derived from retaining group members or territorial neighbors. *Am. Nat.*, **128**, 604–610.

Smith, R.J.F. (1992). Alarm signals in fishes. *Rev. Fish Biol. Fish.*, **2**, 33–63.

Smith, R.J.F. and Smith, M.J. (1989). Predator-recognition behaviour in two species of gobies, *Asterropteryx semipunctatus* and *Gnatholepis anjerensis*. *Ethology*, **81**, 279–290.

Smythe, N. (1977). The function of mammalian alarm advertising: social signals or pursuit invitation? *Am. Nat.*, **111**, 191–194.

Suboski, M.D. and Templeton, J.J. (1989). Life skills training for hatchery fish: social learning and survival. *Fish. Res.*, **7**, 343–352.

Suboski, M.D., Bain, S., Carty, A.E., McQuoid, L.M., Seelen, M.I and Seifert, M. (1990). Alarm reaction in acquisition and social transmission of simulated-predator recognition by zebra danio fish (*Brachydanio rerio*). *J. Comp. Psychol.*, **104**, 101–112.

Thomson, D.A. (1963). *A histological study and bioassay of the toxic stress secretion of the boxfish, Ostracion lentiginosus.* Ph.D. thesis, University of Hawaii, Manoa.

Tinker, S.W. (1978). *Fishes of Hawaii.* Hawaiian Services Ltd., Honolulu.

Trivers, R.L. (1971). The evolution of reciprocal altruism. *Q. Rev. Biol.*, **46**, 35–56.

Verheijen, F.J. (1956). Transmission of a flight reaction amongst a school of fish and the underlying sensory mechanisms. *Experientia*, **12**, 202–204.

Werner, E.E., Gilliam, J.F., Hall, D.J. and Mittelbach, G. (1983). An experimental test of the effects of predation risk on habitat use in fish. *Ecology*, **64**, 1540–1548.

Whoriskey, F.G. (1991). Stickleback distraction displays—sexual or foraging deception against egg cannibalism. *Anim. Behav.*, **41**, 989–995.

Winn, H.E. and Bardach, J.E. (1959). Differential food selection by moray eels and a possible role of the mucous envelope of parrotfishes in reduction of predation. *Ecology*, **40**, 296–298.

Winn, H.E., Marshall, J.A. and Hazlett, B. (1964). Behavior, diel activities, and stimuli that elicit sound production and reaction to sounds in the longspine squirrelfish. *Copeia*, **1964**, 413–425.

Wisenden, B.D., Chivers, D.P. and Smith, R.J.F. (1994). Risk-sensitive habitat use by brook stickleback (*Culaea inconstans*) in areas associated with minnow alarm pheromone. *J. Chem. Ecol.*, **20**, 2975–2983.

Wisenden, B.D., Chivers, D.P., Brown, G.E. and Smith, R.J.F. (1995a). The role of experience in risk assessment: avoidance of areas chemically labelled with fathead minnow alarm pheromone by conspecifics and heterospecifics. *Écoscience*, **2**, 116–122.

Wisenden, B.D., Chivers, D.P. and Smith, R.J.F. (1995b). Early warning in the predation sequence: a disturbance pheromone in Iowa darters (*Etheostoma exile*). *J. Chem. Ecol*, **21**, 1469–1480.

Zaret, T.M. and Kerfoot, W.C. (1975). Fish predation on *Bosmina longirostris*: body size selection versus visibility selection. *Ecology*, **56**, 232–237.

8 *Evading predators*

Jean-Guy J. Godin

8.1 Introduction

Predation is known to affect the behaviour, morphology and life history of prey animals, as well as their populations and communities, over both evolutionary and ecological time scales (reviewed in Endler 1986, 1991, 1995; Sih 1987; Lima and Dill 1990; Milinski 1993; Persson *et al.*, Chapter 12). Predation events typically follow a sequence of consecutive conditional stages, namely, prey encounter, detection, identification, approach (attack), subjugation (capture) and consumption (Endler 1986, 1991; Lima and Dill 1990; see also Fig. 7.1 in Smith, Chapter 7). An individual's risk of predation can thus be defined as its probability of being eaten by a predator during some discrete time period. This can be expressed as P (death) $= 1 - \exp(-\alpha d T)$, where α is the prey's rate of encounter with predators while vulnerable, d the probability of death given an encounter, and T the time spent vulnerable to an encounter (Lima and Dill 1990). Predators vary widely in their ability to detect, capture and consume prey, while prey vary in their ability to detect, avoid and escape predators (Curio 1976; Vermeij 1982; Sih 1987; Lima and Dill 1990; Endler 1991; Fuiman and Magurran 1994; Reimchen 1994; Smith, Chapter 7). Such variation in the risk of predation determines the potential for the evolution of antipredator defences in prey (Vermeij 1982; Reimchen 1994).

In general, the risk of predation of an individual prey increases with successive stages in the above predation sequence (Endler 1986, 1991). Therefore, prey should be selected to interrupt this sequence by means of defences, so as to minimise their risk of predation (Endler 1986; 1991; Sih 1987; Lima and Dill 1990). Any trait that reduces an individual's risk of predation and thus increases its fitness is herein referred to as an antipredator defence or antipredator adaptation, regardless of its evolutionary history (*sensu* Abrams 1990; Reeve and Sherman 1993). Defences may be behavioural, morphological or chemical in nature (e.g. Edmunds 1974; Endler 1986, 1991; Sih 1987; Lima and Dill 1990; Milinski 1993; Smith, Chapter 7).

Because defences acting early in the predation sequence (i.e. pre-attack) tend to be more generalised, more frequent and energetically less costly (per event) than those acting later in the sequence (Endler 1986, 1991), it is not entirely arbitrary to consider defences that occur prior to predator attack separately from those that occur post-attack. Edmunds (1974) made a similar distinction between primary and secondary defences. Defences used by prey fish to avoid being detected and attacked by predators have been reviewed by Smith (Chapter 7). Here, I consider antipredator defences in teleost fishes that occur following predator approach/attack. Such defences, collectively referred to herein as predator evasion defences (*sensu* Weihs and Webb 1984), serve to reduce the prey's probability of being captured given that an attack as occurred ($1 - e_1$, $1 - e_2$ in Fig. 7.1 of Smith, Chapter 7) and, if captured, its probability of being consumed ($1 - e_3$ in Fig 7.1 of Smith, Chapter 7). These probabilistic events constitute part of one of the major components of the prey's overall predation risk, namely d (as defined above).

This chapter provides an overview of the diversity of known or potential defences mitigating capture and ingestion by predators in teleost fishes. Because the fitness benefits and costs associated with particular defences differ, individual prey will often be required to make trade-offs between alternative defence tactics and between antipredator defences and other behavioural activities when faced with a threat of predation (Sih 1987; Lima and Dill 1990; Milinski 1993). Therefore, the general approach followed is a functional one, with the optimality (cost–benefit) paradigm used where appropriate to explain the existence of diverse antipredator defences in fishes. Proximate mechanisms and constraints on behaviour are addressed if they help explain function. Important conceptual issues are identified and recent advances in knowledge and suggestions for future research directions are provided.

8.2 Defences mitigating predator capture of prey

8.2.1 Fleeing

Moving away from an approaching predator (i.e. fleeing) is perhaps one of the most commonly observed antipredator responses in animals (Edmunds 1974; Keenleyside 1979; Ydenberg and Dill 1986; Sih 1987; Lima and Dill 1990, Fuiman and Magurran 1994). Fleeing behaviour in fishes is highly variable, and the particular form of the flight response initiated by a prey fish depends at least on features of the predatory stimulus and the context in which it is presented. If the predator stimulus is detected early and from a sufficient distance away and/or if the prey fish is swimming when attacked, then fleeing may simply involve a gradual increase in the speed of sustained steady swimming (Webb 1982; Weihs and Webb 1984; Roberts 1992). Alternatively, if danger is more imminent (e.g. predator is very close when detected), then flight may be much more rapid. This so-called fast-start or startle response consists of two kinematic phases: an initial, brief (< 0.1 s) preparatory phase (stage 1) involving a unilateral muscular contraction that typically bends the fish into a C-shape, and a subsequent propulsive phase (stage 2) during which the tail bends in the opposite direction to that in stage 1 and the body turns and accelerates forward (see inset of Fig. 8.3). Less commonly, in certain species, the body may bend into a S-shape during stage 1, but does not turn very much during stage 2 allowing the prey to accelerate in line with its original body axis. Stage 2 is typically followed by a phase (stage 3) of sustained steady swimming and diverse turning manoeuvres of varying duration (Webb 1976, 1982, 1986; Eaton and Hackett 1984; Roberts 1992; Domenici and Blake 1993a, b; Pitcher and Parrish 1993). A fast-start evasive response can be elicited by mechanical or visual stimuli naturally associated with an attacking predator, and its stage 1 is usually mediated by the Mauthner neuron that functionally links the sensory input to the motor response (Eaton and Hackett 1984; Roberts 1992). The net displacement of the fleeing prey fish is mainly in the horizontal plane when attacked by fish predators (e.g. Pitcher and Parrish 1993), but mainly along the vertical plane and away from the water surface when attacked by aerial predators (e.g. Seghers 1974; Katzir and Camhi 1993; Litvak 1993). These two general forms of fleeing response (i.e. fast start versus gradual increase in sustained swimming speed) represent the extremes of a continuum of flexible escape responses, all of which tend to displace the prey fish away from the predatory stimulus.

In general, piscivorous fish predators are more likely to abort an attack on prey fish that initiate a rapid flight response compared with prey that do not flee (Neill and Cullen 1974; Major 1978; Webb 1981, 1984a, b, 1986; Webb and Skadsen 1980). Therefore, as a predator evasion tactic, fleeing in fishes can be relatively effective (Table 8.1; Fig. 8.1) compared with the option of not fleeing (i.e. immobility, see Section 8.2.2). However, fleeing is particularly variable in terms of timing, direction, speed and acceleration, and distance travelled during escape (see below). Prey success at evading capture by fleeing is correspondingly highly variable within and between species (Table 8.1; Fuiman 1994; Fuiman and Magurran 1994), and depends on both the aforementioned components of flight in prey, the relative body sizes and shapes of predator and prey (Table 8.1; Fuiman 1994; Fuiman and Magurran 1994), habitat structural complexity (see Section 8.2.3), prey experience with predators (Fuiman and Magurran 1994; Huntingford *et al.* 1994), prey health and condition (Mesa *et al.* 1994) and the type and behaviour of predators (Table 8.1; Fig. 8.1; Fuiman 1994; Fuiman and Magurran 1994). Within the sensory and motor constraints imposed by its body size and/or age (e.g. Noakes and Godin 1988; Fuiman and Magurran 1994), a prey fish that has detected an attacking predator must sequentially decide (1) whether and when to flee, (2) in which direction to flee, (3) how fast to flee and (4) how far to flee. These decisions should determine individual prey success at evading capture, and each is addressed in turn below. Although a large body of information exists on the various components of fleeing for solitary fish under laboratory conditions, little research has been done on the fleeing behaviour of individuals within shoals and on the link between the behaviour of individual members of a shoal and the fleeing response of the shoal as a whole.

Table 8.1 Examples of the relative success of prey fish escaping attacks of piscivorous fish by using a flight response. The mean ratio of predator to prey length, the mean or range of mean probability of surviving a predator attack, and the proportion of individuals that responded to an attacking predator by fleeing are given for each prey species.

Prey species	Predator species	Predator : prey length ratio	Proportion of prey fleeing	Probability of surviving attack	Reference
Fathead minnow[1]	Tiger muskellunge[2]	4.29 : 1	0.26	0.54	Webb 1982
	Tiger muskellunge	2.70 : 1	–	0.20	Moody *et al.* 1983
	Rainbow trout[3]	4.55 : 1	0.90	0.76–0.94	Webb 1982
	Smallmouth bass[4]	4.29 : 1	0.81	0.88	Webb 1982
	Rock bass[5]	2.74 : 1	0.75	0.72–0.86	Webb 1982
Northern anchovy[6]	Clownfish[7]	3.77–15.3 : 1	0.05–0.80	0.05–0.60	Webb 1981
	Northern anchovy	2.5–14.2 : 1	0.16–1.0	0.34–0.93	Folkvord and Hunter 1986
	Northern anchovy	10.0–11.9 : 1	0.17–0.54	0.01–0.03	Booman *et al.* 1991
	Chub mackerel[8]	3.8–31.8 : 1	0.26–0.96	0.14–0.61	Folkvord and Hunter 1986
Bluegill sunfish[9]	Largemouth bass[10]	3.64 : 1	0.82	0.97	Webb 1986
	Tiger muskellunge	4.00 : 1	–	0.48	Moody *et al.* 1983
Largemouth bass	Largemouth bass	4.57 : 1	0.93	0.92–1.00	Webb 1986
Tiger muskellunge	Tiger muskellunge	3.58 : 1	0.96	0.93–1.00	Webb 1986
Atlantic herring[11]	Whiting[12]	4.7 : 1	0.52	0.46	Gallego and Heath 1994
	Whiting	2.9 : 1	0.82	0.60	Gallego and Heath 1994
Hawaiian anchovy[13]	Jack[14]	3.6–9.8 : 1	–	0.62	Major 1978
Banded killifish[15]	White perch[16]	2.1 : 1	–	0.84–0.93	Morgan and Godin 1985

Species latin names: 1 = *Pimephales promelas*, 2 = *Esox* sp., 3 = *Oncorhynchus mykiss*, 4 = *Micropterus dolomieui*, 5 = *Ambloplites rupestris*, 6 = *Engraulis mordax*, 7 = *Amphiprion percula*, 8 = *Scomber japonicus*, 9 = *Lepomis macrochirus*, 10 = *Micropterus salmoides*, 11 = *Clupea harengus*, 12 = *Merlangius merlangus*, 13 = *Stolephorus purpureus*, 14 = *Caranx ignobilis*, 15 = *Fundulus diaphanus*, 16 = *Morone americana*

Timing of flight

When a prey fish has detected an approaching predator, it must continually assess whether or not to flee during the course of the predator's approach. Such a decision has to be made very quickly (typically < 1 s) in cases of ambushing fish predators that suddenly attack using acceleration fast-starts (Webb 1976; Webb and Skadsen

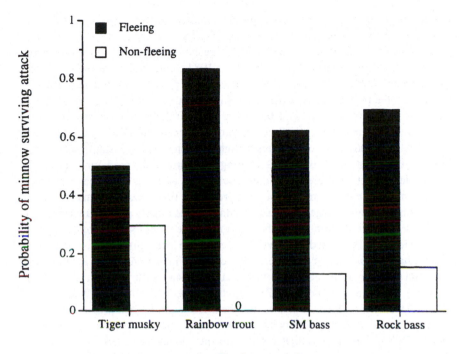

Predator species

Fig. 8.1 Proportions of fathead minnows, *Pimephales promelas*, that survived an attack/chase by different fish predators, depending on whether or not they fled from the approaching predator in a laboratory tank. The predators were the tiger muskellunge, *Esox* sp., rainbow trout, *Oncorhynchus mykiss*, smallmouth (SM) bass, *Micropterus dolomieui*, and rock bass, *Ambloplites rupestris*. Adapted from Webb (1984a).

1980; Harper and Blake 1991) and bird predators striking from above the water surface (Katzir and Intrator 1987; Katzir and Camhi 1993; Litvak 1993). In cases of cursorial (i.e. stalking or pursuit) predators that approach more slowly from some distance away, the prey will likely have more time to decide whether and when to flee. This decision will determine the flight initiation distance (FID), the distance separating prey and predator at the instant the prey flees (*sensu* Ydenberg and Dill 1986), and also its chance of successfully escaping the predator (Dill 1973; Webb 1982, 1984a, 1986; Fuiman 1993; Katzir and Camhi 1993). Proper timing of the flight response is critical; escape responses that are either too early or too late may be fatal (Webb 1976, 1982; Fuiman 1993; Katzir and Camhi 1993). The decision to flee, and associated FID, in prey fish can be understood from both a proximate and functional perspective.

At the proximate level, timing of flight will be determined by the prey's response threshold to the particular predatory stimulus it is exposed to and the physiological latency between the initial receptor neural response to the stimulus and the beginning of the motor response. For visually-mediated escapes, a prey's response threshold can be calculated as the looming threshold (Dill 1974a), or more accurately the apparent looming threshold (Webb 1981), defined as the rate of change of the angle ($d\alpha/dt$) subtended by the approaching predator's cross-sectional silhouette on the prey's retina at the start of its flight response. This threshold angular rate is assumed to be constant for a given individual, irrespective of the body size and attack velocity of the predator. According to Dill's (1974a) looming threshold model, a prey initiates a flight response at the instant its apparent looming threshold (ALT) is exceeded during the course of the predator's approach. His model predicts that prey should escape sooner, and thus at a greater distance away, from larger and/or more rapidly approaching predators than smaller and/or slower ones. Because larger and faster predators presumably pose a greater risk of mortality, such graded timing of the flight response would seem threat-sensitive (*sensu* Helfman 1989) and adaptive. In accordance with this model, certain fish species have shown longer FID (i.e. earlier escapes) in response to an approaching potential predator of increasing body size (Dill 1974a; Hurley and Hartline 1974; Helfman 1989; but see Webb 1982) and attack speed (Dill 1973, 1974a; Fuiman 1993). Available evidence indicates that the likelihood of a fish predator aborting an attack on a prey fish is positively correlated with the prey's ALT (Webb 1982, 1984a, 1986); that is, prey that respond to the predator's attack earlier (lower ALT), and thus from further away (longer FID), are less likely to be pursued by the predator and more likely to survive the encounter than prey that respond later (higher ALT) and at a shorter FID (Dill 1973; Webb 1982, 1984a, 1986; but see Fuiman (1993) for a more complex relationship in fish larvae).

An individual's ALT does not remain constant throughout its lifetime, however, as it can be modified by experience with predators (Benzie 1965; Dill 1974b; Huntingford *et al.* 1994) and by changing sensory capabilities due to growth (Noakes and Godin 1988; Fuiman 1993, 1994; Fuiman and Magurran 1994). Observed inter-individual variation in ALT, and corresponding variation in FID, of prey within the same population may thus be partly attributable to individual differences in body size and prior experience with predators. In contrast, observed variation in ALT among prey species of similar size that were attacked by the same fish predator, and among similar-sized prey individuals of the same species that were attacked by different predator species, are more difficult to explain proximately (Webb 1982, 1986). Differences between prey fish species in the ability to accelerate during escape, in manoeuvrability and in the benefits and costs of flight, and differences in the cross-sectional body shape of fish predators may account for some of the variation in ALT and FID among prey species (Webb 1982, 1986).

Because not all encounters with predators, or all moments during an encounter, are equally dangerous and because fleeing is costly at least in terms of energetic expenditure and lost opportunities to engage in other activities, individual prey should not necessarily flee as soon as they have detected a predator at a distance (Ydenberg and Dill 1986; Lima and Dill 1990). Ydenberg and Dill (1986) developed a functional model of flight that makes general predictions about the timing of fleeing and thus the distance from a predator at which the prey chooses to flee. In their economic (cost–benefit) model, the prey chooses continually between two behavioural options, staying where it is (and perhaps continuing an ongoing activity) or fleeing, as the distance between itself and the predator decreases during the course of the latter's approach/attack, and the distance at which the prey flees (the FID) is determined by a balance between the costs of fleeing and remaining. The costs of fleeing (lines F in Fig. 8.2A) are assumed to increase linearly, and the cost of remaining (namely, risk of capture; lines R in Fig. 8.2A) to decrease proportionately, with increasing distance to the predator. The prey is hypothesised to choose at every distance between itself and the predator the behavioural option (fleeing or remaining) with the lowest cost, and should choose flight over remaining when $R > F$. The optimal flight initiation distance ($D*$) is thus defined as the intersection of the F and R cost curves in Fig. 8.2A. The model predicts that the prey's FID should increase with increasing cost of remaining ($D_2* \rightarrow D_3*$) and decrease with increasing cost of fleeing ($D_2* \rightarrow D_1*$). The maximum FID for a prey animal is constrained by its ability to detect predators. The constraints may be morphological or physiological (e.g. visual acuity) or behavioural (e.g. vigilance level) (Ydenberg and Dill 1986). It is not necessarily implied that decisions about fleeing are made in a cognitive way, but rather that the prey may use simple 'rules-of-thumb', such as the apparent looming threshold (Dill 1974a) described above, to decide when to flee.

Several studies support the qualitative predictions of the economic hypothesis of fleeing in prey animals (reviewed by Ydenberg and Dill 1986; Lima and Dill 1990). The salient results of some of these studies pertaining to fishes, as well as more recent ones, are briefly summarised below. Given that the cost of remaining (= risk of predation) increases theoretically with increasing distance from a refuge (Dill 1973; Lima and Dill 1990), an individual prey's FID should therefore increase when it is increasingly further from the nearest refuge when attacked by a predator. Such a relationship has been observed for banded killifish, *Fundulus diaphanus*, (McLean and Godin 1989) and the cichlid fish *Melanochromis chipokae* (Dill 1990), but not for the armoured (and presumably less vulnerable) ninespine stickleback, *Pungitius pungitius*, and threespine stickleback, *Gasterosteus aculeatus* (McLean and Godin 1989). Also suggestive of a similar relationship is the inverse relationship between FID and cover density (which is negatively correlated with distance to nearest cover) in stream-dwelling brook charr, *Salvelinus fontinalis*, reported by Grant and Noakes (1987). The cost of remaining also presumably increases with increasing predator size and attack speed, but should decrease with increasing crypsis or defensive body armour in prey (Ydenberg and Dill 1986; Lima and Dill 1990). Increasing FID with increasing predator size (Dill 1974a; Hurley and Hartline 1974; Helfman 1989) and attack velocity (Dill 1973, 1974a; Fuiman

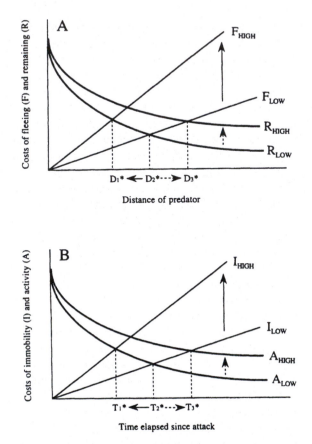

Fig. 8.2 (A) A general economic model of flight initiation distance in prey animals. The marginal cost of remaining (R) when a predator is at a particular distance is assumed to be directly proportional to risk of capture, which increases to a maximum when $D = 0$. Costs of flight (F), such as lost foraging opportunity and energy expenditure, increase linearly with increasing distance of the predator (i.e. time to attack). The prey should choose flight over current activity (i.e. remaining) when $R > F$. The intersection of the R and F curves defines the optimal flight initiation distance (D^*) for a prey being approached by a predator. The model predicts that (1) increasing the cost of flight decreases the optimal flight distance ($D_1^* < D_2^*$) and (2) increasing predation risk (cost of remaining) increases the optimal flight distance ($D_3^* > D_2^*$). Adapted from Ydenberg and Dill (1986). (B) A general economic model of immobility time under predation risk in prey animals. The marginal costs of remaining immobile (I), such as lost foraging and mating opportunities, are assumed to increase linearly with time elapsed since the predator initiated its approach/attack. The cost of remaining active (A), the alternative prey behaviour, should be proportional to the risk of predation, which is assumed to decrease non-linearly with time elapsed since the predator attack. The prey should choose immobility over activity when $I < A$. The intersection of the I and A curves defines the optimal duration of immobility (T^*) for a prey under predation hazard. The model predicts that (1) increasing the cost of immobility (lost opportunities) decreases the optimal immobility time ($T_1^* < T_2^*$) and (2) increasing the cost of activity (predation risk) increases the optimal immobility time ($T_3^* > T_2^*$). Adapted from Ydenberg and Dill (1986).

1993) have been reported for a few fish species. As predicted economically, the FID of non-armoured banded killifish (McLean and Godin 1989) and weakly armoured ninespine stickleback (Benzie 1965) to a potential predatory threat were both greater on average than that of the more heavily armoured threespine stickleback (but see Abrahams 1995). A similar relationship seems to exist within species. In certain populations of the brook stickleback, *Culaea inconstans*, some individuals possess the full pelvic skeleton with spines whereas others totally lack pelvic spines. Although he did not measure FID, Reist (1983) observed that the (more vulnerable) spineless phenotype was more likely to flee from a stalking or pursuing pike, *Esox lucius*, than the spined phenotype. Such phenotype-dependent flightiness may not be general, however, as McLean and Godin (1989) failed to observe any difference in the likelihood of fleeing in response to an approaching predator model between (pelvic) spined and spineless morphs of the ninespine sticklebacks originating from the same population.

In contrast to the above support for the predicted positive relationship between FID and the cost of remaining, there exits little unequivocal evidence for the predicted inverse relationship between FID and cost of flight in fishes. Because an individual necessarily has to abandon its current activity to flee from predators, an important cost of flight is lost opportunities. The economic model predicts that a prey should delay fleeing (i.e. exhibit shorter FID) from an approaching predator with increasing expected benefits from continued foraging, mating, interacting aggressively with conspecifics, or other activities. As predicted by this model, fish that are engaged in foraging (Krause and Godin 1996) or escalated fights (Jakobsson *et al.* 1995) have shorter FID in response to predators than conspecifics not engaged in these activities. Grant and Noakes (1987) observed a negative relationship between FID and short-term feeding rate in individual stream-dwelling brook charr. These relationships may be the result of individuals either actively choosing to delay their flight from a potential predator to gain further benefits from their current activity, as predicted by the economic model, or reduced vigilance for predators (i.e. behavioural constraint) when engaged in such activities (Milinski 1990). The challenge is to be able to discriminate between these two alternative mechanisms, which will require further research on the relationship between the costs of flight and FID.

Furthermore, some studies have reported increased FID in prey fish with increasing prior exposure to predators (e.g. Benzie 1965; Dill 1974b); this relationship is also consistent with the economic model of flight if an individual's estimate of risk is enhanced with past experience with predators (a reasonable assumption).

As demonstrated above, the timing of an individual's flight from an approaching predator can be influenced by its assessment of current risk and the costs of fleeing, both of which are known to vary with group membership size (Godin 1986; Lima and Dill 1990; Magurran 1990; Pitcher and Parrish 1993). Therefore, the FID of prey that live in groups is also expected to vary with group size, according to Ydenberg and Dill's (1986) economic model. However, the particular form of the relationship between group size and FID is difficult to predict, because many variables associated with group living (e.g. dilution of risk, enhanced foraging efficiency) tend to confound the measurement of costs and benefits of flight and remaining as a function of group size (Ydenberg and Dill 1986). To my knowledge, only three studies have quantified predator-induced FID in fish over a range of shoal sizes (reviewed by Godin 1986, Ydenberg and Dill 1986). The particular form of the group size–FID relationship is different in each study, ranging from no effect of group size in free-ranging banded killifish (Godin and Morgan 1985), longer FID in solitary spottail shiners, *Notropis hudsonius*, than in groups but no variation among groups (Seghers 1981), to maximal FID at intermediate-sized groups in banded killifish in the laboratory (J.-G.J. Godin and A.R. Hanson, unpublished data). None of these relationships can be explained simply by the increase in group vigilance (perceptual ability) with increasing shoal size in fishes (Magurran *et al.* 1985; Godin *et al.* 1988), from which one would predict an asymptotically increasing FID with increasing group size. Neither does the observed variability in the group size–FID relationship constitute support per se for the economic hypothesis of flight. To functionally understand the diversity of such relationships, one would need to quantify or control for the benefits and costs of membership in groups of varying size, as was done by Dill and Ydenberg (1987) with water striders, *Aquarius* (= *Gerris*) *remigis*.

Equally complex and diverse should be the relationship between the body size of individual fish and their FID in response to predatory attacks, because the costs of remaining and fleeing will scale with body size owing to size-dependent changes in perceptual and motor abilities (Noakes and Godin 1988; Fuiman and Magurran 1994), vulnerability to predation (McGurk 1993; Pepin 1993; Fuiman 1994; Fuiman and Magurran 1994) and energy requirements (Schmidt-Nielsen 1984) during ontogeny. Unfortunately, virtually nothing is known about the body size–FID relationship in fishes. There are three notable exceptions. Fuiman (1993) observed that the FID of larval herring, *Clupea harengus*, decreased with increasing body length between 8 and 32 mm; he attributed this response to ontogenetic changes in the sensory systems of the larvae. Abrahams (1995) observed a similar inverse relationship for the armoured brook stickleback, but not for the non-armoured fathead minnow, *Pimephales promelas*. Conversely, Grant and Noakes (1987) reported a positive relationship between the body length of brook charr and the distance at which they fled from an approaching human observer in the stream. These authors attributed the apparent greater 'wariness' of larger charr to their probable greater susceptibility to bird predators (i.e. greater benefit of fleeing) and their lower food requirements (i.e. lower opportunity cost of fleeing) compared with smaller charr. Knowledge about how the benefits and costs of fleeing and remaining vary with body size in fishes will not only allow for predictions about how FID should vary with body size, but also contribute to a better understanding of the body-size scaling of vulnerability to predation.

Flight trajectory

The direction in which a fish initially flees from an attacking predator and any subsequent changes in swimming direction during pursuit (collectively referred to as its flight trajectory or escape path) can affect its success at evading capture. If a prey's escape path is linear and directly in line with the predator's attack path, then it will likely be intercepted eventually by the predator because pisciverous predators are generally larger (e.g. Table 8.1), and thus can move faster (e.g. Blake 1983), than their fish prey. Small prey fish, however, are able to effect body turns over smaller angles than their larger predators (Webb 1976, 1981; Blake 1983). Therefore, an individual prey may increase its chances of escape by rapidly turning away from the predator's attack path at unpredictable times and turning angles. Such escape manoeuvres are generally referred to as protean movements, and are assumed to reduce the likelihood of the predator intercepting the fleeing prey (Humphries and Driver 1970; Edmunds 1974).

The kinematic optimality models of predator evasion of Howland (1974) and Weihs and Webb (1984) show that the relative speed of locomotion and relative body turning angle of prey and predator (in two-dimensional space) interact to determine the outcome of the encounter. In particular, Weihs and Webb's (1984) model predicts the optimal prey turning angle that maximises the instantaneous distance separating prey and predator for varying ratios (K) of predator attack speed and prey escape speed. When $K \leq 1$ (i.e. prey is faster or as fast as predator), the prey should flee in direct line with the predator's attack path. When $K > 1$ (i.e. predator faster than prey), however, the fleeing prey should turn away from the predator's attack path and the optimal prey turning angle should be relatively small (i.e. $< 21°$ from the line heading directly away from the predator). Much is known about the kinematics of body turning during fleeing in prey fish in response to attacks from fish predators or simulated fright stimuli, but relatively little information is available on flight trajectories of prey fish under attack from bird predators. Moreover, most studies have been carried out in laboratory tanks which can constrain fish escape manoeuvres. Therefore, the following discussion will necessarily be restricted to predatory fish–prey fish interactions under laboratory conditions. The available data show that the initial direction in which individual prey fish flee, relative to the direction of an attacking fish predator or of the source origin of the fright stimulus, is non-random, but is also not restricted to corresponding turning angles of $< 21°$ as predicted by Weihs and Webb's (1984) kinematic model.

Most commonly during fast-start escape responses, the prey's flight path tends to be away from the source of the predator stimulus along the horizontal plane because the fish's axial musculature contralateral to the stimulus contracts during kinematic stage 1, thus forming a C-shaped body turn (Fig. 8.3). The angle of such a body turn has previously been regarded as preprogrammed at the onset of the escape response and thus the initial flight trajectory to be relatively fixed at about 90° (relative to the initial position of the prey's body axis) and predictable

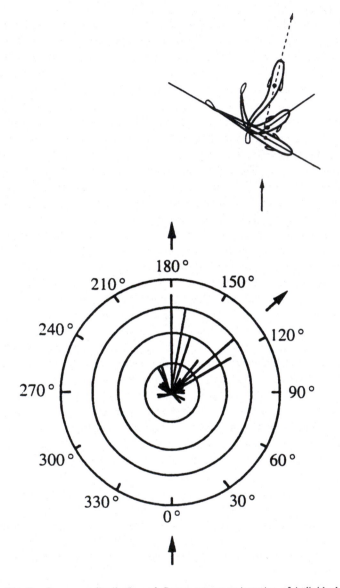

Fig. 8.3   Circular frequency distribution of C-start escape trajectories of individual angelfish, *Pterophyllum eimekei*, ($n = 62$) in response to a mechanical stimulus (arrow at bottom of the graph) applied to the side of the tank. Only responses away from the direction of the stimulus are shown. Responses to left or right are plotted as if the stimulus were always on the right side of the fish. The frequency interval is 10°, and each concentric circle represents a frequency of two fish. Modal escape trajectories occur at 180° and 130° away from the direction of the fright stimulus. The inset illustrates the form of the C-start escape of a fish in response to a directional fright stimulus (arrow). Adapted from Domenici and Blake (1993a).

(Webb 1976, 1984b). In certain species, the body is instead bent into an S-shape during stage 1 and the initial escape path is thus roughly in the same direction as the prey's original body axis (Webb 1976, 1986). More recent studies, however, have shown that the orientations of the prey fish's body at the end of its initial C-shaped body turn (end of kinematic stage 1) and at the beginning of its escape path (end of stage 2), relative to the predator stimulus, are highly variable and to some degree under the prey's control (Webb 1981, 1982, 1986; Eaton and Emberley 1991; Domenici and Blake 1993a, b; Fuiman 1993; Domenici and Batty 1994).

Although fish escape trajectory angles are linearly related to the direction of the predator (or fright) stimulus (Eaton and Emberley 1991; Domenici and Blake 1993a) and generally inversely related to their body size (Webb 1976; Domenici and Blake 1993b; but see Webb 1981; Fuiman 1993 for fish larvae), they vary considerably after the initial turn (at end of kinematic stage 1) away from the stimulus. For example, the majority of similar-sized angelfish, *Pterophyllum eimekei*, escaped away from the direction of an acoustic fright stimulus presented laterally (Domenici and Blake 1993a). Although their escape trajectories extended over a wide range of angles, the fish escaped preferentially at about 130° and 180° away from the direction of the stimulus (Fig. 8.3) and thereby potentially maximised their instantaneous distance from the stimulus source. Similarly, the majority of herring exposed to either an attacking fish predator (Fuiman 1993) or an acoustic fright stimulus (Domenici and Batty 1994) escaped away from the direction of the stimulus. The angles of their escape trajectories were also not randomly distributed, but rather had a broad mode at 130–170° away from the direction of the acoustic stimulus in adult herring (Domenici and Batty 1994) and modes at 140–160° and 0–20° away from the attack path of a fish predator in larval herring (Fuiman 1993). Similar-sized individuals of four different prey fish species exhibited on average initial body turns of about 10–70° (relative to the initial position of their body axis) when escaping from an attacking largemouth bass, *Micropterus salmoides* (Webb 1986). In response to attacks from four different fish predator species, fathead minnows fled more slowly and on average at angles of 57–69° away from the predator's attack path when the latter attacked at relatively low speeds, but escaped at higher speeds and exhibited larger body turns of 81–89° on average when fleeing from faster attacking predators (Webb 1982).

A fish's social environment may also influence its initial escape angle. Coordinated predator evasion manoeuvres, some of which involving similar escape trajectories among individuals, are well documented in shoaling fish (see Section 8.2.4). Furthermore, prior experience with predators can modify prey escape trajectory. For example, wild-caught threespine sticklebacks originating from a population that experiences high predation pressure were significantly more likely to escape at right angles away from the line of 'attack' of a fish predator model than similar-sized sticklebacks from a lower-risk population, which tended to swim towards the approaching predator (Huntingford *et al.* 1994). However, when young sticklebacks from these two populations were separately reared in the laboratory without their father (who often quickly swims after them and retrieves them in its mouth back to the nest), the differences in their escape trajectories noted above disappeared. It would thus appear that interactions with a rapidly accelerating larger fish (their father) during ontogeny sufficiently resembles actual encounters with fish predators to enhance the development of antipredator evasion in sticklebacks from the high-risk population, but interestingly not in those from the low-risk population (Huntingford *et al.* 1994). Because body turning can constrain acceleration in teleost fishes, the particular escape trajectory chosen by an individual may also reflect a behavioural trade-off between the magnitude of the initial fast-start body turn and acceleration performance, with lower acceleration occurring at larger turning angles (Domenici and Blake 1993b).

Collectively, the results of the aforementioned studies strongly suggest that initial prey trajectories during fast-start escapes in fishes are not fixed and predictable, but rather considerably variable and to some extent under behavioural control. Domenici and Blake (1993a) suggested that observed within-species variability in escape trajectories is adaptive, because it prevents predators from learning any fixed single escape trajectory and adjusting their attack path accordingly to intercept the fleeing prey. Unfortunately, there are few studies relating the initial escape trajectory of individual prey fish to their success at escaping an attacking predator. Exceptions include the following two studies. Fuiman (1993) observed no significant difference in the distributions of escape path angles of herring larvae that successfully escaped an attacking fish predator compared with fleeing larvae that were captured. None the less, flight responses were least successful (59%) when larvae fled directly towards the attacking predator, and most successful (79%) when their escape path was directly away (180°) from the line of attack of the predator. Webb (1982) reported that individual fathead minnows tended to be more successful at escaping predatory rainbow trout, *Oncorhynchus mykiss*, and smallmouth bass, *Micropterus dolomieui*, when the angle of their initial escape trajectory was on average 57–69° away from the predator's attack path compared to 81–89°. A limitation inherent to such studies of predator–prey encounter is the difficulty in differentiating between the effect of prey turning angle on their escape success and the effects of other co-variables of the flight response (e.g. timing, escape speed) and of predator behaviour (e.g. attack speed) that might be correlated with prey turning angle. Even less is known about the relationship between the particular type of turning manoeuvres adopted by a prey fish during the more variable kinematic phase 3 of its escape and its success at escaping the predator (see also Section 8.2.4). Future research therefore needs to be directed at quantifying the fitness consequences of fish escape trajectories if we are to functionally understand the observed variation in this component of prey flight behaviour.

Flight speed

A prey fish's chance at escaping an attacking predator in theory depends, at least in part, on their relative linear swimming performance (i.e. acceleration rate and swimming speed) during the encounter (Howland 1974; Weihs and Webb 1984; Webb 1986). In general, selection should favour predators that minimise closure time (the anticipated time to prey interception, *sensu* Dill 1973) and favour prey that maximise this time and thus prolong the encounter (Webb 1976, 1982, 1986; Weihs and Webb 1984). Predator closure time is directly related to the prey's flight initiation distance (FID) from the predator and inversely related to the difference between prey escape speed and predator attack speed (Dill 1973). Therefore, an individual prey can reduce its risk of being captured by increasing its FID (see Section 8.2.1) and by accelerating and swimming faster than the attacking predator. Because maximum acceleration rate, steady swimming speed and distance travelled per unit time tend to increase with increasing body size in fishes (Webb 1976, 1981; Blake 1983; Domenici and Blake 1993b; Fuiman 1993, 1994) and because fish predators are usually larger than their fish prey (Table 8.1), a predator should eventually intercept its fleeing prey, on average, assuming that both are swimming at maximum capacity, have similar geometric body shape and their locomotor trajectories are linear. To the contrary, attack success of fish predators is very variable and generally much less than 100% (Table 8.1), and observed differences in the locomotor performances of predator and prey (and hence closure time) cannot fully explain the outcome of their interaction (Webb 1982). This discrepancy between theory and observation results in part from known submaximal (e.g. Webb 1984b, 1986; Harper and Blake 1991) and variable (e.g. Dill 1974b, 1990; Webb 1986, 1982; Huntingford *et al.* 1994) swimming performances of predator and prey during encounters with one another. The latter observations suggest that there are costs (presumably energetic ones) associated with maximal swimming performance, and that interacting fish predators and prey are capable of optimising their acceleration rates and swimming speeds to maximise their respective fitness. I review below evidence for such behavioural modification of linear swimming performance in prey fish and relate it to the costs and benefits of their behaviour and their success at escaping predators.

In theory, one should expect a positive relationship between an individual's speed of escape from an attacking predator and its anticipated vulnerability to the predator, everything else being equal. Consistent with this proposition, Webb (1982, 1986) and Dill (1990) observed that fish which initiated their escape at short distances (i.e. short FID) from an approaching fish predator stimulus, and thus were presumably at greater risk of mortality to predation (see Section 8.2.1), tended to escape at greater speeds than those that fled from further away (i.e. at longer FID). In contrast, no relationship between FID and escape speed was reported for zebra danios, *Danio rerio*, 'attacked' by a predator model (Dill 1974a). Because predator closure time is inversely related to predator attack speed (see above), an individual's risk of predation theoretically increases with increasing predator attack speed. Therefore, it would be adaptive for prey to increase their escape speed in response to increasing predator attack speed. As expected, a positive linear relationship ($r^2 = 0.40$, $P < 0.001$, $n = 53$) was observed between the escape speed of individual banded killifish and the speed of attack of predatory brook charr in a large circular laboratory tank; average killifish escape speed was significantly greater than average charr attack speed (J.-G.J. Godin and L.M. Taylor, unpublished data). Similarly, individual fathead minnows fled at greater speeds in response to faster attacking fish predators (Webb 1982).

Escape speeds of prey fish may also be modified based on their prior exposure to predators, both over evolutionary and ecological times. For example, Huntingford *et al.* (1994) reported that threespine sticklebacks originating from a high predation risk population fled from an approaching trout predator model at greater speeds, on average, than those originating from a low risk population. Selection presumably favoured higher escape speeds in that environment where sticklebacks frequently encounter fish predators. Prey fish may also learn from prior experience (and successful escapes) the level of danger associated with a particular type of predator and adjust their escape speed accordingly. This is illustrated by individual zebra danios decreasing the speed of their escape response from an innocuous fright stimulus (a looming cinematographic black circle) over repeated exposures to it (Dill 1974b).

Some data indicate that the probability of a fish predator aborting an attack increases with increasing prey acceleration during flight (Webb 1986). It has therefore been suggested that a high level of swimming performance (acceleration, velocity) by prey during flight might reliably signal its high relative escape ability to the predator and thus deter further pursuit and any capture attempt (Webb 1982, 1986; see also Hasson 1991). In this context, it would be interesting to investigate the effects of factors (such as low food ration, parasitism and gravidity) that negatively affect physical condition on the swimming performance of fleeing prey fish (Mesa *et al.* 1994) and their probability of escaping predator attacks.

Flight distance

The distance travelled by a prey during the course of its flight response from an approaching predator will depend mainly on its body length, its escape velocity and acceleration pattern, and the duration of the flight response itself. For a given escape speed, larger fish can travel further than smaller fish over the same time interval (e.g. Webb 1981; Fuiman 1993; Fuiman and Magurran 1994). Escape speed is potentially under the prey's control (see previous section), while flight duration will largely be determined by whether or not the predator pursues the prey and, if it does, for how long (Anholt *et al.* 1987) and by the prey's stamina (Blake 1983), its distance from the nearest refuge at the start of its flight (Dill 1990) and the tortuousness of its flight trajectory (see previous section). The prey has potential behavioural control over the latter two parameters, and perhaps also over predator pursuit time if it can deter the predator through some aspects of its fleeing behaviour (see Hasson 1991). If the prey chooses to stop fleeing before the predator stops pursuing it or before it reaches the relative safety of a refuge, then it will incur an increased risk of capture. Because this cost is far greater than the costs associated with continued flight (e.g. energy expenditure, lost opportunities), the prey should not prematurely end its flight when it has accurately assessed its instantaneous risk of capture to be high. However, underestimation of this risk owing to uncertainty (Lima and Dill 1990; Bouskila and Blumstein 1992; Sih 1992) and/or high energy demands, perhaps caused by parasitism (Milinski 1993; Huntingford *et al.* 1994), in prey may result in shorter flight distances and potentially increased probability of capture.

Ambush or lunging piscivorous predators (e.g. pike) rarely pursue their fish prey beyond the prey's location at the start of its flight (Webb and Skadsen 1980; Webb 1984a, b; Harper and Blake 1991). Although cursorial fish predators (e.g. rainbow trout) may pursue their fish prey, they are reluctant to do so when the attacked prey responds with a fast-start escape (Webb 1981, 1984a, b, 1986). Therefore, most fish predator pursuits and corresponding prey escapes are relatively brief (typically < 2 s) and over relatively short distances (typically < 30 cm) (Major 1978; Webb 1981, 1984b,1986; Moody *et al.* 1983; Wahl and Stein 1988; Harper and Blake 1991; Fuiman 1993; Fuiman and Magurran 1994). Flight distances of fish in response to actual or simulated bird predator attacks are similarly short (Seghers 1974; Godin and Sproul 1988; Katzir and Camhi 1993; Litvak 1993). The true relationship between a prey's probability of successfully escaping an attacking predator and its flight distance is difficult to ascertain, because any prey that is captured while still fleeing necessarily has a truncated flight distance compared with a prey that flees a certain distance and is not captured. Based on the available data, it appears that the distance a prey flees in response to an attacking predator is not as important in determining its escape success as are the timing, initial trajectory and velocity/acceleration of its fleeing response (see previous sections).

8.2.2 Immobility

Prey animals may alternatively cease their current activity and remain immobile when they are threatened by predators, particularly visually guided ones (Edmunds 1974; Suarez and Gallup 1983; Arduino and Gould 1984; Sih 1987; Lima and Dill 1990; Smith 1992, Chapter 7). This is an effective behavioural tactic in reducing encounter rate with predators and the probability of being detected and attacked, especially in prey species that are cryptically coloured (Sih 1987; Lima and Dill 1990; Endler 1991; Smith, Chapter 7). In certain circumstances, immobility may also be adopted by prey as a tactic to reduce the likelihood of being captured *after* they have been approached or attacked by a predator (e.g. Edmunds 1974; Stein 1979; Endler 1986; Sih 1987; Huntingford *et al.* 1994). However, care must be taken to distinguish prey immobility as a behavioural response to an approaching predator from no apparent response (with the prey being relatively immobile), due to perceptual, developmental or physiological constraints (Ydenberg and Dill 1986; Fuiman and Magurran 1994). In fishes, these two alternatives may be distinguished behaviourally by signs of 'arousal' or 'awareness', such as change in rate of movement of the opercula (Metcalfe *et al.* 1987; Knudsen *et al.* 1992), erection of spines and fins (Altbäcker and Csányi 1990; Huntingford *et al.* 1994), subtle body movements (Godin 1986; Krause and Godin 1996) or specific body postures (Smith and Smith 1989), exhibited by otherwise immobile individuals when exposed to a predatory threat.

In fishes, immobility may occur as the initial response to an approaching/attacking predator (e.g. Brown 1984; Tulley and Huntingford 1987; McLean and Godin 1989; Radabaugh 1989; Huntingford *et al.* 1994) or as a secondary response immediately following a preceding defence, such as fleeing a short distance away from the predator (e.g. McPhail 1969; Seghers 1974; Godin and Sproul 1988; Gotceitas and Godin 1991, 1993; Huntingford *et al.* 1994). An individual's decision (*sensu* Dill 1987) to either 'freeze' (= immobility) or remain active (e.g. flee, continue foraging or mating) when threatened, and secondly how long to remain immobile, should depend on the relative fitness benefits and costs of immobility and activity. Because visual predators tend to preferentially attack moving prey (e.g. Stein 1979; Ibrahim and Huntingford 1989; Katzir and Intrator 1987; Krause and Godin 1995; Martel and Dill 1995), activity is risky whereas immobility may reduce the prey's likelihood of being captured in certain circumstances, such as when the predator is very close and escape appears unlikely (McPhail 1969; Webb and Skadsen 1980; Reist 1983; Arduino and Gould 1984; Webb 1984a). A visual predator may lose sight of a previously active, targeted prey if the latter suddenly becomes immobile and more cryptic (Stein 1979; Rahel and Stein 1988; Hastings 1991). However, a major cost of immobility is lost opportunities to obtain and/or defend resources, such as food (Williams and Brown 1991) and mates (Hastings 1991; Magnhagen 1993).

Therefore, the tendency of individual fish to remain immobile when approached or attacked by predators should be inversely related to (1) their vulnerability to predation whilst immobile and (2) the expected benefits of remaining active whilst under threat of predation. The above hypothesised inverse relationship between immobility and vulnerability is supported by a number of studies. For example, McLean and Godin (1989) observed that the non-armoured, and presumably more vulnerable, banded killifish was significantly less likely to remain immobile (i.e. not flee) when 'attacked' by a predator model than the heavily armoured threespine stickleback. Ninespine sticklebacks, which possess smaller and less robust spines and fewer and smaller lateral bony plates than threespine sticklebacks, had an intermediate tendency to remain immobile (Fig. 8.4). McLean and Godin (1989) showed further that individual fish which opted to remain immobile, rather than flee, when attacked were on average significantly closer to a potential refuge (and thus presumably less vulnerable) than individuals that did not remain immobile but fled (see also Grant and Noakes 1987). In contrast, signal blennies, *Emblemaria hypacanthus*, that were approached by predatory fish were more likely to remain immobile on the substratum when they were further away from a shelter in the coral reef rubble (Hastings 1991). The difference between blennies

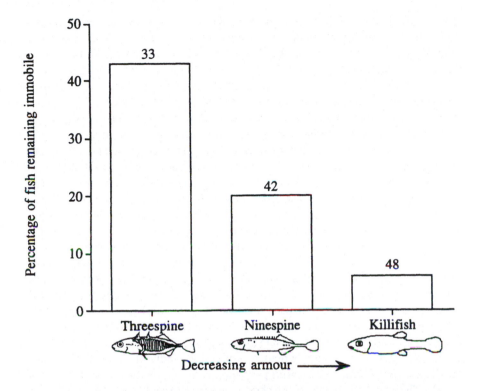

Fig. 8.4 Percentage of individually tested threespine sticklebacks, *Gasterosteus aculeatus*, ninespine sticklebacks, *Pungitius pungitius*, and banded killifish, *Fundulus diaphanus*, which became immobile (i.e. did not flee) when 'attacked' by an underwater predator model. The fish species are ordered in decreasing amount of bony body armour (from left to right along the abscissa). Dorsal and pelvic spines and lateral body plates are illustrated for both stickleback species; the killifish has no such armour. Sample sizes are indicated above the bars. Adapted from McLean and Godin (1989).

and sticklebacks in the relationship between distance from a refuge and the likelihood of opting for immobility when threatened may reflect differences in their vulnerability to predators when away from shelter. Finally, cryptically coloured fish are also more likely to remain immobile when approached by a potential predator than less cryptic conspecifics (Radabaugh 1989; Fuiman and Magurran 1994).

In comparison, little information is available to test the other hypothesis of an inverse relationship between immobility and the benefits of remaining active. One of the major benefits of remaining active whilst under the risk of predation is continued acquisition and defence of resources. For example, hungry animals or those with an energy deficit tend to take more risk to obtain food, because of the greater marginal fitness value of ingested food to them than to well-fed individuals and those with high energy reserves (reviewed in Godin 1990; Lima and

Dill 1990; Milinski 1993; Hart, Chapter 5). Therefore, hungry fish should have a lower tendency to become immobile when approached or attacked by a predator than relatively satiated conspecifics, everything else being equal. To my knowledge, this hypothesis has yet to be tested experimentally with fish under direct predator attack. None the less, Smith's (1981) study of Iowa darters, *Etheostoma exile*, exposed to alarm chemicals associated with predation risk, provides some support for this hypothesis. Darters that had been previously deprived of food responded to a mixture of brine shrimp extract (feeding stimulant) and conspecific skin extract (containing a predator alarm signal) by increasing food searching, but they responded to the same stimulus by remaining immobile when satiated.

If a prey fish has opted for immobility (outside a refuge) in response to an approaching/attacking predator and has survived the encounter, it must still decide how long to remain immobile. Presumably, this decision should depend on the prey's assessment of the marginal costs of remaining immobile or alternatively resuming activity following the unsuccessful predation attempt. These costs are illustrated in a simple economic model of immobility time (Fig. 8.2B), which I derived from Ydenberg and Dill's (1986) economic model of fleeing (see Section 8.2.1). The costs of immobility (e.g. lost foraging and mating opportunities) are assumed to increase linearly with time elapsed since the predator attack (lines *I* in Fig. 8.2B). The costs of resuming activity (e.g. increased predation risk) are assumed to decrease proportionately with increasing time elapsed since the predator attack (lines *A* in Fig. 8.2B), because mobile predators are increasingly likely to leave the immediate vicinity of an unsuccessful prey capture attempt with time (e.g. Thomas 1977). The prey is hypothesised to choose, at every instant of time elapsed since the attack, the behavioural option (remain immobile or resume activity) with the lowest cost. For example, the prey should remain immobile rather than resume activity when $A > I$. The optimal duration of immobility (T^*) is thus defined as the intersection of the *I* and *A* cost curves (Fig. 8.2B). The model predicts that the time a prey remains immobile after being threatened should decrease with increasing cost of immobility ($T_2^* \rightarrow T_1^*$) and increase with increasing cost of resuming activity ($T_2^* \rightarrow T_3^*$).

There exist only limited data to test these predictions in fishes. In support of the first prediction, threespine sticklebacks (Giles 1983, 1987; Godin and Sproul 1988; but see Tierney *et al.* 1993) and juvenile Atlantic salmon, *Salmo salar*, (Gotceitas and Godin 1991) reduced time spent immobile following a simulated predation threat with increasing hunger level or energy deficit (i.e. increasing cost of immobility). With regards to the model's second prediction, the cost of resuming activity following an initial threat is presumably high when the predator is near and/or very dangerous and when prey are more vulnerable to predation while active; under such circumstances, the prey should remain immobile longer. When a fish predator was swimming nearby, lumpfish larvae, *Cyclopterus lumpus*, remained attached (with a ventral adhesive disk) to a surface and immobile significantly longer than in the absence of the predator (Williams and Brown

1991). Gotceitas and Godin (1993) observed that juvenile Atlantic salmon fled a short distance before remaining immobile near the substratum and delayed foraging longer (i.e. remained immobile longer) after being 'attacked' by a stuffed kingfisher than by an unrealistic, bird-shaped predator model (that presumably looked less dangerous to the fish). Similarly, FitzGerald and van Havre (1985) reported an inverse (concave-down) relationship between the membership size of free-ranging shoals of sticklebacks and the time the shoal remained relatively immobile following exposure to an overhead fright stimulus. That is, the fish resumed 'normal' activity sooner when they were in increasingly larger and safer shoals.

In summary, body armour, cryptic coloration and proximity to cover promote immobility as an initial response to approaching predators in fishes. It thus seems that, in general, when a fish appears to be relatively invulnerable, owing to any of the aforementioned traits or factors, it is more likely to freeze in response to an approaching predator than more vulnerable conspecifics. However, there remains a paucity of quantitative information on the efficiency of the immobility tactic in escaping predator capture in fishes. When a prey remains immobile, its chance of surviving an attack depends on predator error, which is typically low ($< 10\%$) for fish predators attacking fish larvae and small juveniles (Fuiman 1994; Fuiman and Magurran 1994). The limited data available suggest that, once an attack has occurred, immobility is a less effective predator evasive tactic than fleeing for relatively small fish (Fuiman 1994; Fuiman and Magurran 1994) and even for larger ones in some circumstances (Fig. 8.1). Also generally lacking are data on other factors which ought to influence a prey's decisions to opt for immobility or not when threatened and how long to remain immobile, because of their known effects on an individual's risk of predation or its expected benefits of remaining active under predation hazard. These factors include the prey's own sociality (solitary vs. group living), the species identity, body size, swimming performance and attack distance of the predator, and the local density and distribution of resources (e.g. food, mates) and competitors. In addition, the general predictions of the economic model of immobility time presented here remain largely untested. The information lacunae identified above represent opportunities for future research which will enhance our understanding of the fitness benefits and costs of immobility as an antipredator tactic in fishes and thus of the circumstances under which it would be favoured by selection.

8.2.3 Refuging

Although refuging can be broadly defined as any strategy that reduces an individual's predation risk (Sih 1987, 1992), this term is generally reserved to describe the behavioural association of a prey with either biotic or abiotic features of its environment that results in a lower risk of predation than otherwise. Among the known abiotic refuges used by fishes when threatened by predators are submerged woody debris (Everett and Ruiz 1993), interstices or crevices within the substratum (Keenleyside 1979; Dill 1990; Hastings 1991; Gotceitas and Brown 1993), objects on the substratum surface (e.g. leaves, shells, Keenleyside 1979), the substratum itself (i.e. burrowing, Colgan 1974; Colgan and Costeloe 1980; Tallmark and Evans 1986; Gotceitas and Brown 1993), air-water interface (Christensen and Persson 1993), water depth (Power 1987; Schlosser 1988; Harvey and Stewart 1991; Angermeier 1992) and non-aquatic habitats (i.e. leaping out of the water, Colgan and Costeloe 1980; Baylis 1982; Sayer and Davenport 1991; Christensen and Persson 1993). Biotic refuges include beds of aquatic vegetation (Werner *et al.* 1983; Christensen and Persson 1993; Kramer *et al.*, Chapter 3, Persson *et al.*, Chapter 12), conspecific or heterospecific shoals (see Section 8.2.4) and mutualistic behavioural associations with other animals such as anemones and echinoderms (Edmunds 1974; Keenleyside 1979).

Individual fish gain relatively greater safety from predation by seeking refuge. In structurally complex habitats such as beds of aquatic vegetation (or simulated vegetation), this benefit is accrued mainly because fish predators in general are less active, visually encounter prey less often, and are less efficient at attacking them than in open water habitats (Savino and Stein 1982, 1989a; Gotceitas and Colgan 1987; Wolf and Kramer 1987; Eklöv and Hamrin 1989; Christensen and Persson 1993; Eklöv and Persson 1995), and partly because prey behaviour is also modified (e.g. increased immobility and crypsis, increased shoaling) by physical features of the refuge in ways that further reduce their risk (Savino and Stein 1982, 1989a). A habitat need not be structurally complex to serve as a refuge. For example, because the attack success of different predators may be adversely affected by the depth at which their prey fish are located, water depth per se may thus serve as a spatial refuge for certain prey fish. The relationship between water depth and risk of predation, however, is not simple and depends on the relative sizes of fish predator and prey and whether the predators are piscivorous fish or wading/diving predators. In freshwater streams, ponds and lakes at least, small fish tend to be safer from larger fish predators in shallow waters (Power 1987; Schlosser 1988; Harvey and Stewart 1991; Angermeier 1992), but larger fish are safer from avian or terrestrial predators in deeper waters (Power 1987; Schlosser 1988; Power *et al.* 1989; Harvey and Stewart 1991).

Refuging behaviour is, however, associated with certain costs. While in a refuge, an individual fish may incur a lost opportunity cost due to missed foraging and mating opportunities outside the refuge and/or to increased resource competition within the refuge (Werner *et al.* 1983; Gilliam and Fraser 1987; Sih 1992; Persson *et al.*, Chapter 12). Moreover, the particular refuge chosen by the prey may not be absolute if it harbours predators other than the one it was evading by seeking refuge in the first instance. For example, by entering a bed of aquatic vegetation to evade an open-water fish predator, a prey fish could place itself at a greater risk of predation from ambushing fish (Savino and Stein 1989a, b) and invertebrate (Reimchen 1994) predators that occur in this habitat.

Threatened fish should therefore be expected to assess the benefits and costs of entering a particular refuge relative to available alternative evasive responses. Several experimental studies (reviewed by Lima and Dill 1990; Milinski 1993; Kramer *et al.*, Chapter 3; Persson *et al.*, Chapter 12) have shown that individual fish are capable of assessing different habitats in terms of their relative expected foraging benefits and/or predation risk (i.e. effectiveness as refuges), and of choosing the habitat among those available that offers either the greatest safety from predation or the best compromise (trade-off) between foraging gains and safety from predation. Once an individual has entered a particular refuge to evade an attacking predator, it must then decide how long to remain there before leaving and emerging into more open space. Presumably the longer a prey stays in the refuge, the more likely the predator will have left the immediate area and the safer it becomes to emerge from the refuge, but at the increasing costs of lost opportunities and possibly increased risk of predation from other predators within the refuge. Sih (1992) developed a Bayesian statistical decision model to explore the effects of the prey's uncertainty about predation risk on its optimal refuge use. The prey continuously must decide whether to stay in the refuge or to emerge into open habitat. His model predicts increasing refuging times in prey with

(1) increasing vulnerability to predation,

(2) decreasing hunger level or increasing condition,

(3) increasing predation risk outside the refuge,

(4) decreasing lost opportunity cost in the refuge relative to outside the refuge, and

(5) decreasing certainty about the ambient risk of predation outside the refuge.

Few experimental studies have explicitly tested the above hypotheses. However, there exist some data that are consistent with Sih's (1992) model. For example, signal blennies spent more time in refuges (crevices within coral reefs) when threatened by potential fish predators at a site with high predator density than at another site with lower predator density (Hastings 1991). European minnows, *Phoxinus phoxinus*, in small shoals spent more time hiding in the gravel substratum or in a weed bed after being attacked by a pike than minnows in larger and safer shoals (Magurran and Pitcher 1987). Similarly, in response to an artificial overhead fright stimulus, mixed-species shoals of sticklebacks in tidal pools fled to the opposite side of the pool and sometimes entered submerged algal mats. Latency time to resume 'normal' activity, which consisted of an unknown proportion of refuging time for some of the shoals, decreased non-linearly with increasing shoal size (FitzGerald and van Havre 1985). In these studies, prey refuging time was proportional to their perceived risk of predation or their vulnerability outside refuges. Refuging time may also be inversely proportional to the expected benefits (i.e. foraging reward rate and marginal value of food) of emerging from the refuge.

Creek chub, *Semotilus atromaculatus,* (Gilliam and Fraser 1987) and European minnows (Pitcher *et al.* 1988) spent less time in a refuge when threatened by fish and avian predators, respectively, as more food was available outside the refuge. Similarly, juvenile Atlantic salmon that had sought cover after being 'attacked' with a kingfisher model emerged from cover to resume foraging sooner when they were hungrier and thus had greater energetic requirements, irrespective of their social status (Gotceitas and Godin 1991).

An individual fish's decisions of whether or not to seek refuge in response to a predation threat and how long to stay in a particular refuge not only influence its risk of mortality due to predation, but also have consequences for predator–prey inter-actions and the dynamics of their respective populations, competitive interactions and community structure (Sih 1987, 1992; Werner 1992; Kramer *et al.*, Chapter 3; Persson *et al.*, Chapter 12). Therefore, knowledge of the factors that influence these particular antipredator behavioural decisions in individual fish is necessary for a more comprehensive understanding of the role of refuges and refuging be-haviour by prey fish in higher-level population and community processes. The behavioural models of Sih (1992) and Hugie and Dill (1994), among others (see Kramer *et al.*, Chapter 3; Persson *et al.*, Chapter 12), provide useful conceptual frameworks for future research on refuge use in particular and habitat use in general in fishes.

8.2.4 Shoaling

Individual fish must often decide whether and when to join, stay or leave social groups or shoals (Pitcher and Parrish 1993). Shoaling will be favoured by selection when the fitness of group members exceed, on average, that of solitary individuals in the population (Pitcher and Parrish 1993). One known fitness benefit of shoaling is a reduction in individual risk of predation (reviewed by Godin 1986; Magurran 1990; Parrish 1992; Pitcher and Parrish 1993). When individual fish are threatened by predators, they commonly form a shoal or join a nearby fish shoal rather than stay alone (Godin 1986; Magurran 1990; Parrish 1992; Pitcher and Parrish 1993; Krause and Tegeder 1994; Tegeder and Krause 1995). Therefore, shoaling may serve as a tactic for individuals to avoid predators and evade capture. Here, I will restrict my discussion to the potential role of shoaling in generating (post-attack) antipredator benefits. Smith (Chapter 7) deals with the role of shoaling behaviour in avoiding predator detection and attack. For a more comprehensive recent re-view of fish shoaling behaviour, including its antipredator function, the reader is referred to Pitcher and Parrish (1993).

Once a predator has detected and attacked a shoal, its members might accrue antipredator benefits through a number of mechanisms. These include the risk di-lution effect, predator confusion effect, differential positioning within the shoal, evasive manoeuvres and transfer of information among shoal members (reviewed by Godin 1986; Magurran 1990; Parrish 1992; Pitcher and Parrish 1993). Both the dilution and confusion effects of shoaling may be viewed as inherent 'passive' benefits of shoaling and, in theory, all members of the shoal benefit equally (Parrish 1992; Pitcher and Parrish 1993). However, the confusion effect may be enhanced overall, and its antipredator benefits unevenly shared, by certain behaviours exhibited by members of the shoal when under attack (see below). Similarly, shoal members may accrue unequal 'active' antipredator benefits through differential positioning within the shoal, evasive manoeuvres and information transfer (Parrish 1992; Pitcher and Parrish 1993).

Dilution of risk

The numerical dilution effect of shoaling is simply the proportional reduction in the probability of a given individual in the group being captured by a predator with increasing shoal size, given that the shoal has been detected and attacked by the predator. Theory states that this conditional probability of predation is the reciprocal of group size, such that the (log) attack rate per individual prey within the group is predicted to be inversely related to (log) group size with a slope of -1 (Bertram 1978; Foster and Treherne 1981). Empirical evidence for this predicted relationship in fishes was first provided by Morgan and Godin (1985; Fig. 8.5). Surprisingly few other studies have provided quantitative support for the risk dilution effect of grouping in fishes (e.g. Wisenden and Keenleyside 1994). Notwithstanding the above theoretical and empirical evidence for risk dilution, Turner and Pitcher (1986) have argued that simple dilution of attack alone cannot be a selective pressure favouring grouping per se, because it applies only to members of a particular group being attacked and not to the remainder of the individuals in the population. They proposed that risk dilution is only one component of the 'attack abatement' (or encounter-dilution) effect of group living. Their attack abatement model takes into account predator search behaviour and encounter probabilities with solitary prey and prey in groups of different sizes in the population, as well as dilution of risk in any given shoal encountered. Interestingly, Pitcher and Parrish (1993) proposed that attack dilution might also confer additional indirect benefits to surviving members of a group in the form of information gained about the attacking predator and wider spread of information about the presence of a hungry predator in the vicinity. Such acquired (socially transmitted) knowledge may be of some benefit in any subsequent encounters with predators. For example, it has recently been shown that individual fish which live in groups may have the opportunity to learn to recognise unfamiliar predators (Mathis *et al.* 1996) and site-associated risks (Chivers and Smith 1995) by observing the predator-mediated behavioural responses of more experienced shoal mates.

Predator confusion

The predator confusion effect is a decrease in the likelihood of predator attack or in predator attack success on a group of prey with increasing group size owing to perceptual confusion in the predator caused by multiple potential targets (Ohguchi

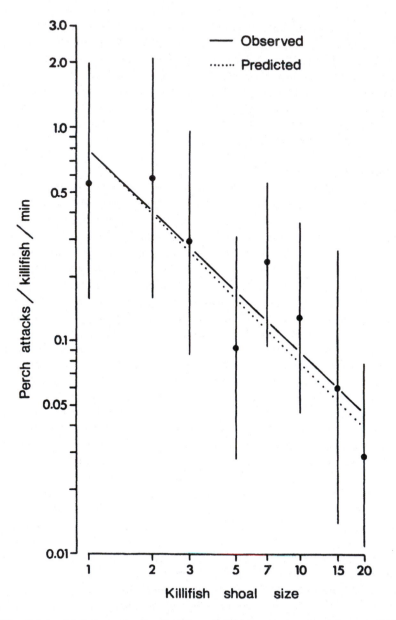

Fig. 8.5   Relationship between the rate of white perch (*Morone americana*) attacks per individual banded killifish (*Fundulus diaphanus*) and killifish shoal size, plotted on logarithmic scales. The observed slope (solid line) is -0.945 ± 0.117 (95% confidence limits). The dotted line represents the expected slope of -1.0, assuming a simple dilution effect. The two slopes are not significantly different ($t = 0.10$, $df = 78$, $P > 0.90$, two-tailed). Means and SD obtained from $\log_{10}$ transformed data are shown. Each mean is based on 10 replicate trials. From Morgan and Godin (1985).

1981; Milinski 1990). Overloading of the visual sensory channels with information originating from many simultaneously moving targets appears to be the neurophysiological basis for the confusion effect (Milinski 1990).

In theory, the relative magnitude of the confusion effect on a predator encountering a group of prey should increase with increasing size of the prey group, the compactness of the prey group (i.e. density of individuals within the group), the movement of the prey within the group, and the uniformity of their appearance (size, behaviour, colour, etc.) (Ohguchi 1981; Landeau and Terborgh 1986; Milinski 1990). Considerable evidence indicates that fish behave in ways that should enhance the perceptual confusion of an attacking predator. For example, when given a choice under experimental conditions, individual fish have been shown to prefer to shoal (1) with a larger group of fish over a smaller group or a solitary fish (Hager and Helfman 1991; Krause and Godin 1994a; Tegeder and Krause 1995) and (2) with fish of similar body size over those of different sizes (Ranta *et al.* 1992; Krause and Godin 1994a). Furthermore, fish shoals under threat of predation typically become more compact (Pitcher and Parrish 1993) and often increase their movements, either individually in the form specific behaviours such as 'skittering' (Magurran and Pitcher 1987) or collectively in the form of coordinated turns and other evasive manoeuvres (Pitcher and Parrish 1993).

Although some studies have demonstrated a decrease in predator attack efficiency with increasing prey fish shoal size (Neill and Cullen 1974; Major 1978; Tremblay and FitzGerald 1979; Landeau and Terborgh 1986; Krause and Godin 1995) and have attributed this phenomenon to predator confusion, other studies (Morgan and Godin 1985; Parrish 1993) have shown no significant effect of fish shoal size on predator attack efficiency. Assuming that a particular predator is confused by a shoal of prey fish, it none the less remains unknown whether all members of the shoal would benefit equally from the confusion effect. Differences in the behaviour, or in other external phenotypic traits, of individuals within the shoal under attack would reduce the strength of the confusion effect. Moreover, if fish predators have evolved tactics to counter perceptual confusion when attacking fish shoals, such as preferentially attacking individuals at the periphery of or stragglers from the shoal (see below), splitting the shoal into smaller groups (Parrish 1989), ramming at grouped prey (Parrish 1993) or hunting cooperatively (Parrish 1992), then this poses some difficulty in demonstrating experimentally the operation of the confusion effect in piscivorous predators. Further empirical studies of the predator confusion effect in fishes are clearly needed before more can be said about its relative importance as an antipredator mechanism of shoaling.

Spatial positioning within the shoal
There is much evidence showing that the fitness costs and benefits of group living are not evenly distributed across space within a group, such that individual group members experience differential fitness returns depending on their positions within

the group (reviewed by Krause 1994). Hamilton's (1971) 'selfish herd' theory states that individuals at the periphery or edge of the group are preferentially attacked by predators over those in the centre of the group. This implies that individuals at the group's periphery are at a higher risk of predation than those in the centre and that central positions are thus safer than peripheral positions. The selfish-herd effect predicts that individuals within a social group should move towards the geometric centre of the group when threatened by a predator. However, peripheral or frontal positions in a group may be better for maximising feeding rate compared with central positions (Krause 1993a, 1994). Therefore, the particular spatial position taken up by an individual at any given time within a social group should depend at least on its current internal state (e.g. hunger, social status) and its assessment of the local risk of predation, and thus will likely reflect a trade-off between foraging gains and the risk of predation or parasitism (Krause 1993a, 1994).

Consistent with the selfish-herd effect, some studies have reported greater predator attack rates on individual fish at the periphery of shoals (McKaye *et al.* 1992) or straggling from shoals (Major 1978; Morgan and Godin 1985; Parrish 1989, 1993) than on conspecifics in other positions within the shoal. Similar supporting evidence from other animal groups is reviewed by Krause (1994). Such predation at the group's margins may depend, however, on the hunting mode of the predator and/or the size and geometry of the prey fish shoal. For example, certain piscivorous fishes attack shoaling fish near the centre of the shoal more frequently than fish at the periphery of the shoal (Parrish 1989; Parrish *et al.* 1989). So central positions within a shoal may not always be relatively safer. Notwithstanding the above evidence, actual per capita predation risks incurred by fish in peripheral and central positions within shoals are not well documented because, as pointed out by Krause (1994), most studies have not related attack and predation rates to the number of prey fish occurring in central and peripheral positions within shoals at the instant of the attacks.

Of particular interest here is whether or not shoaling individuals tend to move towards the geometric centre of the shoal when threatened by a predator, as predicted by Hamilton's (1971) selfish-herd model. This question has been recently addressed experimentally. Krause (1993b) used 'Schreckstoff' (an alarm chemical located in the skin of cyprinid fishes that is released when a fish is wounded by a predator) to frighten single naive European minnows placed in shoals of dace, *Leuciscus leuciscus*, that had been previously habituated to Schreckstoff. Immediately following the introduction of Schreckstoff into the laboratory tank, the minnows moved towards the centre of the shoal whereas the positions of the (Schreckstoff-habituated) dace did not change on average. Furthermore, Krause and Tegeder (1994) showed that individual threespine sticklebacks at the edge of a shoal of conspecifics began the aggregation process sooner and faster than central conspecifics following an overhead fright stimulus simulating an aerial predator, such that individual fish minimised approach time to a conspecific (see also Krause 1994). These findings suggest that individual fish prefer central, and presumably safer, positions in a shoal over peripheral ones when they have assessed their current risk of predation to have increased. However, the tendency for individual fish to move towards the centre of the shoal when threatened by a predator can be affected at least by their current hunger state (Krause 1993a), perceptual constraints (Krause and Tegeder 1994) and parasitism (Krause and Godin 1994b). Additional experimental and theoretical (Krause and Tegeder 1994; Tegeder and Krause 1995) research on the relative positioning of individual fish within shoals before and after a predator attack should prove fruitful in increasing our understanding of the antipredator role of active position choice within a fish shoal.

Shoal evasive manoeuvres

Fish shoals typically become more compact and cohesive when an approaching predator has initially been detected (reviewed by Pitcher and Parrish 1993). In response to an escalation of the predator's approach into an attack, the shoal might exhibit one or more of a number of coordinated evasive behaviours, including 'avoid' (intact shoal rapidly swimming away from the predator), 'fountain effect' (the fleeing shoal splits, passes by the predator's sides and may reassemble behind it) and 'flash expansion' (the shoal suddenly breaks up with individuals rapidly swimming in all directions), among others (reviewed by Pitcher and Parrish 1993). Several of these evasive responses involve synchronised body turns, acceleration and shifting of positions of individuals within the shoal, which may make it difficult for the predator to visually fixate and pursue a single fleeing target (i.e. may enhance the predator confusion effect). However, the relative effectiveness of these various shoal evasive manoeuvres in reducing per capita risk of predation, above that expected from the dilution effect, remains unknown.

Transmission of information among shoal members

Compaction in a shoal under attack facilitates its coordinated evasive manoeuvres, because the closer proximity of shoal members allows for rapid communication of their respective velocity and direction changes as mediated through pressure waves that are detected by the inner ear and lateral line systems and through visual cues from nearest neighbours (Pitcher and Parrish 1993). Information about a predator's attack on the shoal may also be socially transmitted among members of the shoal, such that shoal members that have not themselves detected the approaching predator may still synchronously initiate a flight response in the appropriate direction (e.g. Magurran and Higham 1988; Krause 1993c). Such information transfer can be seen as a wave of flight reaction (alarm) to an attack being propagated from individual to individual across the shoal (e.g. Godin 1986). When the speed of this information transmission is greater than the speed of the approaching predator, all individuals in the shoal potentially gain an antipredator benefit from shoal membership; this phenomenon has been referred to as the Trafalgar effect of group living (Treherne and Foster 1981). To date, only limited empirical evidence exists for the Trafalgar effect in fish shoals (Webb 1980; Godin and Morgan 1985; Godin 1986; Godin *et al.* 1988).

8.2.5 Pursuit deterrence

Fishes have been observed occasionally to exhibit certain behaviours towards predators that appear to be antagonistic and to inhibit their attack or further pursuit (reviewed by Dugatkin and Godin 1992; Smith, Chapter 7). However, little is known about behaviours exhibited by prey fish under attack that result in the predator aborting its attack or pursuit. Behavioural displays that are directed by prey towards predators which inhibit their attack or further pursuit are referred to as deimatic displays (*sensu* Edmunds 1974). Examples of apparent deimatic behaviour in fish elicited when attacked include the 'flash displays' of shoals of silvery fish that result from their sudden and synchronous turning (Pitcher and Parrish 1993), flashing of photophores in some marine fishes (Lane 1967, cited in Smith 1992), rapid body inflation in puffer and porcupine fish (Edmunds 1974) and electrical organ discharges in electric fishes (Belbenoit 1986). However, there is no unequivocal evidence that such displays cause predators to abort their attack.

8.2.6 Deflection and diversion of attack

Predators may be deceived by prey to direct their attacks on certain parts of the prey's body, such that the attack is less successful than otherwise.

Numerous freshwater fishes (particularly characins, cichlids, cyprinids) and marine fishes (particularly wrasses, damselfishes, butterflyfishes) possess false eyespots (ocelli) that typically consist of a dark spot surrounded by one or more brightly coloured rings or possess large black spots without such rings. These eyespots and black spots are mainly located in the caudal region of the body (e.g. McPhail 1977; Zaret 1977; Neudecker 1989; Winemiller 1990). One of the many proposed hypotheses for the evolution of such eyespots or black caudal spots is that they function as eye mimics that either misdirect (deflect) predator attacks towards less vulnerable caudal regions of the prey's body, intimidate the predator or confuse the predator about the identity and/or orientation of the prey and thereby deter attack (Zaret 1977; Neudecker 1989; Meadows 1993).

Except for the experimental studies of McPhail (1977) and Dale and Pappantoniou (1986), available evidence for an antipredator function of false eyespots or caudal spots in general is largely comparative (correlational) or anecdotal (e.g. Neudecker 1989; Winemiller 1990; Meadows 1993). McPhail (1977) experimentally placed a black spot on the caudal peduncle of a characin, *Hyphessobrycon panamensis*, that naturally does not possess any black spots on its body. Compared with unmarked conspecifics, the experimentally marked fish received proportionately more attacks on the caudal region of their body than on the anterior or middle body regions from a natural fish predator, *Ctenolucius beani*, under laboratory conditions. Predator attacks were less successful, albeit not significantly so, on the marked than unmarked prey fish. Similarly, Dale and Pappantoniou (1986) injected goldfish, *Carassius auratus*, with black ink spots on the caudal peduncle to simulate eyespots. They observed that the marked goldfish received more bites directed towards their tail than their head from the eye-picking cutlips minnow, *Exoglossum maxillingua*, compared with unmarked goldfish. The results of these two experimental studies are consistent with the aforementioned 'misdirection' hypothesis. Because fish predator attacks directed towards the posterior end of prey tend to be less successful than attacks directed at mid-body (Webb and Skadsen 1980; Webb 1986), prey fish with caudal eyespots may therefore have a greater chance of escaping an attack than fish without such spots, everything else being equal.

It has also been suggested that both hard-rayed dorsal spines and large body depth redirect predator attack from the prey's mid-body towards the caudal or head regions (Moody *et al.* 1983; Webb 1986). Thus, prey fish with deep bodies and/or with hard-rayed dorsal spines may have a greater chance of escaping an attack than more cylindrically-shaped fish without hard dorsal spines, everything else being equal. The evidence for this proposition is, however, only correlational. Largemouth bass directed proportionately more of their attacks away from the centre of the prey's body mass (i.e. near mid-body region) when attacking laterally-compressed and deeper-bodied fish with hard dorsal spines (bass and sunfish) than more cylindrically-shaped fish lacking hard-rayed dorsal spines (fathead minnow and muskellunge) (Webb 1986). Similarly, tiger muskellunge, *Esox lucius* × *E. masquinongy* hybrid, avoided attacking bluegill sunfish, *Lepomis macrochirus*, in mid-body, where hard-rayed spines are located and the body depth is largest, but preferentially (and successfully) attacked fathead minnows which lack such spines in this particular region of the body (Moody *et al.* 1983). Three pisciv-orous esocid species experienced greater capture success, and achieved propor-tionately more mid-body captures, when attacking fathead minnows and gizzard shad, *Dorosoma cepedianum*, than bluegill sunfish (Wahl and Stein 1988).

These findings suggest that body depth and hard dorsal spines redirect predator attacks away from the mid-body region of prey fish, where attacks are most success-ful. Alternatively, the findings could have resulted from differences in antipredator behaviour between the deep-bodied, hard-spined species and the more cylindric-ally shaped, soft-rayed species, rather than from differences in their body depths or the presence or lack of dorsal spines as such. Manipulative experiments, in which predator behaviour towards experimental fish with dorsal spines clipped or with an induced increase in body depth (e.g. Brönmark and Miner 1992; Brönmark and Pettersson 1994; Nilsson *et al.* 1995) is compared with control fish, are necessary to test the hypothesis that dorsal spines and/or body depth in prey deflect predator attack and thus increase their escape success.

Distraction or diversionary displays in prey that redirect or divert a predator's attack away from themselves are known for invertebrates and vertebrates, particularly birds (Edmunds 1974). However, there exists little evidence for their occurrence in fishes. Perhaps the best known example of a predator distraction display in fishes is the 'rooting diversionary display' exhibited by parental male threespine sticklebacks in some populations in response to an approaching group of conspecifics that may be potential predators of their eggs (reviewed by Foster 1994). This display deceptively diverts the predator's attention away from the male's nest and eggs.

8.3   Defences mitigating predator ingestion of prey

8.3.1   Behavioural defences

Most gape-limited piscivorous predators attempt to swallow their captured fish prey whole. However, prior to ingestion, they typically manipulate the prey in their mouth for a varying period of time, which is largely dependent on the ratio of the cross-sectional diameter of the prey's body to the predator's mouth (gape) width (Reimchen 1994). Such predator manipulation (handling) provides the prey an opportunity to escape (Reimchen 1994). Whilst in the jaws or beak of a vertebrate predator, a fish might be able to dislodge itself from the predator's grasp by vigorously and repeatedly flexing its body. Prey may alternatively escape from the grasp of predators by feigning death (thanatosis). If the predator voluntarily relaxes its grasp on the captured, but immobile, prey, this may provide a chance for escape (Edmunds 1974). Although at least one fish species, the cichlid *Haplochromis livingstoni*, is known to use thanatosis in hunting prey (McKaye 1981), I know of no unequivocal example of thanatosis used as an antipredator defence in fishes. Certain fish make distinctive sounds when captured (Smith, Chapter 7). Such apparent 'distress' sounds might startle the predator so that it is more likely to release its prey. There is, however, no evidence for this possibility.

Although there is evidence that some individuals in fish populations have escaped from the grasp of predators and survived the ordeal (e.g. Smith 1992; Reimchen 1994), it is difficult to ascertain whether such escapes were entirely due to prey behaviour or to morphological and chemical antipredator defences (see below), or to a combination of these different traits.

8.3.2   Morphological defences

Morphological defences that should in theory increase a prey fish's chance of evading predation once captured include large body size and body armour.

Large body size in prey fish may provide a refuge from predation because the prey-handling efficiency of gape-limited piscivorous predators is strongly influenced by the relative body size of the prey. Both manipulation time (Hoyle and Keast 1987, 1988; Forbes 1989; Hart and Connellan 1984; Hughes, Chapter 6) and incidence of prey escapes whilst being manipulated by the predator (Reimchen 1991a, b) increase sharply as the body size (or more accurately the cross-sectional diameter of the body) of the prey approaches the maximum gape of the predator. Therefore, over evolutionary time, such inefficient predation should select for relatively large body size in prey fish within populations (Reimchen 1994) and presumably for foraging behaviours that maximise surplus energy and somatic growth rate in individuals (Hart, Chapter 5; Hughes, Chapter 6).

Even over ecological time (lifetime of individuals), predation pressure may select for facultative changes in body shape in prey fish that reduce predator manipulation efficiency and thereby increase the probability of captured fish escaping. One such aspect of body shape in fishes is body depth. Brönmark and Miner (1992) reported that the maximum body depth of crucian carp, *Carassius carassius*, from sections of two ponds in which pike had been experimentally introduced only 12 weeks earlier was significantly greater than that of carp of the same body length from the corresponding pond sections without pike. In a corollary laboratory experiment in which food abundance, fish density and predator presence were controlled, Brönmark and Miner (1992) showed that the presence of pike alone caused a large increase in body depth of carp compared to treatments without pike. Most interestingly, water-borne chemical cues (presumably metabolites) released by pike have been shown experimentally to be sufficient to induce an increase in the body depth of carp, and this morphological change in body shape is reversible in the absence of pike (Brönmark and Pettersson 1994). Such a predator-induced change in body morphology confers a survival benefit to crucian carp by increasing pike handling time and their avoidance of the deep-bodied phenotype (Nilsson *et al.* 1995). Because of the potential costs associated with large body depth (e.g. energy expenditure through increased drag), such plasticity in a morphological antipredator defence would appear particularly adaptive in environments in which prey fish experience intermittent and varying levels of predation pressure, to which they could respond opportunistically by increasing their body depth whenever the risk of predation is perceived to have increased (Brönmark and Miner 1992; Brönmark and Pettersson 1994). Although predator-induced morphological defences are relatively common in invertebrates (Harvell 1990), they appear to be rare in vertebrates, the above example notwithstanding.

Another potential morphological defence is body armour. Body armour, particularly in the form of hard-rayed spines and bony lateral plates (scutes), is phylogenetically widespread among teleost fishes. Based on descriptions in Nelson (1994), 52.5% of the 426 extant teleost families contain at least one species with some body armour. There exists comparative (correlational) and experimental evidence for a post-capture antipredator function of body armour in fishes, most of it from the sticklebacks (Family Gasterosteidae) (reviewed by Wootton 1984; Reimchen 1994). A number of proximate mechanisms could underlie this documented antipredator benefit of armour (reviewed by Wootton 1984; Reimchen 1994). First, erected dorsal and pelvic spines increase the maximum effective cross-sectional diameter of armoured fish, and thus increase manipulation time of gape-limited predators. Second, such spines can be physically hazardous to the predator when locked erect (i.e. puncture the inside of the mouth and esophagus) and render the prey more difficult to swallow than otherwise. Third, both erected spines and bony lateral plates protect to some degree the prey's skin and tissues from physical damage when captured and manipulated by a predator. If the captured prey happened to escape, then it would have a better chance of surviving than otherwise. These mechanisms are of course not mutually exclusive. The first and second mechanisms in particular render armoured prey less profitable energetically to gape-limited predators and increase the likelihood that they would be rejected once captured. As expected if spines confer an antipredator benefit to prey, piscivorous fishes have been shown to prefer to attack and ingest soft-rayed over spiny-rayed fish; this difference in preference appears to be based on the type of fin rays as such (e.g. Reist 1980; Gillen *et al.* 1981; Eklöv and Hamrin 1989).

8.3.3 Chemical defences

Any release of noxious chemicals from the body of a captured fish could in theory result in a piscivorous predator releasing its prey alive. Therefore, such chemicals could confer antipredator benefits to the possessor either directly (through predators rejecting captured individuals) or indirectly (through predators learning to avoid attacking or pursuing such noxious prey fish, see Smith, Chapter 7). The incidence of toxic or noxious chemicals among fishes is relatively low. Based on descriptions in Nelson (1994), only 12 ($<3\%$) out of all extant teleost families contain at least one species with some type of toxin in the skin, spines or viscera. Interestingly, all of these families (e.g. scorpionfishes, toadfishes, trunkfishes, puffers, blennies, gobies), except one, are largely marine.

There exists little direct experimental evidence for an antipredator function of chemical toxins in fishes. One notable exception is the study of Losey (1972) on the poison-fang blenny, *Meiacanthus atrodorsalis*. This species possesses venomous canine teeth with which they can bite predators. Losey observed that different piscivorous fish predators, upon capturing a blenny, would violently quiver their head, distend their jaws and operculi and release the blenny from their mouth. The presumption was that the captured blenny had bitten the inside of the predator's mouth. Usually, the escaping blenny was not physically harmed by the ordeal. When the poisonous canines were experimentally removed from some blennies, these fish were readily eaten by predatory fish, whereas the intact control blennies were rejected initially.

Another type of chemical released by specialised cells of certain fishes (mainly ostariophysians) when the skin is damaged by a predator is an alarm substance or Schreckstoff (Smith 1992, Chapter 7). Such alarm chemicals do not appear to render prey fish unpalatable or to increase their likelihood of been rejected once captured by the predator (Bernstein and Smith 1983). However, Schreckstoff released from a captured fish may attract other predators that interfere with the predation event, thereby allowing the captured fish to escape (Mathis *et al.* 1995).

It is clear that much more research on the taxonomic distribution of chemical toxins in teleost fishes, and on their effectiveness in conferring either direct and indirect antipredator benefits to prey fish, will be required to further understand their evolution and their relative importance as an antipredator defence.

8.4 Conclusions

Teleost fishes possess diverse behavioural, morphological and chemical traits that can serve as defences mitigating predator detection and attack (Smith, Chapter 7) and predator capture and ingestion (this chapter). Such diversity in known and putative antipredator defences is not surprising given the diverse habitats occupied by fishes (Nelson 1994; Dodson, Chapter 2; Kramer *et al.*, Chapter 3; Persson *et al.*, Chapter 12) and the corresponding diversity of predators they experience in these habitats and during the course of ontogeny as they grow and as their vulnerability changes (e.g. Wootton 1984; Fuiman and Magurran 1994; Reimchen 1994). Evolution of multiple defences and adaptive flexibility in their use is favoured in environments where prey encounter different predators and where predation risk varies over space and time (Schall and Pianka 1980; Sih 1987; Pearson 1989; Lima and Dill 1990). Future research on taxonomically divergent species and on geographically distinct populations of the same species will undoubtedly reveal new antipredator defences hitherto unknown.

This review has also shown that individual fish often have control over which particular defence tactic to adopt instantaneously from a set of available alternatives, and also control over aspects of the chosen defence tactic (e.g. timing and speed of flight, time spent in refuge, position within a shoal), in response to a predator attack. Moreover, evidence indicates that fish are able to assess the fitness-related costs and benefits associated with alternative defence options in making decisions about when and how to respond to an approaching predator. Knowledge of the benefits and costs of alternative antipredator tactics, and how these tactics might conflict with the expression of other behavioural activities (e.g. Sih 1987; Lima and Dill 1990; Werner 1992; Milinski 1993), is essential for an understanding of their evolution and current existence in populations. Although there exists experimental evidence for fitness benefits associated with certain defences, evidence for an antipredator function of other putative defences (particularly chemical ones) is either lacking or at best anecdotal or correlational. Even less is known about the costs of antipredator defences. Therefore, future research should be directed more earnestly at identifying the constraints on the evolution and individual use of defences, at quantifying their associated fitness benefits and costs, and at experimentally testing predictions, derived from optimality (e.g. Ydenberg and Dill 1986) and game-theory (e.g. Hugie and Dill 1994) models, about their relative frequency and duration of use by individuals in a population. This approach should prove useful in improving our current understanding of the variation in antipredator tactics commonly observed among individuals within populations (e.g. Sih 1987; Lima and Dill 1990; Huntingford *et al.* 1994; Reimchen 1994).

In practice, however, it will often be difficult to quantify the contribution of a specific defence to an individual's fitness, partly because of possible correlations between different defence tactics and multiple functions of traits (e.g. Pearson 1989; Brodie *et al.* 1995). Such difficulty may be circumvented by using analysis of covariance, multiple regression analysis or path analysis (e.g. Bulova 1994; Brodie *et al.* 1995), for example, in analysing observed variation in prey success at avoiding and evading predators. Further insights into the function of a particular putative defence may be gained by relating the relative success of prey, who have adopted the defence, at avoiding and evading predators to each consecutive stage of the predation sequence (Pearson 1989; Endler 1986, 1991; Reimchen 1994).

Finally, knowledge of both proximate and functional aspects of predator–prey interactions, and particularly of interindividual variation in antipredator traits, is often required for comprehensive explanations of the patterns of animal abundance and distribution observed at the levels of the population and community (Ives and Dobson 1987; Werner 1992; Kramer *et al.*, Chapter 3; Persson *et al.*, Chapter 12). Such knowledge, in turn, should be helpful in predicting the impact of natural and human-made disturbances to habitat on interacting populations of predators and prey.

Acknowledgements

I thank Joe Brown, Larry Dill, Jens Krause, Andy Sih and Jan Smith for their valuable comments on an earlier version of the chapter. The preparation of this chapter and my own research cited therein have been supported by the Natural Sciences and Engineering Research Council of Canada, for which I am grateful.

References

Abrahams, M.V. (1995). The interaction between antipredator behaviour and antipredator morphology: experiments with fathead minnows and brook sticklebacks. *Can. J. Zool.*, **73**, 2209–2215.

Abrams, P.A. (1990). The evolution of anti-predator traits in prey in response to evolutionary change in predators. *Oikos*, **59**, 147–156.

Altbäcker, V. and Csányi, V. (1990). The role of eyespots in predator recognition and antipredator behaviour of the paradise fish, *Macropodus opercularis* L. *Ethology*, **85**, 51–57.

Angermeier, P.L. (1992). Predation by rock bass on other stream fishes: experimental effects of depth and cover. *Environ. Biol. Fish.*, **34**, 171–180.

Anholt, B.R., Ludwig, D. and Rasmussen, J.B. (1987). Optimal pursuit times: how long should predators pursue their prey? *Theor. Popul. Biol.*, **31**, 453–464.

Arduino, P.J. and Gould, J.L. (1984). Is tonic immobility adaptive? *Anim. Behav.*, **32**, 921–923.

Baylis, J.R. (1982). Unusual escape response by two cyprinodontiform fishes, and a bluegill predator's counter-strategy. *Copeia*, **1982**, 455–457.

Belbenoit, P. (1986). Fine analysis of predatory and defensive motor events in *Torpedo marmorata* (Pisces). *J. exp. Biol.*, **121**, 197–226.

Benzie, V.L. (1965). *Some aspects of the anti-predator responses of two species of stickleback*. D. Phil. thesis, University of Oxford, Oxford.

Bernstein, J.W. and Smith, R.J.F. (1983). Alarm substance cells in fathead minnows do not affect the feeding preference of rainbow trout. *Environ. Biol. Fish.*, **9**, 307–311.

Bertram, B.C.R. (1978). Living in groups: predators and prey. In *Behavioural ecology: an evolutionary approach* (eds. J.R. Krebs and N.B. Davies), pp. 64–96. Blackwell Scientific Publ., Oxford.

Blake, R.W. (1983). *Fish locomotion*. Cambridge University Press, Cambridge.

Booman, C., Folkvord, A. and Hunter, J.R. (1991). Responsiveness of starved northern anchovy *Engraulis mordax* larvae to predatory attacks by adult anchovy. *Fish. Bull.*, **89**, 707–711.

Bouskila, A. and Blumstein, D.T. (1992). Rules of thumb for predation hazard assessment: predictions from a dynamic model. *Am. Nat.*, **139**, 161–176.

Brodie, E.D. III, Moore, A.J. and Janzen, F.J. (1995). Visualizing and quantifying natural selection. *Trends Ecol. Evol.*, **10**, 313–318.

Brönmark, C. and Miner, J.G. (1992). Predator-induced phenotypical change in body morphology in crucian carp. *Science*, **258**, 1348–1350.

Brönmark, C. and Pettersson, L.B. (1994). Chemical cues from piscivores induce a change in morphology in crucian carp. *Oikos*, **70**, 396–402.

Brown, J.A. (1984). Parental care and the ontogeny of predator-avoidance in two species of centrarchid fish. *Anim. Behav.*, **32**, 113–119.

Bulova, S.J. (1994). Ecological correlates of population and individual variation in antipredator behavior of two species of desert lizards. *Copeia*, **1994**, 980–992.

Chivers, D.P. and Smith, R.J.F. (1995). Chemical recognition of risky habitats is culturally transmitted among fathead minnows, *Pimephales promelas* (Osteichthyes, Cyprinidae). *Ethology*, **99**, 286–296.

Christensen, B. and Persson, L. (1993). Species-specific antipredatory behaviours: effects on prey choice in different habitats. *Behav. Ecol. Sociobiol.*, **32**, 1–9.

Colgan, P. (1974). Burying experiments with the banded killifish, *Fundulus diaphanus*. *Copeia*, **1974**, 258–259.

Colgan, P. and Costeloe, N. (1980). Plasticity of burying behavior by the banded killifish, *Fundulus diaphanus*. *Copeia*, **1980**, 349–351.

Curio, E. (1976). *The ethology of predation*. Springer-Verlag, Berlin.

Dale, G. and Pappantoniou, A. (1986). Eye-picking behavior of the cutlips minnow, *Exoglossum maxillingua*: applications to studies of eyespot mimicry. *Ann. N.Y. Acad. Sci.*, **463**, 177–178.

Dill, L.M. (1973). An avoidance learning sub-model for a general predation model. *Oecologia*, **13**, 291–312.

Dill, L.M. (1974a). The escape response of the zebra danio (*Brachydanio rerio*) I. The stimulus for escape. *Anim. Behav.*, **22**, 711–722.

Dill, L.M. (1974b). The escape response of the zebra danio (*Brachydanio rerio*) II. The effect of experience. *Anim. Behav.*, **22**, 723–730.

Dill, L.M. (1987). Animal decision making and its ecological consequences: the future of aquatic ecology and behaviour. *Can. J. Zool.*, **65**, 803–811.

Dill, L.M. (1990). Distance-to-cover and the escape decisions of an African cichlid fish, *Melanochromis chipokae*. *Environ. Biol. Fish.*, **27**, 147–152.

Dill, L.M. and Ydenberg, R.C. (1987). The group size–flight distance relationship in water striders (*Gerris remigis*). *Can. J. Zool.*, **65**, 223–226.

Domenici, P. and Batty, R.S. (1994). Escape manoeuvres of schooling *Clupea harengus*. *J. Fish Biol.* **45** (*Suppl.* A), 97–110.

Domenici, P. and Blake, R.W. (1993a). Escape trajectories in angelfish (*Pterophyllum eimekei*). *J. exp. Biol.*, **177**, 253–272.

Domenici, P. and Blake, R.W. (1993b). The effect of size on the kinematics and performance of angelfish (*Pterophyllum eimekei*) escape responses. *Can. J. Zool.*, **71**, 2319–2326.

Dugatkin, L.A. and Godin, J.-G.J. (1992). Prey approaching predators: a cost–benefit perspective. *Ann. Zool. Fennici*, **29**, 233–252.

Eaton, R.C. and Emberley, D.S. (1991). How stimulus direction determines the trajectory of the Mauthner-initiated escape response in a teleost fish. *J. exp. Biol.*, **161**, 469–487.

Eaton, R.C. and Hackett, J.T. (1984). The role of the Mauthner cell in fast starts involving escape in teleost fish. In *Neural mechanisms of startle behavior* (ed. Eaton, R.C.), pp. 213–266. Plenum Press, New York.

Edmunds, M. (1974). *Defence in animals*. Longman, New York.

Eklöv, P. and Hamrin, S.F. (1989). Predatory efficiency and prey selection: interactions between pike *Esox lucius*, perch *Perca fluviatilis* and rudd *Scardinus erythrophthalmus*. *Oikos*, **56**, 149–156.

Eklöv, P. and Persson, L. (1995). Species-specific antipredator capacities and prey refuges: interactions between piscivorous perch (*Perca fluviatilis*) and juvenile perch and roach (*Rutilus rutilus*). *Behav. Ecol. Sociobiol.*, **37**, 169–178.

Endler, J.A. (1986). Defense against predators. In *Predator-prey relationships* (eds. M.E. Feder and G.V. Lauder), pp. 109–134. University of Chicago Press, Chicago.

Endler, J.A. (1991). Interactions between predators and prey. In *Behavioural ecology: an evolutionary approach*, 3rd edn. (eds. J.R. Krebs and N.B. Davies), pp. 169–196. Blackwell Scientific Publ., Oxford.

Endler, J.A. (1995). Multiple-trait coevolution and environmental gradients in guppies. *Trends Ecol. Evol.*, **10**, 22–29.

Everett, R.A. and Ruiz, G.M. (1993). Coarse woody debris as a refuge from predation in aquatic communities. An experimental test. *Oecologia*, **93**, 475–486.

FitzGerald, G.J. and van Havre, N. (1985). Flight, fright and shoaling in sticklebacks (Gasterosteidae). *Biol. Behav.*, **10**, 321–331.

Folkvord, A. and Hunter, J. R. (1986). Size-specific vulnerability of northern anchovy, *Engraulis mordax*, larvae to predation by fishes. *Fish. Bull.*, **84**, 859–869.

Forbes, L.S. (1989). Prey defences and predator handling behaviour: the dangerous prey hypothesis. *Oikos*, **55**, 155–158.

Foster, S.A. (1994). Evolution of the reproductive behaviour of threespine stickleback. In *The evolutionary biology of the threespine stickleback* (eds. M.A. Bell and S.A. Foster), pp. 381–398. Oxford University Press, Oxford.

Foster, W.A. and Treherne, J.E. (1981). Evidence for the dilution effect in the selfish herd from fish predation on a marine insect. *Nature, Lond.*, **292**, 466–467.

Fuiman, L.A. (1993). Development of predator evasion in Atlantic herring, *Clupea harengus* L. *Anim. Behav.*, **45**, 1101–1116.

Fuiman, L.A. (1994). The interplay of ontogeny and scaling in the interactions of fish larvae and their predators. *J. Fish Biol.*, **45** *(Suppl.* A), 55–79.

Fuiman, L.A. and Magurran, A.E. (1994). Development of predator defences in fishes. *Rev. Fish Biol. Fish.*, **4**, 145–183.

Gallego, A. and Heath, M.R. (1994). Vulnerability of late larval and early juvenile Atlantic herring, *Clupea harengus*, to predation by whiting, *Merlangius merlangus. J. Fish Biol.*, **45**, 589–595.

Giles, N. (1983). Behavioural effects of the parasite *Schistocephalus solidus* (Cestoda) on an intermediate host, the three-spined stickleback, *Gasterosteus aculeatus* L. *Anim. Behav.*, **31**, 1192–1194.

Giles, N. (1987). Predation risk and reduced foraging activity in fish: experiments with parasitized and non-parasitized three-spined sticklebacks, *Gasterosteus aculeatus* L. *J. Fish Biol.*, **31**, 37–44.

Gillen, A.L., Stein, R.A. and Carline, R.F. (1981). Predation by pellet-reared tiger muskellunge on minnows and bluegills in experimental systems. *Trans. Am. Fish. Soc.*, **110**, 197–209.

Gilliam, J.F. and Fraser, D.F. (1987). Habitat selection under predation hazard: test of a model with foraging minnows. *Ecology*, **68**, 1856–1862.

Godin, J.-G.J. (1986). Antipredator function of shoaling in teleost fishes: a selective review. *Naturaliste can.*, **113**, 241–250.

Godin, J.-G.J. (1990). Diet selection under the risk of predation. In *Behavioural mechanisms of food selection* (ed. R.N. Hughes). NATO ASI Series, Vol. G20, pp. 739–769. Springer-Verlag, Berlin.

Godin, J.-G.J. and Morgan, M.J. (1985). Predator avoidance and school size in a cyprinodontid fish, the banded killifish (*Fundulus diaphanus* Lesueur). *Behav. Ecol. Sociobiol.*, **16**, 105–110.

Godin, J.-G.J. and Sproul, C.D. (1988). Risk taking in parasitized sticklebacks under threat of predation: effects of energetic need and food availability. *Can. J. Zool.*, **66**, 2360–2367.

Godin, J.-G.J., Classon, L.J. and Abrahams, M.V. (1988). Group vigilance and shoal size in a small characin fish. *Behaviour*, **104**, 29–40.

Gotceitas, V. and Brown, J.A. (1993). Substrate selection by juvenile Atlantic cod (*Gadus morhua*): effects of predation risk. *Oecologia*, **93**, 31–37.

Gotceitas, V. and Colgan, P. (1987). Selection between densities of artificial vegetation by young bluegills avoiding predation. *Trans. Am. Fish. Soc.*, **116**, 40–49.

Gotceitas, V. and Godin, J.-G.J. (1991). Foraging under the risk of predation in juvenile Atlantic salmon (*Salmo salar* L.): effects of social status and hunger. *Behav. Ecol. Sociobiol.*, **29**, 255–261.

Gotceitas, V. and Godin, J.-G.J. (1993). Effects of aerial and in-stream threat of predation on foraging by juvenile Atlantic salmon (*Salmo salar*). In *Production of juvenile Atlantic* salmon, *Salmo salar, in natural waters.* (eds. R.J. Gibson and R.E. Cutting), *Can. Spec. Publ. Fish Aquat. Sci.*, **118**, 35–41.

Grant, J.W.A. and Noakes, D.L.G. (1987). Escape behaviour and use of cover by young-of-the-year brook trout, *Salvelinus fontinalis. Can. J. Fish. Aquat. Sci.*, **45**, 1390–1396.

Hager, M.C. and Helfman, G.S. (1991). Safety in numbers: shoal size choice by minnows under predatory threat. *Behav. Ecol. Sociobiol.*, **29**, 271–276.

Hamilton, W.D. (1971). Geometry for the selfish herd. *J. theor. Biol.*, **31**, 295–311.

Harper, D.G. and Blake, R.W. (1991). Prey capture and the fast-start performance of northern pike *Esox lucius. J. exp. Biol.*, **155**, 175–192.

Hart, P.J.B. and Connellan, B. (1984). Cost of prey capture, growth rate and ration size in pike, *Esox lucius* L., as functions of prey weight. *J. Fish Biol.*, **25**, 279–292.

Harvell, C.D. (1990). The ecology and evolution of inducible defenses. *Q. Rev. Biol.*, **65**, 323–340.

Harvey, B.C. and Stewart, A.J. (1991). Fish size and habitat depth relationships in headwater streams. *Oecologia*, **87**, 336–342.

Hasson, O. (1991). Pursuit-deterrent signals: communication between prey and predator. *Trends Ecol. Evol.*, **6**, 325–329.

Hastings, P.A. (1991). Flexible responses to predators in a marine fish. *Ethol. Ecol. Evol.*, **3**, 177–184.

Helfman, G.S. (1989). Threat-sensitive predator-avoidance in damselfish-trumpetfish interactions. *Behav. Ecol. Sociobiol.*, **24**, 47–58.

Howland, H.C. (1974). Optimal strategies for predator avoidance: the relative importance of speed and manoeuvrability. *J. theor. Biol.*, **47**, 333–350.

Hoyle, J.A. and Keast, A. (1987). The effect of prey morphology and size on handling time in a piscivore, the largemouth bass (*Micropterus salmoides*). *Can. J. Zool.*, **65**, 1972–1977.

Hoyle, J.A. and Keast, A. (1988). Prey handling time in two piscivores, *Esox americanus vermiculatus* and *Micropterus salmoides*, with contrasting mouth morphologies. *Can. J. Zool.*, **66**, 540–542.

Hugie, D.M. and Dill, L.M. (1994). Fish and game: a game theoretic approach to habitat selection by predators and prey. *J. Fish Biol.*, **45** (*Suppl.* A), 151–169.

Humphries, D.A. and Driver, P.M. (1970). Protean defence by prey animals. *Oecologia*, **5**, 285–302.

Huntingford, F.A., Wright, P.J. and Tierney, J.F. (1994). Adaptive variation in antipredator behaviour in threespine stickleback. In *The evolutionary biology of the threespine stickleback* (eds. M.A. Bell and S.A. Foster), pp. 277–296. Oxford University Press, Oxford.

Hurley, A.C. and Hartline, P.H. (1974). Escape response in the damselfish *Chromis cyanea* (Pisces: Pomacentridae): a quantitative study. *Anim. Behav.*, **22**, 430–437.

Ibrahim, A.A. and Huntingford, F.A. (1989). The role of visual cues in prey selection in three-spined sticklebacks (*Gasterosteus aculeatus*). *Ethology*, **81**, 265–272.

Ives, A.R. and Dobson, A.P. (1987). Antipredator behavior and the population dynamics of single predator–prey systems. *Am. Nat.*, **130**, 431–447.

Jakobsson, S., Brick, O. and Kullberg, C. (1995). Escalated fighting behaviour incurs increased predation risk. *Anim. Behav.*, **49**, 235–239.

Katzir, G. and Camhi, J.M. (1993). Escape response of black mollies (*Poecilia shenops*) to predatory dives of a pied kingfisher (*Ceryle rudis*). *Copeia*, **1993**, 549–553.

Katzir, G. and Intrator, N. (1987). Striking of underwater prey by a reef heron, *Egretta gularis schistacea. J. comp. Physiol.*, **160A**, 517–523.

Keenleyside, M.H.A. (1979). *Diversity and adaptation in fish behaviour.* Springer-Verlag, Berlin.

Knudsen, F.R., Enger, P.S. and Sand, O. (1992). Awareness reactions and avoidance responses to sound in juvenile Atlantic salmon, *Salmo salar* L. *J. Fish Biol.*, **40**, 523–534.

Krause, J. (1993a). Positioning behaviour in fish shoals: a cost–benefit analysis. *J. Fish Biol.*, **43** (*Suppl.* A), 309–314.

Krause, J. (1993b). The effect of 'Schreckstoff' on the shoaling behaviour of the minnow: a test of Hamilton's selfish herd theory. *Anim. Behav.*, **45**, 1019–1024.

Krause, J. (1993c). Transmission of fright reaction between different species of fish. *Behaviour*, **127**, 37–48.

Krause, J. (1994). Differential fitness returns in relation to spatial position in groups. *Biol. Rev.*, **69**, 187–206.

Krause, J. and Godin, J.-G.J. (1994a). Shoal choice in the banded killifish (*Fundulus diaphanus*, Teleostei, Cyprinodontidae): effects of predation risk, fish size, species composition and size of shoals. *Ethology*, **98**, 128–136.

Krause, J. and Godin, J.-G.J. (1994b). Influence of parasitism on the shoaling behaviour of banded killifish, *Fundulus diaphanus. Can. J. Zool.*, **72**, 1775–1779.

Krause, J. and Godin, J.-G.J. (1995). Predator preferences for attacking particular prey group sizes: consequences for predator hunting success and prey predation risk. *Anim. Behav.*, **50**, 465–473.

Krause, J. and Godin, J.-G.J. (1996). Influence of prey foraging posture on flight behavior and predation risk: predators take advantage of unwary prey. *Behav. Ecol*, **7**, 264–271.

Krause, J. and Tegeder, R.W. (1994). The mechanism of aggregation behaviour in fish shoals: individuals minimise approach time to neighbours. *Anim. Behav.*, **48**, 353–359.

Landeau, L. and Terborgh, J. (1986). Oddity and the 'confusion effect' in predation. *Anim. Behav.*, **34**, 1372–1380.

Lane, E.D. (1967). A study of the Atlantic midshipman, *Porichthys porosissimus*, in the vicinity of Port Aransas, Texas. *Contrib. Mar. Sci.*, **12**, 1–53.

Lima, S.L. and Dill, L.M. (1990). Behavioral decisions made under the risk of predation: a review and prospectus. *Can. J. Zool.*, **68**, 619–640.

Litvak, M.K. (1993). Response of shoaling fish to the threat of aerial predation. *Environ. Biol. Fish.*, **36**, 183–192.

Losey, G.S. (1972). Predation protection in the poison-fang blenny, *Meiacanthus atrodorsalis*, and its mimics, *Ecsenius bicolor* and *Runnula laudandus* (Blenniidae). *Pac. Sci.*, **26**, 129–139.

McGurk, M.D. (1993). Allometry of herring mortality. *Trans. Am. Fish. Soc.*, **122**, 1035–1042.

McKaye, K.R. (1981). Death feigning: a unique hunting behavior by the predatory cichlid, *Haplochromis livingstoni* of Lake Malawi. *Environ. Biol. Fish.*, **8**, 81–96.

McKaye, K.R., Mughogho, D.E. and Lovullo, T.J. (1992). Formation of the selfish school. *Environ. Biol. Fish.*, **35**, 213–218.

McLean, E.B. and Godin, J.-G.J. (1989). Distance to cover and fleeing from predators in fish with different amounts of defensive armour. *Oikos*, **55**, 281–290.

McPhail, J.D. (1969). Predation and the evolution of a stickleback. *J. Fish. Res. Board Can.*, **26**, 3183–3208.

McPhail, J.D. (1977). A possible function of the caudal spot in characid fishes. *Can. J. Zool.*, **55**, 1063–1066.

Magnhagen, C. (1993). Conflicting demands in gobies: when to eat, reproduce, and avoid predators. *Mar. Behav. Physiol.*, **23**, 79–90.

Magurran, A.E. (1990). The adaptive significance of schooling as an anti-predator defence in fish. *Ann. Zool. Fennici*, **27**, 51–66.

Magurran, A.E. and Higham, A. (1988). Information transfer across fish shoals under predator threat. *Ethology*, **78**, 153–158.

Magurran, A.E. and Pitcher, T.J. (1987). Provenance, shoal size and the sociobiology of predator-evasion behaviour in minnow shoals. *Proc. R. Soc. Lond. B*, **229**, 439–465.

Magurran, A.E., Oulton, W.J. and Pitcher, T.J. (1985). Vigilant behaviour and shoal size in minnows. *Z. Tierpsychol.*, **67**, 167–178.

Major, P.F. (1978). Predator-prey interactions in two schooling fishes, *Caranx ignobilis* and *Stolephorus purpureus*. *Anim. Behav.*, **26**, 760–777.

Martel, G. and Dill, L.M. (1995). Influence of movement by coho salmon (*Oncorhynchus kisutch*) parr on their detection by common mergansers (*Mergus merganser*). *Ethology*, **99**, 139–149.

Mathis, A., Chivers, D.P. and Smith, R.J.F. (1995). Chemical alarm signals: predator deterrents or predator attractants? *Am. Nat.*, **145**, 994–1005.

Mathis, A., Chivers, D.P. and Smith, R.J.F. (1996). Cultural transmission of predator recognition in fishes: intraspecific and interspecific learning. *Anim. Behav.*, **51**, 185–201.

Meadows, D.W. (1993). Morphological variation in eyespot of the foureye butterflyfish (*Chaetodon capistratus*): implications for eyespot function. *Copeia*, **1993**, 235–240.

Mesa, M.G., Poe, T.P., Gadomski, D.M. and Petersen, J.H. (1994). Are all prey created equal? A review and synthesis of differential predation on prey in substandard condition. *J. Fish Biol.*, **45** (*Suppl.* A), 81–96.

Metcalfe, N.B., Huntingford, F.A. and Thorpe, J.E. (1987). The influence of predation risk on the feeding motivation and foraging strategy of juvenile Atlantic salmon. *Anim. Behav.*, **35**, 901–911.

Milinski, M. (1990). Information overload and food selection. In *Behavioural mechanisms of food selection* (ed. R.N. Hughes). NATO ASI Series, Vol. G20, pp. 721–736. Springer-Verlag, Berlin.

Milinski, M. (1993). Predation risk and feeding behaviour. In *Behaviour of teleost fishes*, 2nd edn. (ed. T.J. Pitcher), pp. 285–305. Chapman and Hall, London.

Moody, R.C., Helland, J.M. and Stein, R.A. (1983). Escape tactics used by bluegills and fathead minnows to avoid predation by tiger muskellunge. *Environ. Biol. Fish.*, **8**, 61–65.

Morgan, M.J. and Godin, J.-G.J. (1985). Antipredator benefits of schooling behaviour in a cyprinodontid fish, the banded killifish (*Fundulus diaphanus*). *Z. Tierpsychol.*, **70**, 236–246.

Neill, S.R. St.J. and Cullen, J.M. (1974). Experiments on whether schooling by their prey affects the hunting behaviour of cephalopod and fish predators. *J. Zool., Lond.*, **172**, 549–569.

Nelson, J.S. (1994). *Fishes of the world,* 3rd edn. John Wiley and Sons, New York.

Neudecker, S. (1989). Eye camouflage and false eyespots: chaetodontid responses to predators. *Environ. Biol. Fish.*, **25**, 143–157.

Nilsson, P.A., Brönmark, C. and Pettersson, L.B. (1995). Benefits of a predator-induced morphology in crucian carp. *Oecologia*, **104**, 291–296

Noakes, D.L.G. and Godin, J.-G.J. (1988). Ontogeny of behavior and concurrent developmental changes in sensory systems in teleost fishes. In *Fish physiology*, Vol. 11 (eds. W.S. Hoar and D.J. Randall), pp. 345–395. Academic Press, New York.

Ohguchi, O. (1981). Prey density and selection against oddity by three-spined sticklebacks. *Adv. Ethol.*, **23**, 1–79

Parrish, J. K. (1989). Re-examining the selfish herd: are central fish safer? *Anim. Behav.*, **38**, 1948–1053.

Parrish, J.K. (1992). Do predators 'shape' fish schools: interactions between predators and their schooling prey. *Neth. J. Zool.*, **42**, 358–370.

Parrish, J.K. (1993). Comparison of the hunting behavior of four piscine predators attacking schooling prey. *Ethology*, **95**, 233–246.

Parrish, J.K., Strand, S.W. and Lott, J.L. (1989). Predation on a school of flat-iron herring, *Harengula thrissina. Copeia*, **1989**, 1089–1091.

Pearson, D.L. (1989). What is the adaptive significance of multicomponent defensive repertoires? *Oikos*, **54**, 251–253.

Pepin, P. (1993). An appraisal of the size-dependent mortality hypothesis for larval fish: comparison of a multispecies study with an empirical review. *Can. J. Fish. Aquat. Sci.*, **50**, 2166–2174.

Pitcher, T.J. and Parrish, J.K. (1993). Functions of shoaling behaviour in teleosts. In *Behaviour of teleost fishes*, 2nd edn. (ed. T.J. Pitcher), pp. 363–439. Chapman and Hall, London.

Pitcher, T.J., Lang, S.H. and Turner, J.A. (1988). A risk-balancing trade off between foraging rewards and predation hazard in a shoaling fish. *Behav. Ecol. Sociobiol.*, **22**, 225–228.

Power, M.E. (1987). Predator avoidance by grazing fishes in temperate and tropical streams: importance of stream depth and prey size. In *Predation: direct and indirect impacts on aquatic communities* (eds. W.C. Kerfoot and A. Sih), pp. 333–351. University Press of New England, Hanover, NH.

Power, M.E., Dudley, T.L. and Cooper, S.D. (1989). Grazing catfish, fishing birds, and attached algae in a Panamanian stream. *Environ. Biol. Fish.*, **26**, 285–294.

Radabaugh, D.C. (1989). Seasonal colour changes and shifting antipredator tactics in darters. *J. Fish Biol.*, **34**, 679–685.

Rahel, F.J. and Stein, R.A. (1988). Complex predator–prey interactions and predator intimidation among crayfish, piscivorous fish, and small benthic fish. *Oecologia*, **75**, 94–98.

Ranta, E., Lindström, K. and Peuhkuri, N. (1992). Size matters when three-spined sticklebacks go to school. *Anim. Behav.*, **43**, 160–162.

Reeve, H.K. and Sherman, P.W. (1993). Adaptation and the goals of evolutionary research. *Q. Rev. Biol.*, **68**, 1–32.

Reimchen, T.E. (1991a). Trout foraging failures and the evolution of body size in stickleback. *Copeia*, **1991**, 1098–1104.

Reimchen, T.E. (1991b). Evolutionary attributes of headfirst prey manipulation and swallowing in piscivores. *Can. J. Zool.*, **69**, 2912–2916.

Reimchen, T.E. (1994). Predators and morphological evolution in threespine stickleback. In *The evolutionary biology of the threespine stickleback* (eds. M.A. Bell and S.A. Foster), pp. 240–276. Oxford University Press, Oxford.

Reist, J.D. (1980). Selective predation upon pelvic phenotypes of brook stickleback, *Culaea inconstans*, by northern pike, *Esox lucius. Can. J. Zool.*, **58**, 1245–1252.

Reist, J.D. (1983). Behavioral variation in pelvic phenotypes of brook stickleback, *Culaea inconstans*, in response to predation by northern pike, *Esox lucius. Environ. Biol. Fish.*, **8**, 255–267.

Roberts, B.L. (1992). Neural mechanisms underlying escape behaviour in fishes. *Rev. Fish Biol. Fish.*, **2**, 243–266.

Savino, J.F. and Stein, R.A. (1982). Predator-prey interaction between largemouth bass and bluegills as influenced by simulated, submersed vegetation. *Trans. Am. Fish. Soc.*, **111**, 255–266.

Savino, J.F. and Stein, R.A. (1989a). Behavioural interactions between fish predators and their prey: effects of plant density. *Anim. Behav.*, **37**, 311–321.

Savino, J.F. and Stein, R.A. (1989b). Behavior of fish predators and their prey: habitat choice between open water and dense vegetation. *Environ. Biol. Fish.*, **24**, 287–293.

Sayer, M.D.J. and Davenport, J. (1991). Amphibious fish: why do they leave water. *Rev. Fish Biol. Fish.*, **1**: 159–181.

Schall, J.J. and Pianka, E.R. (1980). Evolution of escape behavior diversity. *Am. Nat.*, **115**, 551–566.

Schlosser, I.J. (1988). Predation risk and habitat selection by two size classes of a stream cyprinid: experimental test of a hypothesis. *Oikos*, **52**, 36–40.

Schmidt-Nielsen, K. (1984). *Scaling. Why is animal size so important?* Cambridge University Press, Cambridge.

Seghers, B.H. (1974). Geographic variation in the responses of guppies (*Poecilia reticulata*) to aerial predators. *Oecologia*, **14**, 93–98.

Seghers, B.H. (1981). Facultative schooling behavior in the spottail shiner (*Notropis hudsonius*): possible costs and benefits. *Environ. Biol. Fish.*, **6**, 21–24.

Sih, A. (1987). Predators and prey lifestyles: an evolutionary and ecological overview. In *Predation: direct and indirect impacts on aquatic communities* (eds. W.C. Kerfoot and A. Sih), pp. 203–224. University of New England Press, Hanover, NH.

Sih, A. (1992). Prey uncertainty and the balancing of antipredator and feeding needs. *Am. Nat.*, **139**, 1052–1069.

Smith, R.J.F. (1981). Effect of food deprivation on the reaction of Iowa darters (*Etheostoma exile*) to skin extract. *Can. J. Zool.*, **59**, 558–560.

Smith, R.J.F. (1992). Alarm signals in fishes. *Rev. Fish Biol. Fish.*, **2**, 33–63.

Smith, R.J.F. and Smith, M.J. (1989). Predator-recognition behaviour in two species of gobiid fishes, *Asterropteryx semipunctatus* and *Gnatholepis anjerensis. Ethology*, **83**, 19–30.

Stein, R.A. (1979). Behavioral response of prey to fish predators. In *Predator-prey systems in fisheries management* (eds. R.H. Stroud and H. Clepper), pp. 343–352. Sport Fishing Institute, Washington, D.C.

Suarez, S.D. and Gallup, G.G. Jr. (1983). Emotionality and fear in birds: a selected review and reinterpretation. *Bird Behav.*, **5**, 22–30.

Tallmark, B. and Evans, S. (1986). Substrate-related differences in antipredator behavior of two gobiid fish species and the brown shrimp, and their adaptive value. *Mar. Ecol. Prog. Ser.*, **29**, 217–222.

Tegeder, R.W. and Krause, J. (1995). Density dependence and numerosity in fright stimulated aggregation behaviour of shoaling fish. *Phil. Trans. R. Soc. Lond. B.*, **350**, 381–390.

Thomas, G. (1977). The influence of eating and rejecting prey items upon feeding and food searching behaviour in *Gasterosteus aculeatus* L. *Anim. Behav.*, **25**, 52–66.

Tierney, J.F., Huntingford, F.A. and Crompton, D.W.T. (1993). The relationship between infectivity of *Schistocephalus solidus* (Cestoda) and anti-predator behaviour of its intermediate host, the three-spined stickleback, *Gasterosteus aculeatus. Anim. Behav.*, **46**, 603–605.

Treherne, J.E. and Foster, W.A. (1981). Group transmission of predator avoidance in a marine insect: the Trafalgar effect. *Anim. Behav.*, **29**, 911–917.

Tremblay, D. and FitzGerald, G.J. (1979). Social organisation as an antipredator strategy in fish. *Naturaliste can.*, **105**, 411–413.

Tulley, J.J. and Huntingford, F.A. (1987). Age, experience and the development of adaptive variation in anti-predator responses in three-spined sticklebacks (*Gasterosteus aculeatus*). *Ethology*, **75**, 285–290.

Turner, G.F. and Pitcher, T.J. (1986). Attack abatement: a model for group protection by combined avoidance and dilution. *Am. Nat.*, **128**, 228–240.

Vermeij, G.J. (1982). Unsuccessful predation and evolution. *Am. Nat.*, **120**, 701–720.

Wahl, D.H. and Stein, R.A. (1988). Selective predation by three esocids: the role of prey behavior and morphology. *Trans. Am. Fish. Soc.*, **117**, 142–151.

Webb, P.W. (1976). The effect of size on the fast-start performance of rainbow trout *Salmo gairdneri*, and a consideration of piscivorous predator–prey interactions. *J. exp. Biol.*, **65**, 157–177.

Webb, P.W. (1980). Does schooling reduce fast start response latencies in teleosts? *Comp. Biochem. Physiol.*, **65A**, 231–234.

Webb, P.W. (1981). Responses of northern anchovy, *Engraulis mordax*, larvae to predation by a biting planktivore, *Amphiprion percula. Fish. Bull.*, **79**, 727–735.

Webb, P.W. (1982). Avoidance responses of fathead minnow to strikes by four teleost predators. *J. comp. Physiol.*, **147A**, 371–378.

Webb, P.W. (1984a). Body and fin form and strike tactics of four teleost predators attacking fathead minnow (*Pimephales promelas*) prey. *Can. J. Fish. Aquat. Sci.*, **41**, 157–165.

Webb, P.W. (1984b). Chase response latencies of some teleostean piscivores. *Comp. Biochem. Physiol.*, **79A**, 45–48.

Webb, P.W. (1986). Effect of body form and response threshold on the vulnerability of four species of teleost prey attacked by largemouth bass (*Micropterus salmoides*). *Can. J. Fish. Aquat. Sci.*, **43**, 763–771.

Webb, P.W. and Skadsen, J.M. (1980). Strike tactics of *Esox. Can. J. Zool.*, **58**, 1462–1469.

Weihs, D. and Webb, P.W. (1984). Optimal avoidance and evasion tactics in predator–prey interactions. *J. theor. Biol.*, **106**, 189–206.

Werner, E.E. (1992). Individual behavior and higher-order species interactions. *Am. Nat.*, **140**, S5–S32.

Werner, E.E., Gilliam, J.F., Hall, D.J. and Mittlebach, G.G. (1983). An experimental test of the effects of predation risk on habitat use in fish. *Ecology*, **64**, 1540–1548.

Williams, P.J. and Brown, J.A. (1991). Developmental changes in foraging–predator avoidance trade-offs in larval lumpfish *Cyclopterus lumpus*. *Mar. Ecol. Prog. Ser.*, **76**, 53–60.

Winemiller, K.O. (1990). Caudal eyespots as deterrents against fin predation in the neotropical cichlid *Astronotus ocellatus*. *Copeia*, **1990**, 665–673.

Wisenden, B.D. and Keenleyside, M.H.A. (1994). The dilution effect and differential predation following brood adoption in free-ranging convict cichlids (*Cichlasoma nigrofasciatum*). *Ethology*, **96**, 203–212.

Wolf, N.G. and Kramer, D.L. (1987). Use of cover and the need to breathe: the effects of hypoxia on vulnerability of dwarf gouramis to predatory snakeheads. *Oecologia*, **73**, 127–132.

Wootton, R.J. (1984). *A functional biology of sticklebacks*. Croom Helm, London.

Ydenberg, R.C. and Dill, L.M. (1986). The economics of fleeing from predators. *Adv. Study Behav.*, **16**, 229–249.

Zaret, T. M. (1977). Inhibition of cannibalism in *Cichla ocellaris* and hypothesis of predator mimicry among South American fishes. *Evolution*, **31**, 421–437.

9 *Mating systems and sex allocation*

Anders Berglund

9.1 Introduction

Genes that do not endow their possessors with the capacity to reproduce success-fully (i.e. produce offspring that in turn survive to reproduce) will eventually be lost. It is therefore of paramount importance to all organisms to reproduce successfully. The objective of this chapter is to review the multitude of ways through which mating can be achieved in fishes: under what circumstances can we for instance expect promiscuity, polygamy, monogamy, iteroparity, semelparity, alternative strategies, sex change, simultaneous hermaphroditism or asexual reproduction?

In the first part of the chapter I deal with mating systems. Fishes are a highly diverse taxonomic group and therefore ideally suited for discussing such systems in relation to partners (i.e. polygamy versus monogamy) and to time (i.e. breeding cycles, and semelparity versus iteroparity).

Some mating systems allow for alternative reproductive strategies, which is the second major topic of this chapter. Such alternatives may arise when environments vary, when individuals differ in quality, or when two strategies both have higher payoffs when rare (so-called mixed evolutionarily stable strategies). Again, fishes provide good and sometimes unique examples of this phenomenon. Finally, I will consider sex change as an alternative strategy, and briefly discuss sexual and asexual reproduction.

9.2 Mating systems

Mating systems reflect the number of mates an individual acquires per breeding attempt, and the conflicting interests of the two sexes and how they are resolved. Mating systems in which an individual has several mates are termed *polygamous*, and polygamy may be the basic mating system in the animal kingdom ('all animals are basically polygamous' according to Wilson (1975), p. 327). In such species, the pair-bond may not last longer than the fertilisation event. There are three kinds of polygamy:

(1) *polygyny*, where a male mates with several females, but each female mates with one male only,

(2) *polyandry*, which is the reverse: females have several males, whereas males have only one female, and

(3) *promiscuity*, where both males and females mate with several partners (i.e. both polygyny and polyandry operate).

If one male and one female mate only with each other, the system may be labelled *monogamous*. The pair bond may be short or long term. These are not firm categories but rather extremes of a multidimensional continuum. Moreover, as we shall see, different individuals within a population may adopt different mating systems. This is so because mating systems are influenced by numerous factors, and indeed mating systems can be viewed as 'non-adaptive by-products of the adaptive behaviour of individuals to maximise their reproductive success' (T. H. Clutton-Brock, personal communication). Which factors act in any particular case, their relative impact, and the way they may vary over time or space all shape the mating system in that case (Reynolds 1996).

To understand how mating systems have evolved and are maintained has long been a goal in ecology, and fishes provide excellent subjects to address such questions.

9.2.1 Promiscuity: when mates or resources cannot be monopolised

Promiscuity may be an ancestral fish mating system, with external fertilisation of eggs involving many females and males simultaneously. For instance, the large numbers of cod, *Gadus morhua*, or herring, *Clupea harengus*, gathering on breeding grounds to spawn may serve as examples. This mating system probably leaves few options for exercising mate choice or engaging in intrasexual competition at a level above that of sperm competition, and may help explain the lack of sexual dimorphism in such species. Within cod aggregations, however, pairs may form and males may court females (Keenleyside 1979). Many nesting or brooding fishes are also promiscuous, as exemplified by the mouthbrooding Lake Malawi cichlid *Cyrtocara eucinostomus* (McKaye *et al.* 1990), the brooding pipefish *Syngnathus typhle* (Berglund *et al.* 1988), and the nestbuilding garibaldi *Hypsypops rubicundus* (Sikkel 1989).

9.2.2 Polygyny: defending attractive resources, defending females directly or lekking

Polygyny (i.e. one male with several females) is particularly common in marine fishes. Depending on the resource defended by males, polygyny may be categorised as:

(1) *resource-defence polygyny,* where a male defends a good or safe resource that attracts females,

(2) *female-defence polygyny,* where the male is able to defend groups of females directly against other males, and

(3) *leks,* where aggregated, territorial males compete directly for access to females (Emlen and Oring 1977; Krebs and Davies 1993; see also Grant, Chapter 4, for a review of resource defence in general).

There are several examples of resource-defence polygyny in fishes. In the slimy sculpin, *Cottus cognatus,* a freshwater fish with paternal care from North American lakes, males dig a hole under a rock and try to attract females. When suitable rocks are scarce, males with good breeding sites may attract several females and mate polygynously. Other males obviously remain bachelors. In lakes with plenty of suitable rocks, monogamous pairs are the rule (Mousseau and Collins 1987).

Female garibaldi prefer males with algal nests that contain immature eggs (Sikkel 1989); this is another example of resource-defence polygyny where nests with eggs are the resource. In the mottled sculpin, *Cottus bairdi,* males also defend breeding sites, attract females and mate polygynously (Brown and Downhower 1982). In this species, male size is a better predictor of male mating success than is the quality of his nest site. Therefore, this species may be polygynous because females prefer larger males rather than better sites, perhaps because larger males provide better care. In this case, care is a non-depreciable resource (i.e. a male can care for several clutches just as well as he can care for one), and there is no disadvantage for females to mate polygynously.

In the bluehead wrasse, *Thalassoma bifasciatum,* the converse is true, and females choose spawning sites rather than males (Warner 1987). In this coral reef fish, large males hold spawning territories to which females come to mate. The eggs are pelagic and thus require no parental care. If a resident male is removed from his territory, females are still faithful to that particular site. Even if the male in question subsequently moves to an adjacent site, females do not follow him. This resource-defence polygyny has an interesting twist to it: the males do not determine which sites are good and which ones are not, but simply respond to female site preferences by occupying preferred sites (Warner 1990). When males were experimentally displaced, the mating sites remain the same. However, when females were displaced, site use changed dramatically. Therefore, female activity is the cue used by males to determine which sites are worth defending. Interestingly, not all females actually assess the resources themselves, but some copy the decisions of other females (Warner 1988b). In a particular study area, mating sites were recorded to have remained the same for 12 years (corresponding to four generations). When entire local populations were experimentally substituted, new sites came into use, and these sites continued to be in use afterwards. Thus, by means of traditions transmitted across generations, females tell the males what to defend.

Female-defence polygyny is known for instance in two simultaneously hermaphroditic seabasses. In the barred serrano, *Serranus fasciatus,* and the lantern bass, *S. baldwini,* males defend harems of female-acting hermaphrodites, the latter having home ranges so small that a large male can monopolise the space used by several hermaphrodites (Petersen and Fischer 1986; Petersen 1987). In *S. fasciatus*, harems form at intermediate densities, with the dominant individual acting as a male and actually losing the ovarian function completely. The subordinates spawn as females, but retain their functional testes without getting much chance to use them. These 'females' may, however, streak: a subordinate hides near a pair about to release their gametes and joins in at the right time to release sperm. In large group sizes, the dominant male can no longer monopolise all matings, and some high-ranking hermaphrodites may enjoy reproductive success via either their male or female function (Petersen 1990).

Mating systems may be quite flexible and varied within a population, with different individuals adopting different tactics. At the spawning sites of the freshwater dark chub, *Zacco temmincki* (Katano 1990), dominant males try repeatedly to sequester females for pair-spawnings. Dominants also take females from subdominants. In the process these α-males wear out rapidly. During spawning, other individuals dash toward the pair to release sperm, or to eat eggs. Females do not appear to select any male in particular, and they even seem unable to determine how many eggs to shed on any one occasion as the male pushes the female's belly during spawning, forcing the eggs out.

A lek is a place where males aggregate and display to females, who come there solely for the purpose of mating. According to this definition, females should not receive any direct benefits in the process, only genes, but in practice this has never been shown conclusively. Many hypotheses, not always mutually exclusive, for the formation of leks exist (Wiley 1991; Stillman *et al.* 1993), and perhaps the evolutionary origin of the behaviour may vary from species to species. Males may be clumped because

(1) they thus become more visible to females,

(2) females enjoy reduced predation risk at central territories,

(3) inferior males exploit the attractiveness of a male 'hotshot',

(4) the arena is a preferred 'hotspot' for females, and

(5) females are safer from male sexual harassment among territorial males which are able to exclude satellites.

A lek is characterised by very unequal reproductive success among males, with a few high-ranking males obtaining most matings. Therefore, mating on a lek is largely polygynous, although some females may mate with more than one male. Males often show bizarre development of male secondary sexual characteristics, in spite of any direct benefits to females from choosing a particular lekking male. The problem of female choice in non-resource-based systems has been called the 'lek paradox', but may simply be due to the low cost of exerting choice for females (Reynolds and Gross 1990). In fact, lekking species do not even exhibit a more pronounced sexual dimorphism than do non-lekking species (Höglund 1989).

In fishes, lekking occurs for instance in the mouth-brooding cichlid *Cyrtocara eucinostomus* in Lake Malawi at a most amazing scale: up to 50 000 males may display simultaneously on an arena 4 km long that is used exclusively as a mating site. After mating the female, who does all the brooding, leaves the lek with the eggs in her mouth. In this species, males are larger and more colourful than females, but sneaky female mimics also exist. It is unknown whether males have greatly unequal reproductive success, although the mating system suggests this to be the case (McKaye 1983; McKaye *et al.* 1990). Although this species fulfils the usual lek criteria as lekking bird species do, this is not the case for 'lekking' fish species in general (Loiselle and Barlow 1978).

9.2.3 Polyandry: resource-defence by females, low male reproductive rate or parasitism

Polyandry (i.e. one female mating with several males) is rare in fishes. Conditional polyandry occurs in anemone fishes, *Amphiprion* spp., with young individuals being males and older ones changing into females given the right circumstances. A social group consists of one dominant female and several subordinate males, living in and defending one or several sea anemones. The mating system is typically monogamous; the highest-ranking male aggressively suppresses reproduction in the other males. However, when several anemones are included in the female's territory, the dominant male is unable to suppress other males on other anemones and the mating system consequently becomes polyandrous (Moyer and Sawyers 1973). This is analogous to resource-defence polygyny, and can thus be called resource-defence polyandry. I know of no case of male-defence polyandry in fishes.

A type of sequential polyandry exists in the pipefish, *Nerophis ophidion*. The male receives a clutch of eggs from a female and broods it on his body for several weeks, after which he may mate again. Because a female produces on average twice as many eggs during the male brooding cycle as males can typically brood (Berglund *et al.* 1989), she may remate while her first partner is busy brooding.

A very curious mating system is found in deep-sea angler fishes (Norman and Greenwood 1975). Generally, males can be regarded as parasites on the females' gametic investment (sperm contain little else but genes, whereas eggs are nutritious). However, in these fishes, males have taken things one step further: the tiny male attaches himself to a female and turns into a true parasite on her. Most male organs degenerate except for the testes, and nourishment is provided by the female via a placenta-like arrangement. When several males attach to one female, an unusual, albeit perhaps more accidental than adaptive, form of polyandry is generated.

9.2.4 Monogamy: biparental care, resource defence or low population density

To understand the evolution of monogamy, we must understand what keeps either partner from deserting and mating elsewhere. A pair bond may evolve if, for example, biparental care is more rewarding in fitness terms than searching for other mates (Krebs and Davies 1993). Monogamy is not widespread in fishes, but has been well studied where it occurs. Cichlids are especially famous for their diverse mating systems, including monogamy particularly among New World species. In the strictly monogamous Midas cichlid, *Cichlasoma citrinellum*, parents need to stay together, both to defend their territory against other Midas cichlids and to protect their fry from predators, including conspecifics. A single parent is probably unable to do this successfully (McKaye 1977; McKaye and McKaye 1977; Barlow 1984, 1991, 1992; Keenleyside 1991). The selective advantages accruing to individuals that provide biparental care may therefore explain the evolution of monogamy in this species. In addition, both parents feed their young with mucus from their skin, but this is probably of minor value to the offspring (Noakes and Barlow 1973). Males and females care for the offspring in different ways: males primarily defend the territory, while females concentrate directly on the offspring. Overall, parental expenditure appears to be higher in the female than in the male (Rogers 1988). When offered a choice, females preferred large, aggressive and experienced males, whereas males were indiscriminate in their choice of partners (Rogers and Barlow 1991). Monogamy is not a plastic trait in this species. Even when the sex ratio and crowding were experimentally manipulated to promote polygyny, the mating system did not change (Rogers 1987).

The closely related convict cichlid, *Cichlasoma nigrofasciatum*, is less rigid in this respect. Males of this basically monogamous fish can be bigamous under female-biased sex ratios (Keenleyside 1985), or desert their mate (Keenleyside *et al.* 1990; Wisenden 1994). This species otherwise conforms to the Midas cichlid pattern, that is, biparental care and female choice of larger males which are more apt to defend the young (Keenleyside *et al.* 1985). However, females that care for offspring alone seem to do just as well as females aided by males (Keenleyside *et al.* 1990). Similarly, in *C. panamense*, mate desertion was more common in streams with few predators of offspring and many females, whereas monogamy prevailed in streams with many predators and few females (Townshend and Wootton 1985). Obviously, the need for offspring defence and the potential for further matings influenced this mating system.

If the need for biparental care decreases, monogamy may change into polygyny. In 12 coexisting cichlid species in Lake Tanganyika, Africa, low parental reproductive success following the removal of one parent suggested that both parents were needed to guard the offspring in the monogamous species (Gashagaza 1991). Facultative bigamy occurred when two breeding sites were so close together that a male could monitor both at the same time. Polygyny was found in species using safe but rare refuges for raising their young. One parent could easily guard the young in such refuges (e.g. holes, burrows), but only high-quality males were able to defend refuges spacious enough to accommodate several females (i.e. resource-defence polygyny).

Biparental care, however, is not the sole factor that can promote monogamy.

Males may be hindered from forming harems by dominant males, and subdominants will therefore have to resort to monogamy or not mate at all, as in the cichlid *Lamprologus brichardi* (Limberger 1983). Indeed, monogamy in marine fishes is usually not associated with parental care at all, and often takes the form of repeated pair spawnings with the same partner over long periods of time. These species are usually small-bodied, strongly site-attached, non-seasonal or weakly seasonal, and disperse only by means of planktonic eggs or larvae without any parental care (Barlow 1986). These characteristics collectively make it difficult for an individual to find and keep more than one mate. Site-attachment in both sexes, for instance due to resource defence, hinders mate search, in particular if population densities are low or if individuals have low mobility. Variation in the magnitude of these factors may also cause variation in mating systems. In two species of coral-dwelling hawkfishes, *Neocirrhites armatus* and *Oxycirrhites typus*, monogamy prevailed if suitable corals were small or few, whereas multi-female harems were common among large or abundant corals (Donaldson 1989). Likewise, experimental alterations of coral size produced pairs or harems on small corals and multi-male groups on big corals in a coral-dwelling damselfish, *Dascyllus marginatus* (Fricke 1980). Therefore, this marine version of monogamy is perhaps best regarded as harem polygyny, with a harem size of one.

Low population density may also promote monogamy. In the simultaneous hermaphrodite *Serranus fasciatus*, the territorial seabass mentioned in Section 9.2.2, pairs spawn monogamously under low population densities, taking turns to act as males and females (Petersen 1990).

Short inter-clutch intervals have also been hypothesised to contribute to monogamy. In a pipefish, *Corythoichthys intestinalis*, (Gronell 1984), and in two seahorses, *Hippocampus fuscus* and *H. whitei*, (Vincent *et al.* 1992), females take a long time to prepare their eggs before handing them over to the male (males have brood pouches in these species). Under monogamy, a female's preparatory period takes place while the male broods her previous clutch. Once the male has released the young, both members of the pair are ready to mate again. A female cannot search for a new mate while retaining prepared eggs, as these must be released within a short period. This, however, is a classical egg-and-hen dilemma: it is not clear whether monogamy permits the female to have such preparatory periods or whether preparatory periods select for monogamy. Actually, the former explanation may be more likely, because closely related but polygamous pipefish species lack preparatory periods (Berglund *et al.* 1988). Perhaps monogamy evolved for other reasons, such as territory constraints and low mate encounter rates due to low mobility (seahorses actually have a Guinness record for being the slowest swimming fish!), but is presently maintained for the advantages gained from the short inter-brood intervals. Interestingly, this may be the only animal group where 'pure' monogamy can be found. The potential for extra-pair matings is zero, as males only accept eggs from one female at the time. Moreover, a male's paternity confidence is absolute, because the eggs are fertilised within his brood pouch.

Therefore, large benefits to be gained from biparental care, small territory size or low mate encounter rate may all select for monogamy in fishes. As soon as the conditions favouring monogamy deteriorate or are absent, the mating system is expected to return to polygamy.

9.2.5 Monogamy, polygamy and sexual selection

Mating systems and sexual selection clearly influence each other (Reynolds 1996). Monogamy is often viewed as a constraint on the process of sexual selection, resulting in males and females in monogamous species usually being fairly similar in appearance. Polygamy, on the other hand, has been thought of as increasing the opportunity for sexual selection, thereby promoting the evolution of more or less elaborate and even bizarre ornaments in males. This notion is not entirely correct, as females may be choosy regardless of whether they are monogamous or polygamous. Females in monogamous species may select males that are good at parental care or territory defence, whereas females in polygamous species may utilise morphological characters like coloration or tail length as choice criteria (see Dugatkin and FitzGerald, Chapter 10). These two types of choices may be equally selective and costly to the females (indeed, polygamous females may even have lower choice costs than monogamous females, like in lekking or otherwise aggregating species where mate searching costs are low). Therefore, sexual selection need not be stronger under any particular mating system.

A smaller proportion of males mate under polygyny than under monogamy, however. This increases variation in male reproductive success, and hence the potential for sexual selection. Such variation may also exist under monogamy, but may be due to the existence of non-breeders, variance in female quality (fecundity), or extra-pair matings. This variation can in fact be substantial, although usually less than in polygamous species. For example, in the smallmouth bass, *Micropterus dolomieui*, some males fail to establish a territory and some females fail to breed, and this contributes more than twice as much to variation in egg number brooded by males than do differences in female fertility (males guard eggs and larvae in their nests)(Wiegmann *et al.* 1992). A short breeding season may have precluded small males and females from breeding, as small individuals are slower to start breeding than larger ones. The total variation in male mate number is comparable to that found in polygynous species (e.g. Downhower *et al.* 1987; Wiegmann *et al.* 1992); so sometimes sexual selection can potentially be as strong under monogamy as under other mating systems.

Sexual selection (i.e. selection for traits that increase mating success) results from (1) stronger competition for mates within one sex than within the other, the sex competing more intensely for mates being under stronger sexual selection, and (2) from mate choice (see Dugatkin and FitzGerald, Chapter 10). If one sex has a higher potential reproductive rate than the other (i.e. can potentially remate faster), that sex will be in excess and consequently will compete for the other

(Clutton-Brock and Vincent 1991). The less competitive sex is thus a resource in short supply. In other words, the operational sex ratio (males and females ready to mate) is composed of a larger number of the competitive sex than of the other (Clutton-Brock and Parker 1992).

Usually males are able to remate faster than females, and accordingly males are more competitive than females, but occasionally the reverse occurs (defined as sex-role reversal; Vincent *et al.* 1992). The two pipefish species, *Syngnathus typhle* and *Nerophis ophidion*, are examples of this phenomenon (Berglund *et al.* 1989; Rosenqvist 1990; Berglund 1991). Females in these species produce eggs at a faster rate than males can brood them (Berglund *et al.* 1989). However, in the seahorses, *Hippocampus fuscus* and *H. whitei*, which have a reproductive biology similar to the pipefishes, strict monogamy precludes any differences in potential reproductive rates to arise, and these seahorses consequently are not sex-role reversed (Vincent *et al.* 1992). Clearly, a mating system may sometimes constrain sexual selection. The converse is also true. As already mentioned, female choice, which along with male–male competition shapes sexual selection in species with conventional sex roles, may explain polygyny in mottled sculpins (Brown and Downhower 1982). Large males attract several females by providing better care, and this male size advantage, through the action of female choice, could have channelled the mating system into polygyny.

9.3 Mating systems over time

9.3.1 The timing of reproduction

Fish may spawn more or less periodically. In general, seasonality in the larval environment seems to be a poor predictor of both spawning patterns and recruitment back into the population (Robertson 1991). Seasonal spawning periods may coincide with conditions favourable to parent (e.g. Robertson 1990) or offspring (e.g. Ochi 1986) growth and survival in some species (see also Dodson, Chapter 2). For example, Atlantic mackerel, *Scomber scombrus*, and white hake, *Urophycis tenuis*, release enormous numbers of egg during peaks in zooplankton production, thereby presumably increasing offspring growth rates (Lambert and Ware 1984). Thus, offspring may quickly outgrow predators as well as swamp them by their numbers. Alternatively, fish may spread spawning over time to counteract unpredictable, unfavourable environmental conditions, as may be the case in herring, *Clupea harengus*, and capelin, *Mallotus villosus*. In these species, eggs are released in batches throughout the season, rather than all at once (Lambert and Ware 1984).

Reproductive cycles may be of other kinds than seasonal. In the anemone fish, *Amphiprion melanopus*, mating takes place during the first and third quarters of the moon, and hatching peaks near full and new moon (i.e. during high spring tides). Thus, offspring may become effectively dispersed (Ross 1981). In addition, nocturnal hatching may be safer as plankton-eaters are usually less active at night (Ross 1981). In general, lunar spawning patterns may enhance larval settlement possibilities (Robertson *et al.* 1990). Moreover, adults may use the moon as a cue to synchronise spawning should that be advantageous (Robertson 1991).

Spawning in the damselfish, *Stegastes dorsopunicans*, is synchronised among females as well as with the lunar cycle. As a result, males receive larger clutches on days of high female reproductive activity than they would have if females reproduced continuously. Although male damselfish guard their clutch, they eat some of their eggs. In a large clutch, they will eat a smaller proportion of the eggs, which probably selects for female spawning synchrony (Petersen and Hess 1991).

The saddleback wrasse, *Thalassoma duperrey*, exhibits annual, semilunar and diel reproductive cycles. In this pelagically spawning coral reef fish, winter reproduction is thought to reduce the numbers of larvae swept off the reef, as ocean currents are weak at this time of the year. The reproductive peaks at new and full moons may take advantage of tidal currents to maximise within-habitat dispersal. Moreover, spawning at daily high tides may increase the distance between the newly-spawned eggs and reef planktivores (Ross 1983). In general, diel spawning cycles may affect larval dispersal and survival, but little evidence exists (Robertson 1991). Moreover, spawning may occur when it interferes the least with feeding (Conover and Kynard 1984) or when predation on adults is minimal (Thresher 1984).

9.3.2 Semelparity and iteroparity

In fishes that travel great distances to reach their spawning grounds, the number of reproductive events per lifetime may be quite low (see Dodson, Chapter 2). When reproduction occurs only once followed by death, as in most Pacific salmon (genus *Oncorhynchus*), we call it 'big-bang' reproduction or semelparity. However, Atlantic salmon, *Salmo salar*, also swim up rivers to breed but do not necessarily die afterwards, even though some of the Atlantic rivers are much longer and probably more physically challenging than some of the Pacific ones. Why this difference exists between these two closely related salmonid genera is not known, but may be related to fish density and competition on spawning grounds (Fleming and Gross 1989; van den Berghe *et al.* 1989). Certainly, Atlantic salmon may also die after spawning, and in fact populations may differ substantially in the proportions of semelparous and iteroparous fish. Seemingly, this species has adjusted aspects of its life history to the severity of the reproductive migration (Schaffer 1979).

Generally, some types of environmental uncertainty may select for a longer reproductive life span (if, for instance, juvenile mortality varies a lot), and under such conditions semelparity may be a risky option (Stearns 1992). When adult mortality varies more than that of juveniles, semelparity may be favoured. The evidence for this is mixed, however: in 18 fish species reproductive life span correlated, although only weakly, with variation in spawning success ($r^2 = 0.08$,

Mann and Mills 1979). In flatfish, however, no such correlation was found (Roff 1981).

9.4 Alternative reproductive strategies

Previous sections have focused on differences among and within species. However, within species or even within populations, all individuals may not employ the same reproductive strategy. Here, I use the term 'strategy' in the general sense of Krebs and Davies (1993). For more rigorous definitions of the terms 'strategy' and 'tactic', see Gross (1984, 1996). We have already seen (Section 9.2) that some individuals may reproduce monogamously and others polygamously. In addition, competition between males often force some males to exploit alternative ways to achieve fertilisations, commonly termed sneakers, streakers, satellites, peripherals, female mimics, cuckolders, jacks, helpers or accessory males (e.g. Taborsky 1994; Gross 1996). In general, we may expect such alternative strategies to evolve under three circumstances:

(1) when the environment changes either in time or in space, with one strategy yielding higher net fitness payoffs under some conditions than others,

(2) if individuals differ phenotypically, so that inferior individuals are prevented from employing the best strategy, and

(3) if the two strategies are true alternatives at an evolutionary equilibrium, with equal fitness payoffs (Krebs and Davies 1993).

Let us examine these three possibilities in turn.

9.4.1 Changing environments

Guppies, *Poecilia reticulata*, are live-bearing freshwater fish with internal fertilisation. Females are drab compared to the colourful males, but in areas with many predators males are less colourful than in areas with fewer predators. Predation evidently selects against bright males, whereas sexual selection via female choice selects for colourful and conspicuous males in this species (Endler 1980, 1983, 1995; Houde and Endler 1990; Winemiller *et al.* 1990). Not only does male coloration vary with predation intensity, but so does female mate choice based on male colour. Females from bright-male populations prefer bright males, but females from dull-male populations are generally less responsive to colours in males (Breden and Stoner 1987; Houde 1988; Houde and Endler 1990; Endler and Houde 1995). Such female choice has a genetic basis. In addition to colours, courtship behaviour is also affected by predation risk. In the laboratory, fish from predator-safe streams show risk-reckless courtship, even when exposed to predators. Fish from predator-rich sites, on the other hand, are risk-sensitive and modify courtship in response to predators by decreasing the rate of conspicuous sigmoid displays and increasing the rate of sneaky mating attempts (Magurran and Seghers 1990; Godin 1995). Incidentally, sneaking in guppies is not a 'best-of-a-bad-job' tactic (see below). This behaviour is preferentially performed by larger males under predation risk (Reynolds *et al.* 1993). Moreover, females from high-predation sites become less choosy when they have perceived a fish predator nearby (Godin and Briggs 1996), and are less likely to choose males exhibiting the most conspicuous displays (Stoner and Breden 1988), than females from low-predation sites. Therefore, female choice, as well as selective predation itself, will mould reproductive strategies differently in habitats differing in predation risk. Different populations have consequently diverged genetically (Endler 1995); indeed, this may provide a mechanism for the origin of species (Verrell 1991).

In the bluehead wrasse, small individuals are either females or female-coloured males and both may change into terminal-phase males (so-called protogynous hermaphroditism with initial-phase males). Terminal-phase males hold territories. At sites with few individuals, these large males can effectively monopolise matings on their territories, whereas at sites with many individuals most spawnings are group spawnings with non-territorial males (Warner and Hoffman 1980). Hence, population size can influence mating strategies. Accordingly, in the closely related species *Thalassoma lucasanum*, which has the same reproductive biology but denser and larger local populations, the proportion of primary males is higher and sex-change is less common (Warner and Hoffman 1980).

9.4.2 Strategies conditional on phenotype: making the best of a bad job

Alternative reproductive strategies can also be found within a population and among close neighbours. For instance, when males compete intensely for females, some males will lose out in fights or displays (see Dugatkin and FitzGerald, Chapter 10). An option for these losers may be to try to steal fertilisations instead of fighting for them (Taborsky 1994; Gross 1996). The reproductive success of such a 'thief' is likely to be lower than that of a successful competitor, but may be the best option available to the inferior male. When brook lamprey, *Lampetra planeri*, pairs copulate, satellite males can be seen to circle rapidly around the tails of the copulating pair, probably fertilising some of the eggs (Malmqvist 1983). In Arctic charr, *Salvelinus alpinus*, some males guard the female, while others try to sneakmate with her (Sigurjónsdóttir and Gunnarson 1989). Guarders are larger, whereas sneakers are smaller. Even though female charr may be aggressive against the sneakers, these males probably achieve more fertilisations than if they had tried to guard females in the presence of other males.

In the pupfish, *Cyprinodon pecosensis*, males employ three alternative strategies: some are territorial, some are satellites with the same bright blue coloration as the territorials, and others are female-coloured sneakers (Kodric-Brown 1986). Satellites and sneakers have lower reproductive success, suggesting that these are conditional strategies adopted by social subordinates. As females prefer colourful males, satellites enjoy some mating success, but at the same time satellites have to cope with the aggressive behaviour that this coloration evokes in territorial males. When the number of females was experimentally increased, some satellites changed into territorial owners as male–male competition decreased (Kodric-Brown 1988).

In the bluehead wrasse, the initial-phase males are sneakers, but may also group-spawn with females. As mentioned above, sneakers have a much lower reproductive success on small reefs with small populations than on reefs with large populations, as the terminal-phase males more efficiently monopolise matings there. Therefore, sneakers on small reefs (i.e. in small populations) spend less time reproducing and instead invest in growth, but this apparently does not make up for the early losses in reproduction over their lifetime (Warner 1984a). These males are thus sneakers to begin with, making the best of a bad situation, and very unfortunate small reef sneakers at that, having to make the best of that bad situation also!

A potential cost imposed on dominant males is, as evident from above, stolen fertilisations. In addition, dominants surrounded by many peripheral males may be avoided by females. In two species of nest-building labrids, *Symphodus tinca* and *S. ocellatus*, this latter cost is much greater than that of shared paternity (van den Berghe *et al.* 1989). Dominant males obviously are not attractive to females when peripherals are present in these species, but why this is so remains unclear.

9.4.3 Mixed evolutionarily stable strategies

Even without environmental or phenotypic variations, individuals may still adopt different reproductive strategies. This can happen if the fitness payoff of a given strategy is high when few individuals adopt it, but decreases with increasing frequency in the population (i.e. negatively frequency dependent). Two alternative strategies that fulfil this criterion can coexist at some equilibrium point in a population. At this point, the average payoffs of the two strategies are equal (mixed evolutionarily stable strategies (ESS)). In practice, a mixed ESS has never been rigorously demonstrated because it is difficult, perhaps impossible, to show that two fitness sets are equal (Krebs and Davies 1993). None the less, good attempts exist, and some of the best so far actually come from fishes.

If two alternative strategies have different genetic bases, it is difficult to imagine how both can be maintained in the population unless both confer the same fitness. The inferior strategy should vanish due to the selection against it, unless maintained by pleiotrophy. In the pygmy swordtail, *Xiphophorus nigrensis*, male size is under genetic control (Zimmerer and Kallman 1989; Ryan *et al.* 1990). In the presence of large males, small males adopt a sneak-chase behaviour. As females prefer large courting males, it is not presently understood how small males are maintained in the population.

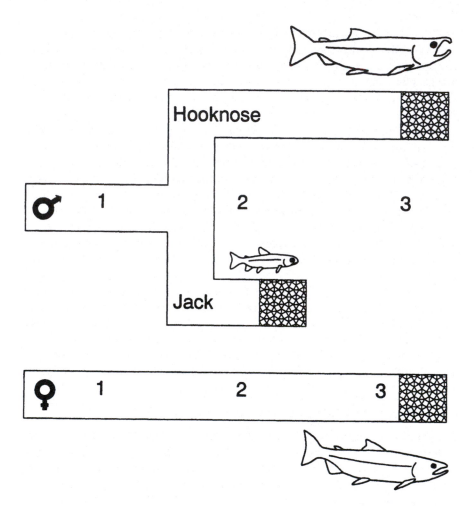

Fig. 9.1   The two life histories in male coho salmon, *Oncorhynchus kisutch*: jacks are sneakers who mature at the age of two years, and hooknoses guard females and mature at three. Both die after breeding, as do females. From Gross (1985).

After spawning in North American rivers, all coho salmon, *Oncorhynchus kisutch*, die. The young spend a year in the stream and then swim to sea where they grow to adulthood. After three years, females return to the river to breed. Males, however, can do either one of two things: if they are large at the age of two years, they may return as 'jacks' to attempt sneaky spawnings. If they are small, they can wait another year at sea, and then mature and return as large 'hooknose' males, which fight for access to spawning females (Fig. 9.1; Gross 1991). Thus, the sneaky strategy is not adopted by those males that are of inferior quality at the age of two. The difference between the two strategies is probably genetic, as well as conditional on growth. The coho female deposits her eggs in a nest she has excavated in the stream bed. The guarding males and the hiding jacks then try to fertilise the eggs as they are spawned, and the closest male may have an advantage in doing so. Large males fight to get close, while small ones hide and sneak. Since intermediate-sized males are no good at either strategy, they are usually further away from the female (incidentally, this may represent a case of disruptive selection).

How can we explain the coexistence of these two strategies in coho salmon? Most likely they represent a mixed ESS, and indeed the jack strategy is negatively density-dependent as required (Gross 1985). When rare, the jack strategy is successful, but as jacks increase in frequency their mating success diminishes. This is because hiding places are in short supply. So, if many jacks go for one and the same female, most of them cannot hide and will have to fight. And fighting is what hooknoses, not jacks, are good at (hooknoses have large, hooked mandibles and canine-like teeth for fighting, and a cartilage back shield for protection), and jacks will lose in these combats. Similarly, when hooknoses increase in frequency, this strategy also becomes less profitable: more and more fights will lower average mating success. Moreover, sperm competition between jacks can create/reinforce the negative frequency dependence: many jacks per hooknose diminishes the average reproductive return of jacks. Therefore, the success of each strategy depends on its relative frequency in the population, with equilibrium occurring at the proportions where fitness is equal for jacks and hooknoses. Evidence suggests that the lifetime reproductive success of the two alternative strategies in coho salmon are apparently equal (Gross 1985). This, however, assumes that survival before going to sea is equal for jacks and hooknoses, that jacks and hooknoses have similar sperm volumes, that sperm of either is equally good at fertilising an egg, and that closeness to the female determines male mating success—obviously much remains to be done before we can conclude that jacks and hooknoses really represent a mixed ESS.

Alternative mating tactics are also common in sunfishes (*Lepomis* spp.): cuckolders parasitise nesting males (Keenleyside 1972; Gross 1982). A mixed ESS may exist in the bluegill sunfish, *L. macrochirus*. Parental males build nests, attract females and mate with them, and subsequently guard the nests. Small cuckolders behave as sneakers, dashing into the nests during spawning to release sperm, whereas larger cuckolders (satellites) employ a female-mimicking behaviour, which deceives the parental males about their actual sex. Satellites have proportionally larger testes than parentals to aid in sperm competition and, since their age distribution is similar to that of parentals, this suggests the presence of a genetic dimorphism (Dominey 1980). Parentals are not free to detect and chase off intruders. If they spend too much time protecting their paternity, the female may leave. Females themselves do not avoid cuckolders.

Life histories in bluegill sunfish are very different for different males (at least in lakes in Ontario, Canada). Parentals mature at seven or eight years of age, and cuckolders at around two. Cuckolders need cover to pursue their strategy, and where cover is abundant competition between cuckolders for matings results in a negatively density-dependent mating success among them. If cover is sparse, cuckolders cannot stay close to the nest. At low cuckolder densities, additional cuckolders increase individual mating success by distracting the parental male. At higher densities, however, competition among cuckolders overrides the distraction effect, rendering success negatively density-dependent (Gross 1982, 1991). As the proportion of parental males decreases, so does the number of nests available to females. Therefore, female choosiness decreases, facilitating parental male control over spawning. Thus, these two male strategies in sunfish work in much the same way as in coho salmon (i.e. are rewarding when uncommon, with a stable equilibrium point). This will occur when the lifetime fitnesses of cuckolders and parentals are equal, that is, when the proportion of cuckolders in the population (which ranged between 11–31%) equals the proportion of eggs they fertilise (which, in approximate agreement, was 11–23%; Gross 1991). If their frequency in the population drops, cuckolders will fertilise proportionately more eggs, but if their frequency increases the reverse will happen: this is what enables the two strategies to coexist in a population.

Very few other good examples of mixed reproductive ESSs are known (but see Shuster and Wade 1991), and indeed we may suspect that such strategies are uncommon in nature (Gross 1996). In the majority of cases, alternative mating strategies can probably be interpreted in terms of either changing environment or doing the best of a bad job.

9.4.4 Sex change and alternative reproductive strategies

Socially inferior males may adopt an alternative strategy other than sneaking; they could change sex and become females! If male–male competition for matings is intense, this may be a small individual's best option. Eventually, after attaining a competitive size, it may pay her to change sex to male. Interestingly, sneakers may exist in parallel to females, as in the bluehead wrasse. Being sneaker or female represent alternative strategies, with their respective proportions depending on population density as described in Section 9.4.1. Within a population, the proportions may represent a mixed ESS, with equal fitness.

In contrast to the mixed ESS case, many good examples of sex change in fishes exist, and the underlying theory is likewise well developed and tested. In general, an animal should change sex when the other sex has a higher reproductive value, providing the cost of doing so is not prohibitive. This idea was formalised in the size-advantage hypothesis (Ghiselin 1969) and its elaborations. This hypothesis states that if the relative reproductive success of the two sexes increases at different rates with size or age, an individual of a particular size or age shall have the sex with the highest success for that size (age). Reproductive success is determined by both survivorship and fecundity, and therefore sexual differences in

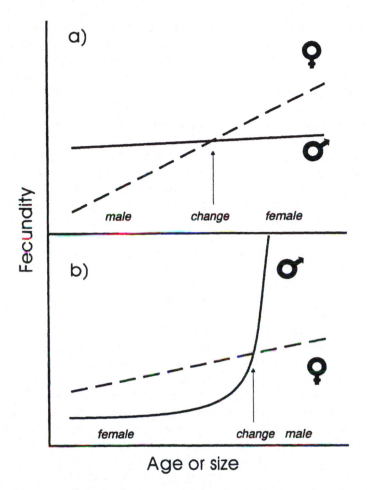

Fig. 9.2 Age-/size-advantage model of sex change in fishes. (a) When large individuals have a higher fecundity as females than as males, sex change from male to female may occur where the lines cross. (b) When male–male competition is intense, a large size may be more important in males than in females; consequently, females may change into males where the curves cross.

either may suffice to select for sex change (Charnov 1982). As can be seen from Fig. 9.2, this can result in a sex change from male to female or vice versa. I will use a few examples to illustrate how this can come about.

To begin, male-to-female change has been carefully examined in only one fish group, the anemonefishes, although it exists in several other groups. In the functionally monogamous *Amphiprion akallopisos*, the reproductive output of a pair depends on female fecundity; a male has enough sperm to ensure fertilisation of clutches of any size. Because larger females are more fecund, it makes good sense for small individuals to be males and for large ones to be females (Fricke 1979). If the female of a pair is removed, the male changes sex into female, and the highest ranking additional male (or an immigrant if no additionals are present) starts playing the male role (Fricke and Fricke 1977). *A. clarkii* shows the same pattern, but in an area with an ample anemone supply, males preferentially emigrated after mate loss rather than change sex (Ochi 1989). In doing so, no time was wasted on the process of sex change. However, anemones are usually too sparse and local predation pressure too intense to make such movements a profitable tactic in most cases.

In general, male-to-female sex change is expected in fishes with little male–male competition. In such species, the potential for polygyny is low. This pattern is common in invertebrates such as shrimps (Charnov 1979) and polychaetes (Berglund 1990), but in sex-changing fish the opposite pattern dominates. Small individuals are females and large ones males; in these species, the polygyny potential is high. The bluehead wrasse also illustrates this phenomenon. On small reefs (where populations are small), fish commonly start out as females and may eventually change into males upon reaching a large size. A large body size is needed for defending a spawning territory; a small male would not stand a chance in the competition. Small primary males instead partake in group spawnings on large reefs (where populations are larger), or they sneak. A female that changes sex to a male will enjoy an enormous potential increase in offspring production. A large male may mate with more than 100 females on a good day! Sex change is socially controlled. When a large male is removed, the largest remaining individual changes sex (if female) and colour into a terminal phase male. This may happen within a day (Warner and Swearer 1991). Sex change in this direction is the rule in labrid fishes. Primary males may or may not exist, and the sexes are usually strikingly dichromatic. Terminal males are territorial and may defend harems, whereas primary males spawn in aggregations or sneak (Roede 1990). Wrasses, however, are not the only fish group with female-to-male change. This practice may be found in angelfishes, parrotfishes and gobies, among others. Commonly, females or nest sites are the resources males compete over, and polygyny is what confers the size-advantage.

Environmental factors may also influence sex change. In the temperate goby, *Coryphopterus nicholsi*, nest sites are an important resource to breeding males. Other fish species (e.g. midshipman, *Porichthys notatus*), various encrusting invertebrates and sea urchins may reduce site availability. This decreases the mating success of smaller male gobies, as they no longer can find nest sites, but increases the success of large males as they now obtain more females. Consequently, an increase in competition between species for nest sites may move the age of sex change upwards in the goby (Breitburg 1987).

The size-advantage hypothesis has been interpreted to state that animals should change sex upon reaching a certain critical size (e.g. Shapiro 1988), and that this size is genetically fixed. However, in cases of social control of sex change, the size at which change occurs depends on the size frequency distribution in the group concerned, and is not fixed to any particular value. The size-advantage hypothesis is perhaps best understood as predicting sex change from reproductive value directly, or from various predictors of reproductive value, and not from only one predictor such as body size (Warner 1988a). Size may indeed be such a predictor, but so may social or environmental factors. Therefore, if the size-advantage hypothesis is taken to mean that sex change is favoured when one sex gains in expected fertility at a faster rate than the other as size or age increases (Fig. 9.2), the hypothesis has so far withstood all attempts of falsification. An additional condition for sex change to evolve is that the cost of changing (in terms of lost time for reproduction, perhaps due to extensive gonadal reconstruction) is not too high. In fact, all fish species known to change sex are external fertilisers with simple sexual organs of the bladder-and-duct type.

By contrast, the proximate cues initiating sex change are poorly understood. Sex change can principally be induced by social suppression or by social induction (Ross 1990). These two mechanisms are not mutually exclusive, and some findings actually suggest a joint action. For example, loss of the dominating female in a group of anemonefish, *Amphiprion bicinctus*, initiates sex change in the highest-ranking male, but the presence of smaller conspecifics is necessary for its completion (Cole and Robertson 1988).

9.4.5 Simultaneous hermaphroditism

If something constrains male size advantage, the typical female-to-male sex change outlined above may not evolve. Instead, individuals above a certain size may benefit from allocating further resources acquired to the female function, as no further benefits will accrue from reproducing as a male. Thus may evolve a simultaneous hermaphrodite, an animal capable of reproducing both as a male and a female. But what can constrain male size advantage such that the potential for polygyny no longer exists? Self-fertilisation, very low population densities and partial parthenogenesis are all possible candidates for such constraints, but a mating behaviour called egg trading seems to be a more common mechanism. Egg trading is an active reciprocation of release of eggs between two individuals, each taking turns spawning as a male and as a female. Eggs are energetically expensive to produce compared with sperm. If a female-acting individual allows its partner to get away with being a cheap male, how can that 'female' ensure that 'she' can be a 'he' next time, thereby enjoying the benefits of acting male? In other words, how is reciprocation ensured in this obvious conflict between the sexes? Egg parcelling (i.e. dividing each clutch of eggs into several small batches released at each spawn) is a possible method (Fischer 1980). Allowing a male-acting fish to fertilise only a small parcel, and then discontinue providing eggs unless the other individual reciprocates by acting female, reduces any loss to a female-acting fish in case of cheating. Such safeguards against cheating has been found in the black hamlet, *Hypoplectrus nigricans*, a coral reef fish that spawns as reciprocating pairs without subsequent parental care. Deserting is not favoured, because the time to find a new partner exceeds the time until next egg parcel delivery. The spawning behaviour of parcelling out the eggs is coupled to a gradual maturation of eggs for spawning throughout the daily spawning period (Fischer and Hardison 1987). This will constrain both male mating success and male size advantage.

Egg trading also allows individual hamlets to allocate most of their resources to the female function. Both members of a pair thereby gain a fecundity advantage compared with a monogamously reproducing couple with separate sexes (Fischer 1981). Allocation to either sexual function generally seems to be determined by the level of sperm competition (high competition—high allocation to the male function), as well as by fertilisation efficiency (low efficiency—high male allocation; Petersen 1991). The currency in the trade is certainly eggs, not sperm. There is no egg surplus after the mating season, and egg fertilisation rate does not drop over the season as would be expected if sperm was a limiting resource (Fischer 1987).

Desertion, instead of reciprocation, may in some cases (e.g. *Serranus tigrinus*, Pressley (1981)), also be prevented by mating shortly before nightfall, after which shelter probably must be sought. If so, this will reduce mate search and consequently promote reciprocation (Warner 1984b).

Small population size may facilitate egg trading by decreasing opportunities to cheat and desert. This has already been described for *Serranus fasciatus* (Sections 9.2.2 and 9.2.4), where egg trading occurs at low densities, and harems exist at higher densities. Polygyny potential obviously increases with population density, and females can no longer defend themselves from non-reciprocating male-acting fish. In other words, a male size advantage arises a soon as conditions allow male–male competition over females. At very low population densities the difficulties of finding any mate at all may promote simultaneous hermaphroditism, as in deep-sea tripod-fishes: whoever you meet you can mate with, and if you meet none, you may be able to mate with yourself! Simultaneous hermaphroditism is quite common among deep-sea fishes (Mead *et al.* 1964).

The borderline between sex change and simultaneous hermaphroditism is not sharp. Some sex changers may eventually spawn reciprocally, each taking turn being male and female (e.g. the polychaete *Ophryotrocha puerilis*, Berglund (1986)), and simultaneous hermaphrodites may exhibit sequential changes in sex allocation (e.g. blue-banded goby, *Lythrypnus dalli*, St Mary (1994)).

9.4.6 Sexual and asexual reproduction

For egg-trading to evolve in simultaneous hermaphrodites, female-acting individuals needed safeguards against cheating, as noted above. Without such safeguards, 'females' have little chance of defending themselves against 'males' who parasitise their somatic investment in eggs. As a reciprocating hermaphroditic pair has a much higher fecundity than a male–female pair, egg-trading fish circumvent a major disadvantage of sexual reproduction (i.e. that females forsake their fecundity by allowing males to contribute genes without having to contribute other resources to the next generation). This strategy, however, is vulnerable to cheating individuals. But if females cannot defend themselves against males, why don't they reproduce asexually? This is indeed a fundamental problem in biology, which remains largely unresolved. Anyhow, the fact that females usually reproduce sexually rather than parthenogenetically suggests some advantages of sex, large enough to compensate for its two-fold larger cost (two-fold because parthenogenetically reproducing females produce only daughters who all have offspring of their own, whereas sexually reproducing females produce 50% sons that in turn produce no offspring of their own. See also Stearns (1987) and Michod and Levin (1988)). One hypothesis that has gained some empirical support is the Red Queen hypothesis. If a species is locked into a coevolutionary arms race with biological enemies (predators or parasites), the production of genetically variable offspring by means of sex may allow the species to escape those enemies. The hypothesis assumes that enemies will disproportionally attack the commonest phenotype, so that less common ones may survive. Parthenogenetic species, on the other hand, will fall victim to their enemies as only genetic copies of females are produced, with a low genetic variation by necessity.

The topminnow genus *Poeciliopsis* contains sexual and asexual species. The asexual ones, comprising entirely females, were formed by hybridisation between two sexual topminnow species, and are triploid. Sperm from one of the ancestral species is required to initiate reproduction, but contributes nothing genotypically or phenotypically to the offspring (Vrijenhoek 1979). Still, males of the ancestral species benefit from spawning with the asexual females, as females of their own species copy the mate choice of the asexual females (Schlupp *et al.* 1994). These fishes are infected by trematode larvae (*Uvulifer* sp.) and develop the black spot disease. The Red Queen hypothesis assumes that asexuals should be more parasitised than sexuals, and that variation in parasite level should be higher in sexuals. This was indeed the case found in an assemblage of one sexual and two asexual topminnow species. When the sexual species was inbred following a population bottleneck, parasite level rose, suggesting that genetic variability decreases susceptibility (Lively *et al.* 1990).

All known asexual fish species require the physical stimulus of sperm to induce embryogenesis, like the topminnow (Moore 1984). However, these species are not ideal subjects for testing the Red Queen hypothesis, as existing species assemblages are bound to enjoy some sexual advantage. If not, the asexual species would make up an increasingly larger proportion of the species assemblage with time, so that males would eventually become rare. As both sexual and asexual females depend on males, all these taxa would eventually go extinct.

9.5 Conclusions

Fishes exhibit diverse reproductive strategies. Mating systems are most commonly polygamous. Polygamy can be of three kinds. Promiscuity is widespread and may be the ancestral fish mating system. Polygyny (i.e. one male with several female) is common when a male holds an attractive resource, when he can monopolise females, and on leks. Polyandry (i.e. one female with several males) is rare, but occurs under special circumstances in anemone fishes, in a pipefish and in deep-sea angler fishes with parasitic males. A species may be monogamous if biparental care is necessary, if mutual resource defence is necessary, or if population size is low. Monogamy is imposed on basically polygamous species and is usually not fixed. Relaxation of any of the above conditions promotes a switch to polygamy.

Mating systems influence and are influenced by sexual selection. Under certain circumstances, monogamy may relax sexual selection, but this need not be so. If individuals vary in quality, there will be competition for high quality partners, and this may be a more important source of variation in reproductive success than the number of partners acquired.

Mating is timed to coincide with conditions favourable for offspring or parent survival. If mating involves a strenuous migration, big-bang reproduction may evolve as in Pacific salmon: reproduce once and die! Why Atlantic salmon, with equally strenuous migrations, sometime survive reproduction is unclear.

Fishes provide extraordinarily good examples of alternative reproductive strategies: sneakers, streakers, satellites, peripherals, mimics and jacks flourish, and mostly represent strategies conditional upon environmental changes or resorted to by low-quality individuals. Some of the best examples of mixed evolutionarily stable strategies, with equal fitness payoffs for each strategy, come from fishes: jacks and hooknoses in coho salmon, and cuckolders and parental males in bluegill sunfishes, may be equally successful. However, many of the assumptions behind the fitness calculations are in need of verification, as are the genetic bases of the different strategies.

Sex change may be an alternative to an inferior alternative strategy: why be a male sneaker when you can be a female? Upon reaching a large size, a change into a harem-holding male may increase reproductive success tremendously. The converse (changing from male to female) occurs when the potential for polygyny is low, that is, when male–male competition is unimportant. Then, the higher fecundity associated with large female body size renders larger individuals better suited to function as females.

Some species are simultaneous hermaphrodites, producing eggs and sperm at more or less the same time. This requires safeguards by the female-acting (egg-producing) individual. Reciprocation is needed, so that a female-acting individual may play male next time, as eggs are more costly to produce than sperm. One such safeguard is egg-trading; by only allowing the 'male' to fertilise small batches of eggs and refusing new batches until reciprocation has occurred, individuals will be relatively safe against cheating and profit from the apparently cooperative system. This mating system usually collapses if there is a potential for polygyny, and egg-trading is replaced with harem polygyny. Very low population densities may also allow simultaneous hermaphroditism to be stable, but this is poorly investigated.

Asexually reproducing fishes may support the Red Queen hypothesis for the prevalence of sexual reproduction. In one study, asexuals were more parasitised than sexuals, as were inbred sexuals when compared to outbred sexuals. Therefore, genetic variability in offspring resulting from sexual reproduction may have great advantages when coping with parasites.

Acknowledgements

I thank Ingrid Ahnesjö, Elisabet Forsgren, Lotta Kvarnemo, Gunilla Rosenqvist, Staffan Ulfstrand, Amanda Vincent, the referees and the editor for constructive criticisms on the manuscript.

References

Barlow, G.W. (1984). Patterns of monogamy among teleost fishes. *Arch. Fischereiwis.*, **35**, 75–123.

Barlow, G.W. (1986). A comparison of monogamy among freshwater and coral-reef fishes. In *Indo-Pacific fish biology. Proceedings of the Second International Conference on Indo-Pacific Fishes,* Tokyo Natinal Museum, Ueno Park, Tokyo, July 29- August 3, (1985) (eds. T. Uyeno, R. Arai, T. Taniuchi and K. Matsuura), pp. 767–775. The Ichtyological Society of Japan, Tokyo.

Barlow, G.W. (1991). Mating systems among cichlid fishes. In *Cichlid fishes—behaviour, ecology and evolution* (ed. M.H.A. Keenleyside), pp. 173–190. Chapman and Hall, London.

Barlow, G.W. (1992). Is mating different in monogamous species? The Midas cichlid fish as a case study. *Amer. Zool.*, **32**, 91–99.

Berglund, A. (1986). Sex change by a polychaete: effects of social and reproductive costs. *Ecology*, **67**, 837–845.

Berglund, A. (1990). Sequential hermaphroditism and the size-advantage hypothesis: an experimental test. *Anim. Behav.*, **39**, 426–433.

Berglund, A. (1991). Egg competition in a sex-role reversed pipefish: subdominant females trade reproduction for growth. *Evolution*, **45**, 770–774.

Berglund, A., Rosenqvist, G. and Svensson, I. (1988). Multiple matings and paternal brood care in the pipefish *Syngnathus typhle*. *Oikos*, **51**, 184–188.

Berglund, A., Rosenqvist, G. and Svensson, I. (1989). Reproductive success of females limited by males in two pipefish species. *Am. Nat.*, **133**, 506–516.

Breden, F. and Stoner, G. (1987). Male predation risk determines female preference in the Trinidad guppy. *Nature, Lond.*, **329**, 831–833.

Breitburg, D.L. (1987). Interspecific competition and the abundance of nest sites: factors affecting sexual selection. *Ecology*, **68**, 1844–1855.

Brown, L. and Downhower, J.F. (1982). Polygamy in the mottled sculpins (*Cottus bairdi*) of southwestern Montana (Pisces: Cottidae). *Can. J. Zool.*, **60**, 1973–1980.

Charnov, E.L. (1979). Natural selection and sex change in a pandalid shrimp: test of a life history theory. *Am. Nat.*, **113**, 715–734.

Charnov, E.L. (1982). *The theory of sex allocation*. Princeton University Press, Princeton.

Clutton-Brock, T.H. and Parker, G.A. (1992). Potential reproductive rates and the operation of sexual selection. *Q. Rev. Biol.*, **67**, 437–455.

Clutton-Brock, T. H. and Vincent, A. (1991). Sexual selection and the potential reproductive rates of males and females. *Nature, Lond.*, **351**, 58–60.

Cole, K.S. and Robertson, D.R. (1988). Protogyny in the Caribbean reef goby, *Coryphopterus personatus*: gonad ontogeny and social influences on sex-change. *Bull. Mar. Sci.*, **42**, 317–333.

Conover, D.O. and Kynard, B.E. (1984). Field and laboratory observations of spawning periodicity and behavior of a northern population of the Atlantic silverside, *Menidia menidia*. *Environ. Biol. Fish.*, **11**, 161–171.

Dominey, W.J. (1980). Female mimicry in male bluegill sunfish—a genetic polymorphism?. *Nature, Lond.*, **284**, 546–548.

Donaldson, T.J. (1989). Facultative monogamy in obligate coral-dwelling hawkfishes (Cirrhitidae). *Environ. Biol. Fish.*, **26**, 295–302.

Downhower, J.F., Blumer, L.S. and Brown, L. (1987). Opportunity for selection: an appropriate measure for evolutionary variation in the potential for selection? *Evolution*, **41**, 1395–1400.

Emlen, S.T. and Oring, L.W. (1977). Ecology, sexual selection and the evolution of mating systems. *Science*, **197**, 215–223.

Endler, J.A. (1980). Natural selection on color patterns in *Poecilia reticulata*. *Evolution*, **34**, 76–91.

Endler, J.A. (1983). Natural and sexual selection on color patterns in poeciliid fishes. *Environ. Biol. Fish.*, **9**, 173–190.

Endler, J.A. (1995). Multiple-trait coevolution and environmental gradients in guppies. *Trends Ecol. Evol.*, **10**, 22–29.

Endler, J.A. and Houde, A.E. (1995). Geographic variation in female preferences for male traits in *Poecilia reticulata*. *Evolution*, **49**, 456–468.

Fischer, E.A. (1980). The relationship between mating system and simultaneous hermaphroditism in the coral reef fish, *Hypoplectrus nigricans* (Serranidae). *Anim. Behav.*, **28**, 620–633.

Fischer, E.A. (1981). Sexual allocation in a simultaneously hermaphroditic coral reef fish. *Am. Nat.*, **117**, 64–82.

Fischer, E.A. (1987). Mating behavior in the black hamlet—gamete trading or egg trading? *Environ. Biol. Fish.*, **18**, 143–148.

Fischer, E.A. and Hardison, P.D. (1987). The timing of spawning and egg production asconstraints on male mating success in a simultaneously hermaphroditic fish. *Environ. Biol. Fish.*, **20**, 301–310.

Fleming, I.A. and Gross, M.R. (1989). Evolution of adult female life history and morphology in a Pacific salmon (coho: *Oncorhynchus kisutch*). *Evolution*, **43**, 141–157.

Fricke, H.W. (1979). Mating system, resource defence and sex change in the anemonefish *Amphiprion akallopisos*. *Z. Tierpsychol.*, **50**, 313–326.

Fricke, H.W. (1980). Control of differing mating systems in a coral reef fish by one environmental factor. *Anim. Behav.*, **28**, 561–569.

Fricke, H.W. and Fricke, S. (1977). Monogamy and sex change by aggressive dominance in coral reef fish. *Nature, Lond.*, **266**, 830–832.

Gashagaza, M.M. (1991). Diversity of breeding habits in lamprologine cichlids in Lake Tanganyika. *Physiol. Ecol. Japan*, **28**, 29–65.

Ghiselin, M.T. (1969). The evolution of hermaphroditism among animals. *Q. Rev. Biol.*, **44**, 189–208.

Godin, J.-G.J. (1995). Predation risk and alternative mating tactics in male Trinidadian guppies (*Poecilia reticulata*). *Oecologia*, **103**, 224–229.

Godin, J.-G.J. and Briggs, S.E. (1996). Female mate choice under predation risk in the guppy. *Anim. Behav.*, **51**, 117–130.

Gronell, A.M. (1984). Courtship, spawning and social organization of the pipefish, *Corythoichthys intestinalis* (Pisces: Syngnathidae) with notes on two congeneric species. *Z. Tierpsychol.*, **65**, 1–24.

Gross, M.R. (1982). Sneakers, satellites and parentals: polymorphic mating strategies in North American sunfishes. *Z. Tierpsychol.*, **60**, 1–26.

Gross, M.R. (1984). Sunfish, salmon, and the evolution of alternative reproductive strategies and tactics in fishes. In *Fish reproduction: strategies and tactics* (eds. G.W. Potts and R.J. Wootton), pp. 55–73. Academic Press, London.

Gross, M.R. (1985). Disruptive selection for alternative life histories in salmon. *Nature, Lond.*, **313**, 47–48.

Gross, M.R. (1991). Evolution of alternative reproductive strategies: frequency-dependent sexual selection in male bluegill sunfish. *Phil. Trans. R. Soc. Lond. B*, **332**, 59–66.

Gross, M.R. (1996). Alternative reproductive strategies and tactics: diversity within sexes. *Trends Ecol. Evol.*, **11**, 92–98.

Höglund, J. (1989). Size and plumage dimorphism in lek-breeding birds: a comparative analysis. *Am. Nat.*, **134**, 72–87.

Houde, A.E. (1988). Genetic difference in female choice between two guppy populations. *Anim. Behav.*, **36**, 510–516.

Houde, A.E. and Endler, J.A. (1990). Correlated evolution of female mating preferences and male color patterns in the guppy *Poecilia reticulata*. *Science*, **248**, 1405–1408.

Katano, O. (1990). Dynamic relationships between the dominance of male dark chub, *Zacco temmincki*, and their acquisition of females. *Anim. Behav.*, **40**, 1018–1034.

Keenleyside, M.H.A. (1972). Intraspecific intrusioms into nests of spawning longear sunfish (Pisces: Centrarchidae). *Copeia*, **1972**, 272–278.

Keenleyside, M.H.A. (1979). *Diversity and adaptation in fish behaviour.* Springer-Verlag, Berlin.

Keenleyside, M.H.A. (1985). Bigamy and mate choice in the biparental cichlid fish *Cichlasoma nigrofasciatum*. *Behav. Ecol. Sociobiol.*, **17**, 285–290.

Keenleyside, M.H.A. (1991). Parental care. In *Cichlid fishes—behaviour, ecology and evolution* (ed. M.H.A. Keenleyside), pp. 191–208. Chapman and Hall, London.

Keenleyside, M.H.A., Bailey, R.C. and Young, V.H. (1990). Variation in the mating system and associated parental behaviour of captive and free-living *Cichlasoma nigrofasciatum* (Pisces, Cichlidae). *Behaviour*, **112**, 202–221.

Keenleyside, M.H.A., Rangeley, R.W. and Kuppers, B.U. (1985). Female mate choice and male parental defense behaviour in the cichlid fish *Cichlasoma nigrofasciatum*. *Can. J. Zool.*, **63**, 2489–2493.

Kodric-Brown, A. (1986). Satellites and sneakers: opportunistic male breeding tactics in pupfish (*Cyprinodon pecosensis*). *Behav. Ecol. Sociobiol.*, **19**, 425–432.

Kodric-Brown, A. (1988). Effects of sex-ratio manipulation on territoriality and spawning success of male pupfish, *Cyprinodon pecosensis*. *Anim. Behav.*, **36**, 1136–1144.

Krebs, J.R. and Davies, N.B. (1993). *An introduction to behavioural ecology*, 3rd edn. Blackwell Scientific Publ., Oxford.

Lambert, T.C. and Ware, D.M. (1984). Reproductive strategies of demersal and pelagic spawning fish. *Can. J. Fish. Aquat. Sci.*, **41**, 1565–1569.

Limberger, D. (1983). Pairs and harems in a cichlid fish, *Lamprologus brichardi*. *Z. Tierpsychol.*, **62**, 115–144.

Lively, C.M., Craddock, C. and Vrijenhoek, R.C. (1990). Red Queen hypothesis supported by parasitism in sexual and clonal fish. *Nature, Lond., 344, 864–866.*

Loiselle, P.V. and Barlow, G.W. (1978). Do fishes lek like birds? In *Contrasts in behavior: adaptations in the aquatic and terrestrial environment* (eds. E. Reese and F.G. Lighton). John Wiley and Sons, New York.

McKaye, K.R. (1977). Competition for breeding sites between the cichlid fishes of Lake Jiloá, Nicaragua. *Ecology*, **58**, 291–302.

McKaye, K.R. (1983). Ecology and breeding behavior of a cichlid fish, *Cyrtocara eucinostomus*, on a large lek in Lake Malawi, Africa. *Environ. Biol. Fish.*, **8**, 81–96.

McKaye, K.R. and McKaye, N.M. (1977). Communal care and kidnapping of young by parental cichlids. *Evolution*, **31**, 674–681.

McKaye, K.R., Louda, S.M. and Stauffer, J.R.J. (1990). Bower size and male reproductive success in a cichlid fish lek. *Am. Nat.*, **135**, 597–613.

Magurran, A.E. and Seghers, B.H. (1990). Risk sensitive courtship in the guppy (*Poecilia reticulata*). *Behaviour*, **112**, 194–201.

Malmqvist, B. (1983). Breeding behaviour of brook lampreys *Lampetra planeri*: experiments on mate choice. *Oikos*, **41**, 43–48.

Mann, R.H.K. and Mills, C.A. (1979). Demographic aspects of fish fecundity. *Symp. Zool. Soc. Lond.*, **44**, 161–177.

Mead, G.W., Bertelson, E. and Cohen, D.M. (1964). Reproduction among deep-sea fishes. *Deep Sea Res.*, **11**, 569–596.

Michod, R.E. and Levin, B.R. (1988). *The evolution of sex*. Sinauer Associates Inc., Sunderland, MA.

Moore, W.S. (1984). Evolutionary ecology of unisexual fishes. In *Evolutionary genetics of fishes* (ed. B.J. Turner), pp. 329–398. Plenum Press, New York.

Mousseau, T.A. and Collins, N.C. (1987). Polygyny and nest site abundance in the slimy sculpin (*Cottus cognatus*). *Can. J. Zool.*, **65**, 2827–2829.

Moyer, J.T. and Sawyers, C.E. (1973). Territorial behaviour of the anemonefish *Amphiprion xanthurus* with notes on the life history. *Jpn. J. Ichthyol.*, **20**, 85–93.

Noakes, D.L.G. and Barlow, G.W. (1973). Ontogeny of parent-contacting behaviour in young *Cichlasoma citrinellum* (Pisces, Cichlidae). *Behaviour*, **46**, 221–255.

Norman, J.R. and Greenwood, P.H. (1975). *A history of fishes*, 3rd edn. Benn, London.

Ochi, H. (1986). Growth of the anemone fish *Amphiprion clarkii* in temperate waters, with special reference to the influence of settling time on growth of 0-year olds. *Mar. Biol.*, **92**, 223–229.

Ochi, H. (1989). Mating behavior and sex change of the anemonefish, *Amphiprion clarkii*, in the temperate waters of southern Japan. *Environ. Biol. Fish.*, **26**, 257–275.

Petersen, C.W. (1987). Reproductive behaviour and gender allocation in *Serranus fasciatus*, a hermaphroditic reef fish. *Anim. Behav.*, **35**, 1601–1614.

Petersen, C.W. (1990). The relationships among population density, individual size, mating tactics, and reproductive success in a hermaphroditic fish, *Serranus fasciatus*. *Behaviour*, **113**, 57–80.

Petersen, C.W. (1991). Sex allocation in hermaphroditic sea basses. *Am. Nat.*, **138**, 650–667.

Petersen, C.W. and Fischer, E.A. (1986). Mating system of the hermaphroditic coral-reef fish, *Serranus baldwini*. *Behav. Ecol. Sociobiol.*, **19**, 171–178.

Petersen, C.W. and Hess, H.C. (1991). The adaptive significance of spawning synchronization in the Caribbean damselfish *Stegastes dorsopunicans* (Poey). *J. Exp. Mar. Biol. Ecol.*, **151**, 155–167.

Pressley, P.H. (1981). Pair formation and joint territoriality in a simultaneous hermaphrodite: the coral reef fish *Serranus tigrinus*. *Z. Tierpsychol.*, **56**, 33–46.

Reynolds, J.D. (1996). Animal breeding systems. *Trends Ecol. Evol.*, **11**, 68–72.

Reynolds, J.D. and Gross, M.R. (1990). Costs and benefits of female mate choice: is there a lek paradox? *Am. Nat.*, **136**, 230–243.

Reynolds, J.D., Gross, M.R. and Coombs, M.J. (1993). Environmental conditions and male morphology determine alternative mating behaviour in Trinidadian guppies. *Anim. Behav.*, **45**, 145–152.

Robertson, D.R. (1990). Differences in the seasonalities of spawning and recruitment of some small neotropical reef fishes. *J. Exp. Mar. Biol. Ecol.*, **144**, 49–62.

Robertson, D.R. (1991). The role of adult biology in the timing of spawning of tropical reef fishes. In *The ecology of fishes on coral reefs* (ed. P.F. Sale), pp. 356–386. Academic Press, New York.

Robertson, D.R., Petersen, C.W. and Brawn, J.D. (1990). Lunar reproductive cycles of benthic-brooding reef fishes: reflections of larval biology or adult biology? *Ecol. Monogr.*, **60**, 311–329.

Roede, M.J. (1990). Growing knowledge and understanding of dichromatism and sex reversal in Labridae. *Bijdr. Dierkd.*, **60**, 225–232.

Roff, D.A. (1981). Reproductive uncertainty and the evolution of iteroparity—why don't flatfish put all their eggs in one basket? *Can. J. Fish. Aquat. Sci.*, **38**, 968–977.

Rogers, W. (1987). Sex ratio, monogamy and breeding success in the Midas cichlid (*Cichlasoma citrinellum*). *Behav. Ecol. Sociobiol.*, **21**, 47–51.

Rogers, W. (1988). Parental investment and division of labor in the Midas cichlid (*Cichlasoma citrinellum*). *Ethology*, **79**, 126–142.

Rogers, W. and Barlow, G.W. (1991). Sex differences in mate choice in a monogamous biparental fish, the Midas cichlid (*Cichlasoma citrinellum*). *Ethology*, **87**, 249–261.

Rosenqvist, G. (1990). Male mate choice and female–female competition for mates in the pipefish *Nerophis ophidion*. *Anim. Behav.*, **39**, 1110–1115.

Ross, R.M. (1981). Experimental evidence for stimulation and inhibition of sex change in the saddleback wrasse *Thalassoma duperrey*. *Pac. Sci.*, **35**, 275.

Ross, R.M. (1983). Annual, semilunar, and diel reproductive rhythms in the Hawaiian labrid *Thalassoma duperrey*. *Mar. Biol.*, **72**, 311–318.

Ross, R.M. (1990). The evolution of sex-change mechanisms in fishes. *Environ. Biol. Fish.*, **29**, 81–93.

Ryan, M.J., Hews, D.K. and Wagner, W.E. Jr. (1990). Sexual selection on alleles that determine body size in the swordtail *Xiphophorus nigrensis*. *Behav. Ecol. Sociobiol.*, **26**, 231–237.

Schaffer, W.M. (1979). The theory of life-history evolution and its application to Atlantic salmon. *Symp. Zool. Soc. Lond.*, **44**, 307–326.

Schlupp, I., Marler, C. and Ryan, M.J. (1994). Benefit to male sailfin mollies of mating with heterospecific females. *Science*, **263**, 373–374.

Shapiro, D.Y. (1988). Behavioral influences on gene structure and other new ideas concerning sex change in fishes. *Environ. Biol. Fish.*, **23**, 283–297.

Shuster, S.M. and Wade, M.J. (1991). Equal mating success among male reproductive strategies in a marine isopod. *Nature, Lond.*, **350**, 608–610.

Sigurjónsdóttir, H. and Gunnarson, K. (1989). Alternative mating tactics of arctic charr, *Salvelinus alpinus*, in Thingvallavatn, Iceland. *Environ. Biol. Fish.*, **26**, 159–176.

Sikkel, P.C. (1989). Egg presence and developmental stage influence spawning-site choice by female garibaldi. *Anim. Behav.*, **38**, 447–456.

St Mary, C.M. (1994). Sex allocation in a simultaneous hermaphrodite, the blue-banded goby (*Lythrypnus dalli*): the effects of body size and behavioral gender and the consequences for reproduction. *Behav. Ecol.*, **5**, 304–313.

Stearns, S.C. (1987). *The evolution of sex and its consequences*. Birkhäuser Verlag, Basel.

Stearns, S.C. (1992). *The evolution of life histories*. Oxford University Press, Oxford.

Stillman, R., Clutton-Brock, T.H. and Sutherland, W.J. (1993). Black holes, mate retention and the evolution of ungulate leks. *Behav. Ecol.*, **4**, 1–6.

Stoner, G. and Breden, F. (1988). Phenotypic differentiation in female preference related to geographic variation in male predation risk in the Trinidad guppy (*Poecilia reticulata*). *Behav. Ecol. Sociobiol.*, **22**, 285–292.

Taborsky, M. (1994). Sneakers, satellites, and helpers: parasitic and cooperative behavior in fish reproduction. *Adv. Study Behav.*, **23**, 1–100.

Thresher, R.E. (1984). *Reproduction in reef fishes*. T. F. H. Publications, Neptune City, NJ.

Townshend, T.J. and Wootton, R.J. (1985). Variation in the mating system of a biparental cichlid fish, *Cichlasoma panamense*. *Behaviour*, **95**, 181–197.

van den Berghe, E.P., Wernerus, F. and Warner, R.R. (1989). Female choice and the mating cost of peripheral males. *Anim. Behav.*, **38**, 875–884.

Verrell, P.A. (1991). Illegitimate exploitation of sexual signalling systems and the origin of species. *Ethol. Ecol. Evol.*, **3**, 273–283.

Vincent, A., Ahnesjö, I., Berglund, A. and Rosenqvist, G. (1992). Pipefishes and seahorses: are they all sex role reversed? *Trends Ecol. Evol.*, **7**, 237–241.

Vrijenhoek, R.C. (1979). Factors affecting clonal diversity and coexistence. *Amer. Zool.*, **19**, 787–797.

Warner, R.R. (1984a). Deferred reproduction as a response to sexual selection in a coral reef fish: a test of the life historical consequences. *Evolution*, **38**, 148–162.

Warner, R.R. (1984b). Mating behavior and hermaphroditism in coral reef fishes. *Amer. Sci.*, **72**, 128–136.

Warner, R.R. (1987). Female choice of sites versus mates in a coral reef fish, *Thalassoma bifasciatum*. *Anim. Behav.*, **35**, 1470–1478.

Warner, R.R. (1988a). Sex change in fishes: hypotheses, evidence, and objections. *Environ. Biol. Fish.*, **22**, 81–90.

Warner, R.R. (1988b). Traditionality of mating-site preferences in a coral reef fish. *Nature, Lond.*, **335**, 719–721.

Warner, R.R. (1990). Male versus female influences on mating-site determination in a coral reef fish. *Anim. Behav.*, **39**, 540–548.

Warner, R.R. and Hoffman, S.G. (1980). Local population size as a determinant of mating system and sexual composition in two tropical marine fishes (*Thalassoma* spp.). *Evolution*, **34**, 508–518.

Warner, R.R. and Swearer, S.E. (1991). Social control of sex change in the bluehead wrasse, *Thalassoma bifasciatum* (Pisces: Labridae). *Biol. Bull.*, **181**, 199–204.

Wiegmann, D.D., Baylis, J.R. and Hoff, M.H. (1992). Sexual selection and fitness variation in a population of smallmouth bass, *Micropterus dolomieiu* (Pisces: Centrarchidae). *Evolution*, **46**, 1740–1753.

Wiley, R.H. (1991). Lekking in birds and mammals: behavioral and evolutionary issues. *Adv. Study Behav.*, **20**, 201–291.

Wilson, E.O. (1975). *Sociobiology. The new synthesis.* Harvard University Press, Cambridge.

Winemiller, K.O., Leslie, M. and Roche, R. (1990). Phenotypic variation in male guppies from natural inland populations: an additional test of Haskins' sexual selection/predation hypothesis. *Environ. Biol. Fish.*, **29**, 179–191.

Wisenden, B.D. (1994). Factors affecting mate desertion by males in free-ranging convict cichlids (*Cichlasoma nigrofasciatum*). *Behav. Ecol.*, **5**, 439–447.

Zimmerer, E.J. and Kallman, K.D. (1989). Genetic basis for alternative reproductive tactics in the pygmy swordtail, *Xiphophorus nigrensis. Evolution*, **43**, 1298–1307.

10 *Sexual selection*

Lee Alan Dugatkin and Gerard J. FitzGerald

10.1 Introduction

Sexual selection refers to the process responsible for the evolution of secondary sexual or epigamic traits. As Darwin (1871) noted, this process 'depends on the advantage which certain individuals have over other individuals of the same sex and species, in exclusive relation to reproduction'. Unlike traits that have evolved via natural selection, epigamic traits enhance an animal's mating success, even though they may reduce its chances for survival. Classic examples of such traits are the peacock's, *Pavo cristatus*, bright colours and elaborate tail, both of which Darwin (1871) believed evolved as a consequence of female mate choice.

Darwin recognised two distinct processes of sexual selection: intrasexual selection, usually male–male competition, and intersexual selection or mate choice, usually female choice. Sexual selection by intrasexual competition is accepted by most biologists, while sexual selection by female choice, until about 15 years ago, remained suspect (Halliday 1983; Bradbury and Andersson 1987; Andersson 1994). An excellent, though somewhat popularised, historical review of sexual selection and a discussion of modern issues is given by Cronin (1991).

Here, we review the recent literature on sexual selection in fishes and suggest directions for future research. Fishes are particularly useful for the experimental examination of ideas generated by models of sexual selection, because a substantial body of evidence now exists that male traits, as well as female mating preferences, have evolved via sexual selection and because many species are of convenient size to allow manipulations in the laboratory. In addition, many species of fish have large geographic distributions, allowing interpopulational studies that may help distinguish between various models of sexual selection (Houde 1993). The existence of many closely-related species (species flocks) in one area, such as the African Great Lakes cichlids (Dominey 1984) and poeciliids (Ryan and Wagner 1987), and of different populations of the same species in close proximity to one another (Magurran *et al.* 1995) also offers opportunities to explore the role of sexual selection in speciation.

We begin by evaluating the evidence that males and females compete for mates and for resources that the opposite sex requires (e.g. breeding territories, nesting materials, the best nest sites, and refuges from predators). We proceed by examining how individuals obtain such resources and discuss some of the alternative behaviours used to avoid paying the costs associated with obtaining resources. We then address mechanisms of mate choice, the consequences of choice for an individual's reproductive success, as well as the strategies and tactics used by rejected individuals to circumvent the choice of others and obtain matings. Some suggestions for future research topics round out the chapter.

Because the literature on sexual selection overlaps with that on other aspects of reproductive behaviour, we encourage readers to consult Chapter 9 by Berglund, on mating systems and reproductive strategies, and Chapter 11 by Sargent, on parental care.

10.2 Intrasexual selection

Many models of sexual selection (e.g. Darwin, 1871; Fisher 1958; Kirkpatrick 1982; Andersson 1994) predict a positive correlation between exaggerated traits and reproductive success in males, even though a negative correlation may exist between these traits and male viability (Fig. 10.1). For example, the most brightly coloured males may obtain the most matings, but these males may be more vulnerable to predators than dull males (as in the guppy, *Poecilia reticulata*; Seghers 1973, Houde 1988a, b; Houde and Endler 1990; Endler 1995). Unfortunately, it has proven difficult to test this quantitative prediction in the field because of large individual differences in the ability to garner resources. For instance, bright males may be able to allocate more resources to both survival and reproduction, leading to a positive correlation between sexually-selected traits and viability among individuals. Another problem in testing quantitative predictions of intrasexual selection theory is that some traits, such as body size, are clearly involved in both male combat and female choice. Despite these problems, it is useful to examine the links between such factors as the development of weapons, sexual differences in body size, colour patterns, and the various behavioural strategies and tactics (*sensu* Gross 1984, 1996) that individuals use to settle contests for key resources.

10.2.1 Colour

Colours in fishes are of two basic types: pigments (biochromes) and structural colours (Endler 1991). Pigments are coloured compounds that are found primarily in chromatophores—cells located mainly in the dermis and epidermis of the skin, eyes and various organs. Chromatophores are highly irregular in shape, but usually appear as a central core giving out many branched processes. This shape permits the rapid colour changes that are seen in some fish species. Structural colours are those produced by light reflecting from structures (e.g. guanine), rather than pigments, such as the silvery colours of many pelagic fishes.

Sexual differences in colour patterns are common in fishes and colour patterns

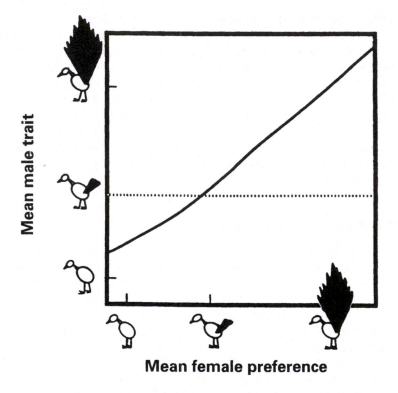

Mean female preference

Fig. 10.1   A Fisherian-like 'null model' of sexual selection. Natural selection operates against, and sexual selection operates in favour of, the exaggerated male trait. The solid curve shows the value at which these forces equilibrate and the dashed line shows the degree of the male trait that maximises survival. From Kirkpatrick (1987).

play a key role in sexual selection. However, interpretation of the significance of colour patterns is complicated by the ability of many fish to rapidly change colour in response to changing environmental conditions, especially in relation to reproduction. Adult males are often brightly coloured, or conspicuously marked, whereas juveniles of both sexes are usually less colourful and more cryptic. Of particular interest here is whether differences in colour among males play a role in intraspecific agonistic interactions that occur when males compete for breeding sites or mates.

During the breeding season, males of some populations of the threespine stickleback, *Gasterosteus aculeatus*, develop a bright red nuptial coloration, bright blue eyes and a dark blue back. This colour pattern is due to the presence of three carotenoids: xanthophyll, a carotene ester and astaxanthin (Wootton 1976). There is, however, considerable variation in the expression of the colour both within and among populations of the species (Reimchen 1989; Frischknecht 1993; Bakker and

Mundwiler 1994). Because it occurs only during the reproductive season and is not displayed by females, red colour appears to be a sexually-selected trait. However, whether male–male competition or female choice has been the primary selective agent in the evolution of coloration in sticklebacks is controversial (FitzGerald 1993).

Tinbergen (1951) and Collias (1990) found that red colour in males 'releases' aggressive behaviour. In these studies, territorial males attacked models of red males more often than grey ones. Other authors, however, have not only failed to confirm the finding of more attacks on red models, but have reported more attacks on grey models (e.g. Muckenstrum 1967, 1969; Peeke *et al.* 1969; Rowland 1982; Rowland and Sevenster 1985).

The role of red in male–male competition appears to vary among populations. For freshwater populations of threespine sticklebacks, bright males usually dominate dull ones (Bakker 1994), perhaps because red colour has an intimidating effect on rivals. This hypothesis, however, remains untested. In contrast, bright males of anadromous populations do not have any advantage in dyadic fights (FitzGerald and Kedney 1987). For a discussion of possible reasons for discrepancies among studies, see FitzGerald (1993) and Bakker (1994).

10.2.2 Weapons

Although many animals have developed ornaments for use in sexual competition, there are surprisingly few examples of this in fishes (Huntingford and Turner 1987). One well-documented exception is the hooked nose (kype) of the adult male coho salmon, *Oncorhynchus kisutch*, which appears to have evolved exclusively for use in male–male combats for mates. The 'hook' comprises an exaggerated curved snout and enlarged teeth that develop only during the breeding season. Other males, called 'jacks', forego the development of this structure. These males compete with their fiercer and larger rivals by remaining relatively cryptic on the breeding grounds and 'sneaking' fertilisations (Gross 1985).

Why weaponry used primarily for male–male combat is apparently rare in fishes merits further investigation. Is the aquatic medium per se less conducive to the evolution of weaponry? Is the lack of weaponry related to fish body shape and morphology (e.g. lack of appendages)? Is the prevalence of male parental care in fishes related to the lack of weaponry? The comparative approach (Harvey and Pagel 1991) would be a fruitful avenue for exploring some of these questions.

10.2.3 Body size and correlated morphological traits

The lack of weaponry in male fishes should not, however, be taken as evidence against male–male competition. Large body size, and traits correlated with it, may confer an advantage in mate competition. For instance, in Atlantic salmon, *Salmo salar*, body size, cardiac–somatic index and relative kype size are all positively correlated with each other and with male dominance rank (Järvi 1991), resulting in dominant males mating more frequently than subordinates. Apparently, intrasexual selection has favoured the evolution of a status-signalling system, whereby males signal their dominance rank by their relative kype size. Dominant males having this conspicuous secondary sexual characteristic are thus able to mate more frequently (Järvi 1991).

10.2.4 Body size and fighting ability

Virtually all models of the evolution of fighting behaviour assume that size (one measure of 'resource holding potential') is positively correlated with the probability of victory in aggressive interactions (Parker 1974; Parker and Rubenstein 1981). In fishes, larger males typically defeat smaller ones, but the minimal size differential which determines the outcome of a fight appears to vary among species (e.g. Barlow *et al.* 1986; Rowland 1989; Dufresne *et al.* 1990). Factors other than size may also play a role in determining the outcome of a fight. As examples, the length of time that male Midas cichlids, *Cichlasoma citrinellum*, were allowed to familiarise themselves with a territory played a role in whether body weight or aggressiveness determined the duration of a combat (Barlow *et al.* 1986), while the 'value' of the resource affected whether smaller pumpkinseed sunfish, *Lepomis gibbosus*, had any chance of defeating larger individuals (Dugatkin and Ohlsen 1990; Dugatkin and Biederman 1991). For more on studies examining the factors that influence the outcome of aggressive interactions among males, see Beacham and Newman (1987), Goldschmidt and Bakker (1990), Enquist *et al.* (1990) and Beaugrand and Beaugrand (1991).

The assumption that a male's ability to defeat a conspecific during an aggressive interaction leads to increased mating opportunities remains to be directly tested in fishes. A direct test of this assumption will also set the stage for untangling intra- and intersexual selection. Consider a simple experiment with three treatments, for example. In all treatments, two males engage in agonistic interactions. In treatment I, after males have fought and a clear winner emerges, place both males in a tank which has a (receptive) female in a Plexiglas tube (thus minimising female choice). Then simply note which male spends more time near the female. In Treatment II, allow females to watch the males fight. Subsequent to this, a female choice test is done in which the female is placed in a small aquarium located between two end tanks, each containing one of the males, and time spent near the males is recorded as a measure of mating preference. Treatment III is similar to Treatment II, except that a female who did not see the males fight is placed in the central aquarium. Treatment III thus controls for behaviours that may be correlated with winning aggressive interactions. A fourth treatment could place the female choice experiment before the male–male interaction phase and examine how choice may influence aggressive interaction. We suggest that experiments along these lines will help us better understand the role that aggressive interactions play in intra- and intersexual selection in fishes.

10.2.5 Costs of fighting

The loss of a single fight can have an important effect on lifetime reproductive success, especially in short-lived fishes (see Enquist and Leimar (1990) for a general treatment of aggressive behaviour in situations where lifetime reproductive success depends on the result of a few aggressive interactions). Even brief fights, lasting only a few seconds, can have high physiological costs, especially for the loser (Chellappa and Huntingford 1988; Haller and Wittenberger 1988). Chellappa and Huntingford (1988) measured glycogen levels in the livers of male threespine sticklebacks following a brief fight. They found that losers had less glycogen than winners, and liver glycogen levels and fight duration were negatively correlated for both the winners and losers. Participation in a brief fight left both fish, but particularly the defeated one, with depleted liver glycogen, which in turn may impair their ability to rapidly respond to conspecifics (e.g. in the context of mating opportunities) and predators. In addition, in many fishes, once an individual has lost one fight, it is more likely to lose future fights (Frey and Miller 1972; Francis 1983; Beaugrand and Zayan 1985; Beacham and Newman 1987; Bakker 1994).

10.2.6 Alternatives to fighting

Given the (often large) costs of fighting, it is not surprising that fish employ alternative strategies and tactics to ensure their reproduction (Gross 1984, 1996; Berglund, Chapter 9). These alternatives include cuckoldry, forced copulation and egg stealing. In addition, sex change might also be considered a life-history strategy for avoiding the costs of fighting.

Cuckoldry

In many species, one finds male reproductive morphs that are distinct in structural, physiological, endocrinological and behavioural traits (Bass 1993). The co-occurrence of alternative forms of salmon and sunfish is well known (Gross and Charnov 1980; Gross 1985; see Fig. 10.2), and the different morphs may have a genetic basis (Gross 1991). In other species, alternative tactics may be facultative (e.g., poeciliids: Ryan 1990; Kodric-Brown 1992; Reynolds *et al.* 1993; Godin 1995).

In species in which individuals can employ more than one tactic, it is of interest to know how the use of a specific behavioural tactic to ensure reproduction is affected by prevailing environmental and social situations. For example, both physical and social aspects of the environment may influence the use of alternative mating tactics. Individual male guppies exhibit two distinct types of mating behaviour: sigmoid displays used in courtship and gonopodial thrusts used in circumventing female choice. Reynolds *et al.* (1993) found that the relative use of sigmoid displays and thrusting by male guppies depended on their body size, and more specifically larger males displayed less often than smaller males at high light

(a)

(b)

(c)

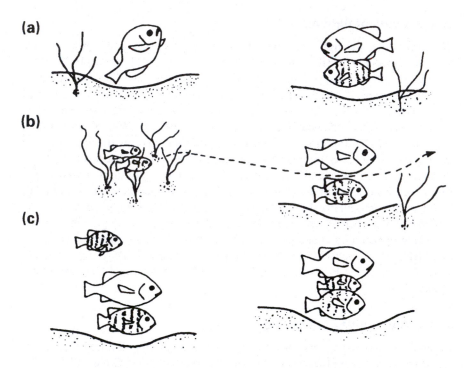

Fig. 10.2   Alternative reproductive strategies in bluegill sunfish, *Lepomis macrochirus*. (a) Parental strategy—parental males build nests and provide care to their young. (b) Sneaker strategy—sneaker males (shown hiding in the vegetation) dart into nests of parental males and shed sperm during oviposition. (c) Satellite strategy—satellite males mimic female appearance (stippled in this diagram) and release sperm while pairing with parental males and true females. From Gross (1984).

intensity, although they did not compensate with increased thrusting. In contrast, at low light intensity, display frequency was unrelated to male size, but thrusting behaviour correlated positively with the relative length of the male's intromittent organ (i.e. the gonopodium).

Egg stealing

Another alternative, or at least a supplement, to courtship is egg stealing. Male threespine sticklebacks often steal eggs from rival males and carry them back to their own nests (van den Assem 1967; Sargent 1989; Rico *et al.* 1992). Rohwer (1978) reported that stolen eggs attract females to such nests where the thief can then obtain matings. Indeed, in at least eight fish species, females prefer males with eggs to those with empty nests (Turner 1993). Although little is known about the costs involved in an attempt to steal eggs from another nest, evidence indicates that egg stealing may be adaptive for both males and females. From the male's perspective, not only are females more attracted to nests with more eggs, but males are able to discriminate between their own eggs and stolen eggs, and adopted eggs are defended less and suffer higher mortality (Sargent 1989). From the female perspective, any mechanism that increases egg number seems to be favoured as it causes an increase in paternal care and egg survival (Sargent 1989, Chapter 11).

Sex changes

A somewhat different way to avoid competition for a breeding space or partner is to change sex (Warner *et al.* 1975). The size-advantage model (Ghiselin 1969; Warner *et al.* 1975; Berglund, Chapter 9) predicts body size change in relation to social and ecological factors that maximise lifetime fitness. In general, the size-advantage model has been successful in predicting switch points in sex-changing species (Warner 1988; Berglund, Chapter 9). For example, an individual bluehead wrasse, *Thalassoma bifasciatum*, born female, may turn into a male when large enough to compete successfully as a territory holder. In other species, such as the anemonefish, *Amphiprion clarkii*, individuals are born males and change into females (e.g. Ochi 1989). For more information on hypotheses regarding sex change in fishes, we refer the reader to Shapiro (1979), Warner (1988) and Berglund (Chapter 9).

Fishes have particularly labile sex-determining mechanisms (Francis 1992). In the context of intrasexual selection, sex change can affect, as well as be affected by, the degree of aggression displayed within a group. Francis and Barlow (1993) have recently found that sex-biased size differences among individuals are not based on endogenous factors, but rather that the sex of an individual is determined by its relative size at maturity. They argue that 'this mode of sex determination, which may be common among teleosts, is a heterochronic variant of post-maturational sex change, one in which some individuals are deflected from a default female trajectory before maturation, as a result of social signals' (Francis and Barlow 1993, p. 10673). Such signals, received as juveniles, affect the sex ratio of adults and thus indirectly the intensity of intrasexual selection.

10.2.7 Female–female competition

Most studies of intrasexual selection in fishes have focused on males, because it is they who most often possess exaggerated epigamic characters and are the most aggressive sex. Females, however, may also compete for limited food, space and mates. When such competition is for food alone, it should be expressed through the evolution of more efficient feeding and food conversion processes, rather than elaborate secondary sexual characteristics.

Females might be expected to compete for males under two conditions: if operational sex ratios are female-biased and/or if male quality varies so that high-quality males are in short supply (Berglund *et al.* 1993; Vincent *et al.* 1994; Berglund, Chapter 9). Female-biased operational sex ratios can occur when females can remate faster than males. This is likely if males invest substantially in offspring, if this investment enhances female reproductive success, and if the investment is at least partly depreciable (i.e. if a male invests with one female, this decreases his chances of mating with another female).

Competition among females for mates has been documented in coho salmon, an anadromous fish that breeds in freshwater streams (van den Berghe and Gross 1989). After fertilisation, the eggs are guarded by females against cannibalistic conspecifics for an average of 8 days (i.e. until her death). The duration of female guarding is positively correlated with body size, probably because larger females have greater energy reserves. van den Berghe and Gross (1989) postulate that females compete for high-quality territories to insure egg development, and this leads to significant variation in female body size and kype size (a secondary sexual character used for fighting, see Section 10.2.2). The biggest females were estimated to have had a 23-fold increase in fitness through increased egg biomass, the ability to acquire a high-quality territory, and success in nest defence.

Competition among female pipefish, *Syngnathus typhle*, occurs during shortages of receptive males (Berglund *et al.* 1989; Berglund 1991). Males prefer large females and small females cease egg production when housed with larger ones. Berglund *et al.* (1989) argued that small females were probably channelling stored energy into somatic growth, in order to become better competitors in the following breeding season. However, this hypothesis assumes that small females were willing to sacrifice present reproductive opportunity for future opportunities. An equally plausible alternative hypothesis is that large females were simply suppressing the reproduction of smaller individuals.

Female saddleback wrasse, *Thalassoma duperrey*, a diandric protogynous hermaphrodite, may change sex to reduce intrasexual competition (Ross 1981). When two females were kept together in an outdoor tank, the larger individual of a pair changed sex, whereas isolated individuals remained females. In threesomes, only the largest females changed sex and inhibited the sex change in the other females.

10.3 Intersexual selection

Mate choice has been operationally defined as any behaviour shown by members of one sex that leads to their being more likely to mate with certain members of the opposite sex than with others (Halliday 1983). At least three hypotheses have been put forward to explain the evolution of female mate preferences (Kirkpatrick and Ryan 1991; Andersson 1994). The *Direct Selection* hypothesis supposes that females receive direct fitness benefits from their preference for a particular male trait. Such benefits include, but are not limited to, reduced mate searching time, food, and decreased predation risk. The *Runaway Selection* hypothesis argues that, because female preference and male traits can become genetically correlated via linkage disequilibrium, slight female preferences for even arbitrary male traits can cause the trait and the preference to 'run away' toward extreme values. The *Good Genes* hypothesis claims that females choose males with genes that increase viability. The most popular of the good genes models is the Hamilton and Zuk (1982) parasite-resistance model, which proposes that females choose males based on their resistance to parasites (expressed via elaborate traits). Unfortunately, aside from the studies of Reynolds and Gross (1992) and Reynolds and Côté (1995), experiments to date have not been able to distinguish between these three hypotheses for the evolution of female preference in fishes.

10.3.1 Benefits and choice criteria

There are any number of benefits, both short-term and long-term, that animals may obtain by choosing certain mates over others. When choosing a mate, an individual may use as its criterion of mate quality either the putative benefit itself (e.g. ability to defend a nest) or some feature of the behaviour or appearance of prospective mates that is a predictor of the benefits they offer (e.g. large size—which may indicate an ability to defend a nest; see Price *et al.* (1993) for a general treatment of sexual selection when females benefit directly from their choice of mates).

In fishes, choice may be for signs of good parental abilities, resources, high social status, good physical condition and effective courtship displays. Of course, these factors are not mutually exclusive and females may use more than one cue when choosing among potential mates.

Choice for resources or males?

In resource-based polygamous systems, male access to females is determined by control of a resource essential to female reproduction (e.g. oviposition site; see Grant, Chapter 4 and Berglund, Chapter 9). In such systems, male fitness is assumed to be positively correlated with both the quality of the resource held and his ability to defend it against competing males. Male–male competition may be an important selective pressure in the evolution of those male characteristics that facilitate acquisition and defence of such resources (see Section 10.2). Female choice of such male characteristics may also be important in enhancing male fitness. Unfortunately, the importance of male characteristics (in addition to, or independent of, resource quality) for female mate choice cannot be readily determined in most breeding systems, especially if differences in secondary sexual characteristics among males are so small that female choice would be difficult to demonstrate.

A common approach used in field studies investigating factors affecting variation in male reproductive success is to record various territory and oviposition site characteristics, as well as male phenotypic characteristics thought likely to be related to reproductive success. Multiple regression analysis can then be employed to determine how much of the overall variation in male mating success is accounted for by the set of characteristics and by each independent variable separately. For example, Kodric-Brown (1983) quantified differences in the microtopography of territories, degree of nuptial coloration, size and agonistic behaviour among male pupfish, *Cyprinodon pecosensis*, breeding in lakes in New Mexico (USA). Both territory and male characteristics explained statistically significant proportions of the observed variation in male reproductive success. Topographic complexity accounted for two thirds of the variance in mating success and coloration for an additional 13%, whereas male size and agonistic behaviour accounted for only 1.8% and thus were less important indicators of female choice. Kodric-Brown (1983) argued that egg mortality is high because male pupfish provide little or no parental care. Females should therefore choose habitats most favourable for the development and hatching of eggs. The story in pupfish is more complex, however, because the observed high positive correlation between the bright blue nuptial coloration of breeding males and reproductive success, independent of territory quality, suggests that females also assess male quality. Bright males are better at excluding conspecific egg predators from oviposition sites (Kodric-Brown 1978). Consequently, by choosing a bright male, females minimise their egg loss to cannibals.

Although informative, correlational studies typically suffer from the problem of confounding variables. In the pupfish study, for example, it was impossible to determine whether the enhancement of male reproductive success was caused by nuptial colour per se or associated courtship and aggressive behaviour, because the intensity of breeding coloration was positively correlated with aggressive behaviour. Similarly, it is difficult to determine whether females were choosing the breeding site per se or the male. Several qualities are usually associated with particular male/site combinations and the identification of the actual basis for choice is not trivial. For example, large and aggressive males may be associated with high-quality resources. Are females independently assessing resources, or do they simply assume that it is the larger, aggressive males who have secured the best sites? Identifying which of these two components of female choice occurs in any given case is crucial to an understanding of the operation of sexual selection. If females choose mating sites, selection will favour those males who can secure sites through male–male competition. However, a choice based on some intrinsic aspects of males may lead to extreme development of secondary sexual characteristics unrelated to combat.

Although difficult, it is possible in some species to examine whether female mate choice is based on characteristics of the male or the territory it defends. For instance, female choice in the bluehead wrasse is probably based on territory quality, as males provide no parental care but do display elaborate coloration and courtship behaviour (Warner 1987). Coloration and courtship could, however, result from the operation of mechanisms other than female choice. By removing males from their sites and allowing other males to replace them, Warner (1987) demonstrated that females were primarily selecting sites and argued that the presence of conspicuous males may simply indicate the site is safe for females. Similar conclusions were reached by Jones (1981) for *Pseudolabrus celidotus*, another species of wrasse.

The relative importance of male quality and site quality may differ among species. Downhower *et al.* (1983) found that female mottled sculpins, *Cottus bairdi*, choose mates on the basis of male characteristics rather than site quality. Unlike male wrasses, male sculpins guard the eggs and larger males are better guarders. Other studies, using a combination of experimental and observational approaches (e.g. river bullhead, *Cottus gobio*; Bisazza and Marconato 1988), indicate that females may choose mates using both site quality and male quality. Undoubtedly, the relative importance of site and male characteristics varies among species, and merits further examination.

Choice for parental abilities
In species with extensive paternal care (see Sargent, Chapter 11), females may choose males based on their ability to act as good parents. Females may use cues such as body size (e.g. Downhower and Brown 1980; Côté and Hunte 1989, 1993), male courtship behaviour (Knapp and Kovach 1991) or intensity of body coloration (e.g. Milinski and Bakker 1990) as reliable indicators of male health and vigour, because bigger males may be best able to defend nests filled with offspring and because less colourful males may be less likely to endure the rigours of parental care.

Rather than testing correlates of parental abilities, we can imagine studies that directly test the hypothesis that females are choosing males based on the ability to provide parental care. For example, one could manipulate the number of eggs that survive when a female spawns with a given male, and test whether such a female expresses a preference for males in treatments with high egg survival. Furthermore, it might be worthwhile examining female mate choice for male parental care in a species in which females 'parcel out' eggs rather than shed all their eggs at once. In the black hamlet, *Hypoplectrus nigricans*, and the chalk bass, *Serranus tortugarun*, both simultaneous hermaphrodites, individuals parcel out eggs and stop interacting with a partner if that partner 'cheats' by continually adopting the role of male, and hence produces the cheaper of the two types of gametes (Fischer 1988). Whether females in other species 'test' for male parental ability in a similar fashion remains unknown.

Good genes and good health
If females are choosing males based on 'good genes' (Kodric-Brown and Brown 1984), the criterion for such choice must be some phenotypic expression of these genes, rather than the genes themselves. The question of whether animals choose their mates on the basis of the quality of their genotypes is perhaps the most controversial issue in the sexual selection literature (see Kodric-Brown (1990) for more on this debate in the context of fish mate choice and Kirkpatrick and Ryan (1991) and Andersson (1994) for general reviews). One point of contention in this debate is how animals might detect variations in the genetic quality of their potential mates and whether sufficient additive genetic variance exists for the trait in question.

Zahavi's (1975) handicap principle provides one possible mechanism by which females might be able to recognise males with 'good genes'. Zahavi's basic idea is that males which survive to mate, even though 'handicapped' by extravagant secondary sexual traits, are good quality males. Although Zahavi's original model has been criticised by population geneticists (Kirkpatrick 1986), it has received recent theoretical support (Pomiankowski 1987; Grafen 1990a, b). Hamilton and Zuk (1982) have argued that a special type of handicap may be the driving force behind the evolution of exaggerated traits. They suggest that coevolutionary arms races between parasites and their hosts might maintain additive genetic variance in fitness components, because populations are rarely at equilibrium. Assuming that parasites have debilitating effects on their hosts and that parasite resistance is heritable, the Hamilton–Zuk model predicts that those species with the most sexual dimorphism should harbour the most parasites, while within species the brightest males should harbour the fewest parasites.

Tests of the Hamilton–Zuk model using fishes have yielded mixed results. Ward (1988) examined parasite load and brightness in 24 species of British freshwater fish and found a positive correlation between the two factors. In contrast, Chandler and Cabana (1991) scored a much larger number of North American freshwater fish for parasites and sexual dichromatism, and found no association between the two variables.

Intraspecific tests have provided better support for the Hamilton–Zuk hypothesis. Kennedy *et al.* (1987) found that female guppies, *Poecilia reticulata*, preferred males with fewer parasites, and that this preference was associated with a higher courtship display rate by less parasitised males. McMinn (1990) concluded that the sigmoid display rate is the major cue used by female guppies to distinguish between parasitised and non-parasitised males. In the best controlled experiment examining parasites and mate choice in guppies to date, Houde and Torio (1992) found that the orange colour spots of males infected with parasites were paler than the colours of full siblings, which were not parasitised, and that female were likely using these colour differences in preferring the non-parasitised individuals. These three studies indicate that parasites play a role in female mate choice in the guppy, but the traits by which females distinguish parasitised from non-parasitised males may differ across populations.

The results of two studies testing the intraspecific prediction of the Hamilton and Zuk model with threespine sticklebacks are more ambiguous. Milinski and Bakker (1990) found that females from a freshwater population preferred the brighter (more intense red) of two males in a standard laboratory choice test. When males were experimentally infected with a ciliate parasite, they lost some of their colour and were rejected by the females. Milinski and Bakker (1990) argued that the intensity of the red coloration is a revealing handicap, because colour intensity correlated positively with male physical condition. If bright males have fewer parasites because they are more resistant, and if resistance is heritable, females choosing bright males will produce sons which will inherit their father's resistance alleles. Milinski and Bakker (1990), however, did not report whether natural populations of sticklebacks are ever exposed to levels of the parasite they used in their experiments. In contrast, FitzGerald *et al.* (1993) found no association between parasite load, degree of nuptial colour, health and mating success for an anadromous population of threespine sticklebacks (see Frischknecht (1993) for further information on the relationship between red coloration and male courtship vigour in sticklebacks). The apparent discrepancy between the results of Milinski and Bakker (1990) and FitzGerald *et al.* (1993) may be due to differences in the virulence of the parasites infecting their respective fish populations (see Folstad *et al.* (1994) for more on the relationship between parasite virulence and fish resistance).

One recent finding linking parasites and sexual ornamentation, which was not envisaged by Hamilton and Zuk (1982), is that of Wedekind (1992). He found that female roach, *Rutilus rutilus*, used two different characteristics of a male's breeding ornamentation to discriminate between males infested with nematode worms and those infected with *Diplozoon* sp. The more heavily a fish was infected by *Diplozoon*, the fewer breeding tubercles it had, and the more nematodes a fish harboured, the fewer breeding tubercles were on its flank. This provides an opportunity for females to mate with males that are resistant to a parasite which they themselves are susceptible to. That is, females that are more resistant to *Diplozoon*, but susceptible to nematodes may base their mate choice on a male's resistance to nematodes, while the opposite may be true for females that are more resistant to nematodes, but susceptible to *Diplozoon*.

Good genes models need not, however, be based on the Hamilton and Zuk (1982) parasite hypothesis. One prediction of good genes models is that females should respond to male secondary sexual characteristics that are phenotypically plastic, that is, traits whose expression depends on the environment. The phenotypic expression of dietary carotenoids in male fish provides a means for testing this hypothesis. In contrast to the difficulty in disentangling whether coloration evolved via male–male competition or female choice in sticklebacks, it seems clear in guppies that coloration evolved via female choice (as male–male competition is minimal; but see Kodric-Brown (1992)). One hypothesis is that a female's choice is based, at least in part, on male carotenoid pigments, because such pigments are honest indicators (Kodric-Brown and Brown 1984) of male foraging ability; that is, males that have bright carotenoid pigmentation are the best foragers. Assuming that foraging ability is a heritable trait, selection may favour females that choose such males. Using a split-brood design, Kodric-Brown (1989) examined female mate choice in guppies when males were fed either a diet high in astaxanthin and canthaxin or a diet lacking these carotenoids. Results indicated that males fed the carotenoid-enhanced diet had red and orange spots that averaged 2.5 times brighter, spent significantly more time near the female in visual trials, and had a higher success than their siblings raised on the carotenoid-free diet (Kodric-Brown 1989). See also Nicoletto (1991, 1992) for more on condition-dependence and female mate choice in the guppy.

Perhaps the strongest support yet for the good genes hypothesis in fishes comes from Reynolds and Gross' (1992) study on female mate choice for male body size in the guppy. In a three-generation experiment, they found that

(1) females preferred larger males,

(2) a significant father–son heritability existed for body size,

(3) larger fathers produced sons and daughters with higher growth rates, and

(4) daughters with higher growth rates had higher reproductive output.

Thus, female mate choice for larger body size may in fact select for good genes in males. Reynolds and Gross (1992) noted, however, that, while their results are consistent with good genes models, other factors such as sensory bias (*sensu* Ryan 1990) may have been involved in the evolutionary origin of this preference.

Female choice and sensory biases

A somewhat different perspective on the initial emergence of female choice is put forth by Ryan (1990) in his 'sensory exploitation' hypothesis. Ryan and Keddy-Hector (1992) compared this hypothesis to other models of the evolution of female choice, and noted that sensory exploitation 'suggests that females prefer male traits that elicit the greatest amount of stimulation from the sensory system'. In other words, selection favours male traits that exploit sensory biases in females. These biases may have evolved in response to selection in other contexts. Ryan and Keddy-Hector (1992, p. S24) noted that 'The sensory exploitation hypothesis differs from the good genes hypothesis with regard to the explanation of the evolution of traits: in the sensory exploitation theory, the sensory biases already in place, regardless of the reason, explain the initial evolution of the trait in the species under study, whereas the good genes theory offers the explanation that good gene selection in the species under study causes females to prefer certain traits'. In their review of the literature, Ryan and Keddy-Hector (1992) found 16 studies in which female choice in fishes was based on male visual traits. Females always preferred males with a greater than average degree of the trait, providing some support for the sensory exploitation hypotheses in fishes (Ryan and Keddy-Hector 1992). For example, Basolo (1990) found that in the *Xiphophorus* genus, which contains the green swordtail, *Xiphophorus helleri*, and the swordless platyfish, *X. maculatus*, a swordless tail is the primitive state (but see Meyer *et al.* (1994) for a molecular-based phylogeny that appears to cast serious doubt on this conclusion, and Basolo (1995) for a recent interpretation). Yet when normally swordless male platyfish have a sword sown on to their tail, females show a strong preference for them compared to their swordless conspecifics.

Female choice and social cues

Most models for the evolution of female mate choice assume that social cues play no role in how a female decides upon which male with whom to mate (see Laland

(1994) and Kirkpatrick and Dugatkin (1994) for exceptions). This assumption, however, remains virtually untested, and when tested has shown to be false. For example, female guppies copy each other's choice of mates, when such an opportunity arises (Dugatkin 1992). In this case, a female's choice is influenced by the preference expressed by other females. Experimental evidence suggests that females copy each other's choice of mates by remembering the identity of a male chosen by another female and preferring such males in subsequent mate-choice tests (Dugatkin 1992).

If female guppies copy each other's choice of mates, it seems reasonable to ask how important copying is in the entire decision process. That is, other work (e.g. Kodric-Brown 1985; Kennedy *et al.* 1987; Houde 1988a, b; Reynolds and Gross 1992; Nicoletto 1991, 1992; Endler and Houde 1995) has shown that female guppies prefer a suite of morphological and behavioural traits that may be population specific. How important is mate-choice copying compared with these other variables in the mating decision process? One way to address this question is to test whether copying can cause a female to reverse her choice of mates. Dugatkin and Godin (1992) examined this by allowing a female to choose between two males in a standard mate-choice test. In the control treatment, the female was allowed to choose between the same males 30 minutes later. The experimental treatment was identical except that just before the female was asked to choose between the males a second time, she observed another female choosing the male that she did not choose previously. Compared to the control treatment, females in the experimental treatment reversed their choice of males more often. Thus, in the guppy, mate copying may play a significant role in the overall decision-making process of a female (see Dugatkin (1996a, b) for more on this question). In addition, mate copying in the guppy seems to be age dependent, as younger females are more likely to copy the choice of older females than vice versa (Dugatkin and Godin 1993).

Why should females copy each other's choice of mates? One hypothesis is that if assessing or searching for males is very costly, mate-choice copying allows females to avoid such costs (Losey *et al.* 1986; Wade and Pruett-Jones 1990; Pruett-Jones 1992). Population genetic models of mate-choice copying, however, indicate that cost-reducing mechanisms are not required for the evolution of female mate-choice copying (Kirkpatrick and Dugatkin 1994), and a recent study by Briggs *et al.* (1996) did not reveal an increased tendency to mate-choice copy in female guppies when predation risk was increased experimentally, which would have rendered direct mate choice potentially costly.

A second type of social information used in female mate choice is displayed in some nest-building species of fishes, in which the number of eggs already in a nest provides a female with a socially-mediated cue she incorporates into her mating decisions (Ridley and Rechten 1981; Constantz 1985; Marconato and Bisazza 1986; Unger and Sargent 1988; Jamieson and Colgan 1989; Goldschmidt *et al.* 1993). Females may prefer to spawn with males that already have eggs in their nests, because eggs may indicate that the male is of 'high' quality (e.g. able to defend a nest against predators). Another possibility is that females choose such nests to spawn into because of increased courtship rates of males guarding nests with many eggs, rather than because of the presence of the eggs per se (Jamieson and Colgan 1989). However, this hypothesis has been criticised by Goldschmidt and Bakker (1990) and Goldschmidt *et al.* (1993). We believe their debate could be resolved if the Jamieson and Colgan (1989) hypothesis is considered as a proximate explanation, rather than an ultimate (evolutionary) explanation, of the behaviour observed. An alternative evolutionary explanation to 'high quality' is the 'risk dilution' (Rohwer 1978) or 'egg-survival' (Jamieson 1995) hypothesis, which argues that females choose nests with eggs because their eggs are less likely to be eaten by egg predators or the guardian male.

Females may also use social cues obtained as juveniles when choosing between potential mates later in life. In the Midas cichlid, two colour morphs, 'normal' and 'gold', coexist within Nicaraguan populations (Barlow 1976) and there is assortative mating amongst colour morphs (McKaye and Barlow 1976). One potential factor involved in establishing assortative mating in this species is that adult females prefer the colour morph they experienced while schooling during the 'subadult' stage. However, whether subadult schools assort by colour in nature is unknown.

10.3.2 Circumventing female choice

'Sneaky males' (i.e. males that do not court females and attempt forced copulations) have been reported in guppies (e.g. Reynolds *et al.* 1993; Godin 1995), salmon (e.g. Gross 1985), and bluegill sunfish (e.g. Gross and Charnov 1980), among other species (see Gross (1996) and Berglund, Chapter 9). In populations where cuckoldry is a frequent event, it will be difficult for females to select males solely for their 'good genes', as they have little control over which male(s) will fertilise their eggs. Female wrasses (*Symphodus tinca* and *S. ocellatus*), however, do discriminate against nesting males with sneakers in the vicinity (van den Berghe *et al.* 1989). When sneakers were removed, the females increased their spawning rate. Because sneakers had no effect on the fertilisation rate of the females, on egg mortality, or on the quality of subsequent paternal care, van den Berghe *et al.* (1989) suggested that female wrasse might be choosing mates based on some measure of their genetic quality.

10.3.3 Male mate choice

The extent to which one sex alone exercises mate choice, or both sexes choose, is likely dependent on the extent to which comparable investment in parental care is made by both sexes. When both sexes make substantial investment, the reproductive success of each will depend on the quality of parental care performed by their partner. In guppies, where males provide only sperm, it is clear that females are the choosy sex (Kodric-Brown 1985). However, in threespine sticklebacks, both males and females invest heavily in reproduction, and males show mate choice by directing more courtship to the larger of two females (Rowland 1982; Sargent *et al.* 1986). This choice is likely adaptive because bigger females are usually more fecund. For a brief review of male mate choice in fishes, we direct the reader to Sargent *et al.* (1986).

10.3.4 Correlation between female preference and preferred male trait

If additive genetic variance for both female mate choice (Bakker and Pomiankowski 1995) and the male trait which females prefer (Houde 1992) exists within a population, then both Fisherian and good genes models of sexual selection predict a positive genetic correlation between these two traits (see Kirkpatrick and Ryan (1991) and Andersson (1994) for reviews). Although this linkage disequilibrium has been the centrepiece of many models of sexual selection, until very recently little data were available on whether such a correlation exists or not (Kirkpatrick and Ryan 1991; Andersson 1994). If such a correlation does in fact exist, then artificial (or natural) selection on the male trait should cause a corresponding shift in female preference. Work on sexual selection in fishes has led the way in providing information on this question (but also see Bakker and Pomiankowski (1995) for other taxa). Selection experiments on sticklebacks (Bakker 1993) and guppies (Breden and Hornaday 1994; Houde 1994) have provided limited evidence for a correlation between female preference and the preferred male trait. In both the stickleback and guppy experiments, male colour pattern was selected and corresponding changes in female preference were noted. Houde (1994) found a correlated change in two of four selected lines, while Breden and Hornaday (1994) found no correlation in any of their four lines. Using a different design, Bakker (1993) found a strong correlation between male redness and female choice for red (see Breden *et al.* (1994) and Pomiankowski and Sheridan (1994) for an exchange on the interpretation of these three studies). Fishes seem to be an ideal taxonomic group for addressing linkage disequilibrium because of the ease with which female preference can be measured, and the relatively short generation time characteristic of many species. Future work on this subject will no doubt prove illuminating.

10.4 Conclusions

A number of new perspectives and new techniques can be now be used to examine questions of sexual selection in fishes. Technologically, the use of genetic markers to determine paternity should provide the means to better understand variance in male reproductive success and the traits that females prefer in males (e.g. Rico *et al.* 1992). In addition, molecular genetics can be employed to examine the role that sexual selection may play in speciation (Meyer *et al.* 1994; Magurran *et al.*

1995). A cognitive ethological approach to sexual selection in fishes may also prove useful (Dugatkin and Wilson 1994). This relatively new discipline focuses on the evolution of cognitive abilities in animals (Bekoff and Jamieson 1990). Understanding the rules which animals use to choose their mates, as well as how those rules are generated (both in the proximate and ultimate sense), can only help us better understand both proximate and ultimate questions regarding sexual selection. Lastly, we hope that the new approaches being developed for conducting comparative studies of behaviour (Harvey and Pagel 1991) will be used by those studying sexual selection in fishes. These approaches will undoubtedly help us unravel the wide array of different mechanisms apparent in mate choice across fish taxa.

Sexual selection appears to be the 'hottest' topic in behavioural ecology today (Gross 1994). Notwithstanding the hundreds of papers a year being published, much work remains to be done before we have a thorough understanding of the process of sexual selection. The development of this field in the past 15 years has been a game of 'catch-up' between theory and empirical studies. In the early 1980s, the-oreticians developed a family of models of sexual selection (e.g. O'Donald 1980; Lande 1980; Kirkpatrick 1982; Andersson 1994), but little controlled experimental work existed to test between alternative models. In the mid- to late 1980s, theoretical work subsided somewhat and an explosion of empirical work appeared. Currently, empirical and theoretical work are both moving ahead at full steam. Interplays between theoreticians and empiricists interested in sexual selection are more common, and rather than having theory but no data, or data but no theoretical construct to work within, the situation today is moving toward the development of empirical tests of the predictions of alternative models. We view this trend as a major step in the right direction and note that some of the most promising work in this area is being done on fishes.

Acknowledgements

We thank Dana Dugatkin, Jean-Guy Godin, Jim Grant and John Reynolds for their comments on the manuscript.

References

Andersson, M. (1994). *Sexual selection*. Princeton University Press, Princeton.

Bakker, T.C.M. (1993). Positive genetic correlation between female preference and preferred male ornament in sticklebacks. *Nature, Lond.,*, **363**, 255–257.

Bakker, T.C.M. (1994). Evolution of aggressive behaviour in threespine stickleback. In *The evolutionary biology of the threespine stickleback* (eds. M.A. Bell and S.A. Foster), pp. 345–380. Oxford University Press, Oxford.

Bakker, T.C.M. and Mundwiler, B. (1994). Female mate choice and male red coloration in a natural three-spined stickleback (*Gasterosteus aculeatus*) population. *Behav. Ecol.*, **5**, 74–80.

Bakker, T.C.M. and Pomiankowski, A. (1995). The genetic basis of female mate preferences. *J. Evol. Biol.*, **8**, 129–171.

Barlow, G.W. (1976). The Midas cichlid in Nicaragua. In *Investigations of the ichthyofauna of Nicaraguan lakes* (ed. T.B.Thorson), pp. 332–358. University of Nebraska Press, Lincoln.

Barlow, G.W., Rogers, W. and Fraley, N. (1986). Do Midas cichlid win through prowess or daring? It depends. *Behav. Ecol. Sociobiol.*, **19**, 1–8.

Basolo, A.L. (1990). Female preference predates the evolution of the sword in swordfish. *Science*, **250**, 808–811.

Basolo, A.L. (1995). A further examination of a pre-existing bias favouring a sword in the genus *Xiphophorus*. *Anim. Behav.*, **50**, 365–375.

Bass, A.H. (1993). From brains to behaviour—hormonal cascades and alternative mating tactics in teleost fishes. *Rev. Fish Biol. Fish.*, **3**, 181–186.

Beacham, J. and Newman, J. (1987). Social experience and the formation of dominance hierarchies in pumpkinseed sunfish. *Anim. Behav.*, **35**, 1560–1563.

Beaugrand, J.P. and Beaugrand, M. (1991). Prior residency and the stability of dominance relationships in pairs of green swordtail fish *Xiphophorus helleri*. *Behav. Proc.*, **24**, 169–175.

Beaugrand, J.P. and Zayan, R. (1985). An experimental model of aggressive dominance in *Xiphophorus helleri*. *Behav. Proc.*, **10**, 1–52.

Bekoff, M. and Jamieson, D. (eds.) (1990). *Interpretation and explanation in the study of animal behavior*. Westview Press, Boulder.

Berglund, A. (1991). Egg competition in a sex-role reversed pipefish: subdominant females trade reproduction for growth. *Evolution*, **45**, 770–774.

Berglund, A., Rosenqvist, G. and Svensson I. (1989). Reproductive success of females limited by males in two pipefish species. *Am. Nat.*, **133**, 506–516.

Berglund, A., Magnhagen, C., Bisazza, A., Köning, B. and Huntingford, F.A. (1993). Female–female competition over reproduction. *Behav. Ecol.*, **4**, 184–186.

Bisazza, A. and Marconato, A. (1988). Female mate choice, male–male competition, and parental care in the river bullhead, *Cottus gobio*. *Anim. Behav.*, **36**, 1352–1360.

Bradbury, J.W. and Andersson, M.B. (eds.) (1987). *Sexual selection: testing the alternatives*. John Wiley and Sons, New York.

Breden, F., Gerhardt, H.C. and Butlin, R. K. (1994). Female choice and genetic correlations. *Trends Ecol. Evol.*, **9**, 343.

Breden, F. and Hornaday, K. (1994). Test of indirect models of selection in the Trinidad guppy. *Heredity*, **73**, 291–297.

Briggs, S.E., Godin, J.-G.J. and Dugatkin, L.A. (1996). Mate-choice copying under predation risk in the Trinidadian guppy (*Poecilia reticulata*). *Behav. Ecol.*, **7**, 151–157.

Chandler, M. and Cabana, G.S. (1991). Sexual dichromatism in North American freshwater fish: do parasites play a role? *Oikos*, **60**, 322–328.

Chellappa, S. and Huntingford, F.A. (1988). Depletion of energy reserves during reproductive aggression in male three-spined stickleback, *Gasterosteus aculeatus* L. *J. Fish Biol.*, **35**, 315–316.

Collias, N.E. (1990). Statistical evidence for aggressive response to red by male three-spined sticklebacks. *Anim. Behav.*, **39**, 401–403.

Constantz, G. (1985). Alloparental care in the tessellated darter, *Etheostoma olmstedi* (Pisces: Percidae). *Environ. Biol. Fish.*, **14**, 175–183.

Côté, I.M. and Hunte, W. (1989). Self-monitoring of reproductive success: nest switching in the redlip blenny (Pisces: Blenniidae). *Behav. Ecol. Sociobiol.*, **24**, 403–408.

Côté, I.M. and Hunte, W. (1993). Female redlip blennies prefer older males. *Anim.Behav.*, **46**, 203–205.

Cronin, H. (1991). *The ant and the peacock*. Cambridge University Press, Cambridge.

Darwin, C. (1871). *The descent of man and selection in relation to sex*. Appleton, London.

Dominey, W.J. (1984). Effects of sexual selection and life history on speciation: species flocks in African cichlids and Hawaiian *Drosophila*. In *Evolution of fish species flocks* (eds. A.A. Echelle and L. Kornfield), pp. 231–249. University of Maine Press, Orono.

Downhower, J.F. and Brown, L. (1980). Mate preference of female mottled sculpins, *Cottus bairdi. Anim. Behav.*, **28**, 728–734.

Downhower, J.F., Brown, L., Pederson, R. and Staples, G. (1983). Sexual selection and sexual dimorphism in mottled sculpins. *Evolution*, **37**, 96–103.

Dufresne, F., FitzGerald, G.J. and Lachance, S. (1990). Age and size-related differences in reproductive success and reproductive costs in threespine sticklebacks (*Gasterosteus aculeatus*). *Behav. Ecol.*, **1**, 140–147.

Dugatkin, L.A. (1992). Sexual selection and imitation: females copy the mate choice of others. *Am. Nat.*, **139**, 1384–1389.

Dugatkin, L.A. (1996a). Interface between culturally based and genetic preferences: female mate choice in *Poecilia reticulata. Proc. Natl. Acad. Sci. USA*, **93**, 2770–2773.

Dugatkin, L.A. (1996b). Copying and mate choice. In *Social learning in animals: the roots of culture* (eds. C.M. Heyes and B.G. Galef, Jr.), pp. 85–105. Academic Press, New York.

Dugatkin, L.A. and Biederman, L. (1991). Balancing asymmetries in resource holding power and resource value in the pumpkinseed sunfish. *Anim. Behav.*, **42**, 691–692.

Dugatkin, L.A. and Godin, J.-G.J. (1992). Reversal of female mate choice by copying in the guppy (*Poecilia reticulata*). *Proc. R. Soc. Lond. B.*, **249**, 179–184.

Dugatkin, L.A. and Godin, J.-G.J. (1993). Female mate copying in the guppy, *Poecilia reticulata*: age-dependent effects. *Behav. Ecol.*, **4**, 289–292.

Dugatkin, L.A. and Ohlsen, S. (1990). Contrasting asymmetries in value expectation and resource holding power: effects on attack behavior and dominance in the pumpkinseed sunfish (*Lepomis gibbosus*). *Anim. Behav.*, **39**, 802–804.

Dugatkin, L.A. and Wilson, D.S. (1994). Fish behavior, partner choice experiments and cognitive ethology. *Rev. Fish Biol. Fish.*, **3**, 368–372.

Endler, J.A. (1991). Variation in the appearance of guppy color patterns to guppies and their predators under different visual conditions. *Vision Res.*, **31**, 587–608.

Endler, J.A. (1995). Multiple-trait coevolution and environmental gradients in guppies. *Trends Ecol. Evol.*, **10**, 22–29.

Endler, J.A. and Houde, A.E. (1995). Geographic variation in female preferences for male traits in *Poecilia reticulata. Evolution*, **49**, 456–468.

Enquist, M. and Leimar, O. (1990). The evolution of fatal fighting. *Anim. Behav.*, **39**, 1–9.

Enquist, M., Leimar, O., Ljungberg, T. Mallner, Y. and Segerdahl, N. (1990). A test of the sequential assessment game: fighting in the cichlid fish *Nannacara anomala. Anim. Behav.*, **40**, 1–14.

Fischer, E.A. (1988). Simultaneous hermaphroditism, Tit-for Tat, and the evolutionary stability of social systems. *Ethol. Sociobiol.*, **9**, 119–136.

Fisher, R.A. (1958). *The genetical theory of natural selection*, 2nd edn. Dover, New York.

FitzGerald, G.J. (1993). Seeing red, turning red. *Rev. Fish Biol. Fish.*, **3**, 286–292.

FitzGerald, G.J. and Kedney, G. (1987). Aggression, fighting, and territoriality in sticklebacks: three different phenomena. *Biol. Behav.*, **12**, 186–195.

FitzGerald, G.J., Fournier, M. and Morrissette, J. (1993). Sexual selection in an anadromous population of threespine sticklebacks: no role for parasites. *Evol. Ecol.*, **8**, 348–356.

Folstad, I., Hope, A.M., Karter, A. and Skorping, A. (1994). Sexually selected color in male sticklebacks: a signal of both parasite exposure and parasite resistance? *Oikos*, **69**, 511–515.

Francis, R.C. (1983). Experiential effects of agonistic behavior in the paradise fish. *Z. Tierpsychol.*, **85**, 292–313.

Francis, R.C. (1992). Sexual lability in teleosts: developmental factors. *Q. Rev. Biol.*, **67**, 1–18.

Francis, R.C. and Barlow, G.W. (1993). Social control of primary sex differentiation on the Midas cichlid. *Proc. Natl. Acad. Sci. USA*, **90**, 10673–10675.

Frey, D.F. and Miller, R.J. (1972). The establishment of dominance relationships in the blue gourami. *Behaviour*, **42**, 8–62.

Frischknecht, M. (1993). The breeding colouration of male three-spined sticklebacks (*Gasterosteus aculeatus*) as an indicator of energy investment in vigour. *Evol. Ecol.*, **7**, 439–450.

Ghiselin, M.T. (1969). The evolution of hermaphroditism among animals. *Q. Rev. Biol.*, **44**, 189–201.

Godin, J.-G.J. (1995). Predation risk and alternative mating tactics in male Trinidadian guppies (*Poecilia reticulata*). *Oecologia*, **103**, 224–229.

Goldschmidt, T. and Bakker, T.C.M. (1990). Determinants of reproductive success of male sticklebacks in the field and in the laboratory. *Neth. J. Zool.*, **40**, 664–687.

Goldschmidt, T., Bakker, T.C.M. and Feuth-de Bruijn, E. (1993). Selective copying in mate choice of female sticklebacks. *Anim. Behav.*, **45**, 541–547.

Grafen, A. (1990a). Biological signals as handicaps. *J. Theor. Biol.*, **144**, 517–546.

Grafen, A. (1990b). Sexual selection unhandicapped by the Fisher process. *J. Theor. Biol.*, **144**, 473–476.

Gross, M.R. (1984). Sunfish, salmon and the evolution of alternative reproductive strategies and tactics in fish. In *Fish reproduction: strategies and tactics* (eds. G.W. Potts and R.J. Wootton), pp. 55–76. Academic Press, London.

Gross, M.R. (1985). Disruptive selection for alternative life histories in salmon. *Nature, Lond.*, **313**, 47–48.

Gross, M.R. (1991). The evolution of alternative strategies: frequency-dependent sexual selection in male bluegill sunfish. *Phil. Trans. R. Soc. Lond. B*, **332**, 59–66.

Gross, M.R. (1994). The evolution of behavioural ecology. *Trends Ecol. Evol.*, **9**, 358–360.

Gross, M.R. (1996). Alternative reproductive strategies and tactics: diversity within the sexes. *Trends Ecol. Evol.*, **11**, 92–98.

Gross, M.R. and Charnov, E.L. (1980). Alternative male life histories in bluegill sunfish. *Proc. Natl. Acad. Sci. USA*, **77**, 6937–6940.

Haller, J. and Wittenberger, C. (1988). Biochemical energetics of hierarchy formation in *Betta splendens*. *Physiol. Behav.*, **43**, 447–450.

Halliday, T.R. (1983). The study of mate choice. In *Mate choice* (ed. P. Bateson), pp. 3–33. Cambridge University Press, Cambridge.

Hamilton, W.D. and Zuk, M. (1982). Heritable true fitness and bright birds: a role for parasites? *Science*, **218**, 384–387.

Harvey, P.H. and Pagel, M.D. (1991). *The comparative method in evolutionary biology*. Oxford University Press, Oxford.

Houde, A.E. (1988a). The effects of female choice and male–male competition on the mating success of male guppies. *Anim. Behav.*, **36**, 888–897.

Houde, A.E. (1988b). Genetic difference in female choice in two guppy populations. *Anim. Behav.*, **36**, 510–516.

Houde, A.E. (1992). Sex-linked heritability of a sexually selected character in a natural population of guppies, *Poecilia reticulata* (Pisces: Poeciliidae). *Heredity*, **69**, 229–235.

Houde, A.E. (1993). Evolution by sexual selection: what can population comparisons tell us? *Am. Nat.*, **141**, 796–803.

Houde, A.E. (1994). Effect of artificial selection on male colour patterns on mating preference of female guppies. *Proc. R. Soc. Lond. B*, **256**, 125–130.

Houde, A.E. and Endler, J.A. (1990). Correlated evolution of female mating preference and male color pattern in the guppy, *Poecilia reticulata*. *Science*, **248**, 1405–1408.

Houde, A.E. and Torio, A.J. (1992). Effect of parasitic infection on male color pattern and female choice in guppies. *Behav. Ecol.*, **3**, 346–351.

Huntingford, F.A. and Turner, A.K. (1987). *Animal conflict*. Chapman and Hall, London.

Jamieson, I. (1995). Do female fish prefer to spawn in nests with eggs for reasons of mate choice copying or egg survival? *Am. Nat.*, **145**, 824–832.

Jamieson, I.G. and Colgan, P.W. (1989). Eggs in the nests of males and their effect on mate choice in the three-spined stickleback. *Anim. Behav.*, **38**, 859–865.

Järvi, T. (1991). The effects of male dominance, secondary sexual characteristics and female mate choice on the mating success of the Atlantic salmon *Salmo salar*. *Ethology*, **84**, 123–132.

Jones, G.P. (1981). Spawning-site choice by female *Pseudolabrus celidotus* (Pisces: Labridae). *Behav. Ecol. Sociobiol.*, **8**, 129–142.

Kennedy, C.E.J., Endler, J.A., Poynton, S.L. and McMinn, H. (1987). Parasite load predicts mate choice in guppies. *Behav. Ecol. Sociobiol.*, **21**, 291–295.

Kirkpatrick, M. (1982). Sexual selection and the evolution of female choice. *Evolution*, **36**, 1–12.

Kirkpatrick, M. (1986). The handicap mechanism of sexual selection does not work. *Am. Nat.*, **127**, 222–240.

Kirkpatrick, M. (1987). Sexual selection by female choice in polygynous animals. *A. Rev. Ecol. Syst.*, **18**, 43–70.

Kirkpatrick, M. and Dugatkin, L.A. (1994). Sexual selection and the evolutionary effects of mate copying. *Behav. Ecol. Sociobiol.*, **34**, 441–449.

Kirkpatrick, M. and Ryan, M.J. (1991). The evolution of mating preferences and the paradox of the lek. *Nature, Lond.*, **350**, 33–38.

Knapp, R.A. and Kovach, J.T. (1991). Courtship as an honest indicator of male parental quality in the bicolor damselfish, *Stegastes partitus*. *Behav. Ecol.*, **2**, 295–300.

Kodric-Brown, A. (1978). Establishment and defence of breeding territories in a pupfish. *Anim. Behav.*, **26**, 818–834.

Kodric-Brown, A. (1983). Determinants of male reproductive success in pupfish. *Anim. Behav.*, **31**, 128–137.

Kodric-Brown, A. (1985). Female preference and sexual selection for male colouration in the guppy (*Poecilia reticulata*). *Behav. Ecol. Sociobiol.*, **17**, 199–205.

Kodric-Brown, A. (1989). Dietary carotenoids and male reproductive success: an environmental component to female choice. *Behav. Ecol. Sociobiol.*, **25**, 309–323.

Kodric-Brown, A. (1990). Mechanisms of sexual selection: insights from fishes. *Ann. Zool. Fennici*, **27**, 87–100.

Kodric-Brown, A. (1992). Male dominance can enhance mating success in guppies. *Anim. Behav.*, **44**, 165–167.

Kodric-Brown, A. and Brown, J.H. (1984). Truth in advertising: the kinds of traits favored by sexual selection. *Am. Nat.*, **124**, 309–323.

Laland, K.N. (1994) Sexual selection with a culturally transmitted mating preference. *Theor. Popul. Biol.*, **45**, 1–15

Lande, R. (1980). Sexual dimorphism, selection selection and adaptation in polygenic characters. *Evolution*, **34**, 292–307.

Losey, G.S. Jr., Stanton, F. G., Telecky, T.M., Tyler, W. A. III and the Zoology 691 Graduate Seminar Class. (1986). Copying others, an evolutionarily stable strategy for mate choice: a model. *Am. Nat.*, **128**, 653–664.

McKaye, K.R. and Barlow, G.W. (1976). Competition between color morphs of the Midas cichlid, *Cichlasoma citrinellum*, in Lake Jiloà, Nicaragua. In *Investigations of the ichthyofauna of Nicaraguan lakes* (ed. T.B.Thorson), pp. 465–475. University of Nebraska Press, Lincoln.

McMinn, H. (1990). Effects of the nematode parasite *Camallanus cotti* on sexual and non-sexual behaviors in the guppy (*Poecilia reticulata*). *Amer. Zool.*, **30**, 245–249.

Magurran, A.E., Seghers, B.H., Shaw, P.W. and Carvalho, G.R. (1995). The behavioral diversity and evolution of guppy, *Poecilia reticulata*, populations in Trinidad. *Adv. Study Behav.*, **24**, 155–202.

Marconato, A. and Bisazza, A. (1986). Males whose nests contain eggs are preferred by female *Cottus gobio*. *Anim. Behav.*, **34**, 1580–1582.

Meyer, A., Morrissey, J. and Schartl, M. (1994). Recurrent origin of a sexually selected trait in *Xiphophorus* fishes inferred from a molecular phylogeny. *Nature, Lond.*, **368**, 539–542.

Milinski, M. and Bakker, T.C.M. (1990). Female sticklebacks use male coloration in mate choice and hence avoid parasitized males. *Nature, Lond.*, **344**, 330–333.

Muckenstrum, B. (1967). L'épinoche et les leurres, complexité de la réaction. *C. R. Acad. Sci. Paris, Series D*, **264**, 745–748.

Muckenstrum, B. (1969). La signification de la livrée nuptiale de l'épinoche. *Rev. Comp. Anat.*, **3**, 39–64.

Nicoletto, P.F. (1991).The relationship between male ornamentation and swimming speed in the guppy, *Poecilia reticulata*. *Behav. Ecol. Sociobiol.*, **38**, 365–370.

Nicoletto, P.F. (1992). Female sexual response to condition-dependent ornaments in the guppy, *Poecilia reticulata*. *Anim. Behav.*, **46**, 441–450.

Ochi, H. (1989). Mating behavior and sex-change of the anemone fish *Amphiprion clarkii*, in the temperate waters of southern Japan. *Environ. Biol. Fish.*, **26**, 257–275.

O'Donald, P. (1980). *Genetic models of sexual selection*. Cambridge University Press, Cambridge.

Parker, G.A. (1974). Assessment strategy and the evolution of fighting behaviour. *J. Theor. Biol.*, **47**, 223–243.

Parker, G.A. and Rubenstein, D.I. (1981). Role assessment, reserve strategy and the acquisition of information in asymmetric animal contests. *Anim. Behav.*, **29**, 221–240.

Peeke, H.V.S., Wyers, E.J. and Herz, M.J. (1969). Waning of the aggressive response in male models in the three-spined stickleback (*Gasterosteus aculeatus*). *Anim. Behav.*, **17**, 224–228.

Pomiankowski, A. (1987). Sexual selection: the handicap principle does work—sometimes. *Proc. R. Soc. Lond. B*, **231**, 123–145.

Pomiankowski, A. and Sheridan, L. (1994). Reply from A. Pomiankowski and L. Sheridan. *Trends Ecol. Evol.*, **9**, 343.

Price, T., Schluter, D. and Heckman, N.E. (1993). Sexual selection when the female benefits directly. *Biol. J. Linn. Soc.*, **48**, 187–211.

Pruett-Jones, S.G. (1992). Independent versus non-independent mate choice: do females copy each other? *Am. Nat.*, **140**, 1000–1009.

Reimchen, T.E. (1989). Loss of nuptial color in threespine sticklebacks (*Gasterosteus aculeatus*). *Evolution*, **43**, 450–460.

Reynolds, J.D. and Côté, I.M. (1995). Direct selection on mate choice: female redlip blennies pay more for better mates. *Behav. Ecol.*, **6**, 175–181.

Reynolds, J.D. and Gross, M.R. (1992). Female mate preference enhances offspring growth and reproduction in a fish, *Poecilia reticulata*. *Proc. R. Soc. Lond. B.*, **250**, 57–62.

Reynolds, J.D., Gross, M.R. and Coombs, M.J. (1993). Environmental conditions and male morphology determine alternative mating behaviour in Trinidadian guppies. *Anim. Behav.*, **45**, 145–152.

Rico, C., Kunhlein, U. and FitzGerald, G.J. (1992). Male reproductive tactics in the threespine stickleback—an evaluation by DNA fingerprinting. *Mol. Ecol.*, **1**, 79–87.

Ridley, M. and Rechten, C. (1981). Female sticklebacks prefer to spawn with males whose nests contain eggs. *Behaviour*, **16**, 152–161.

Rohwer, S. (1978). Parental cannibalism of offspring and egg raiding as a courtship strategy. *Am. Nat.*, **112**, 429–440.

Ross, R.M. (1981). Experimental evidence for stimulation and inhibition of sex change in the Hawaiian reef fish *Thalassoma duperrey*. *Proc. 4th Intl. Coral Reef Symp.*, **2**, 575–580.

Rowland, W.J. (1982). The effects of male nuptial coloration on stickleback aggression: a re-examination. *Behaviour*, **80**, 118–126.

Rowland, W.J. (1989). Mate choice and the supernormality effect in female sticklebacks (*Gasterosteus aculeatus*). *Behav. Ecol. Sociobiol.*, **24**, 433–438.

Rowland, W.J. and Sevenster, P. (1985). Sign stimuli in three-spined sticklebacks (*Gasterosteus aculeatus*): a re-examination of some classic experiments. *Behaviour*, **93**, 241–257.

Ryan, M.J. (1990). Sexual selection, sensory systems and sensory exploitation. *Oxf. Surv. Evol. Biol.*, **7**, 157–195.

Ryan, M.J. and Keddy-Hector, A. (1992). Directional patterns of female mate choice and the role of sensory biases. *Am. Nat.*, **139**, S4–S35.

Ryan, M.J. and Wagner, W.E. (1987). Asymmetries in mating preference between species: female swordtails prefer heterospecifics. *Science*, **236**, 595–597.

Sargent, R.C. (1989). Allopaternal care in the fathead minnow, *Pimephales promelas*: stepfathers discriminate against their adopted eggs. *Behav. Ecol. Sociobiol.*, **25**, 379–385.

Sargent, R.C., Gross, M.R. and van den Berghe, E.P. (1986). Male mate choice in fishes. *Anim. Behav.*, **34**, 545–550.

Seghers, B.H. (1973). *An analysis of geographic variation in the antipredator adaptations of the guppy, Poecilia reticulata*. Ph.D. thesis, University of British Columbia, Vancouver.

Shapiro, D.Y. (1979). Social behavior, group structure and the control of sex reversal in hermaphroditic fish. *Adv. Study Behav.*, **10**, 43–102.

Tinbergen, N. (1951). *The study of instinct*. Clarendon Press, Oxford.

Turner, G.F. (1993). Teleost mating behaviour. In *Behaviour of teleost fishes*, 2nd edn. (ed. T.J. Pitcher), pp. 307–326. Chapman and Hall, London.

Unger, L.M. and Sargent, R.C. (1988). Alloparental care in the fathead minnow, *Pimephales promelas*: females prefer males with eggs. *Behav. Ecol. Sociobiol.*, **23**, 27–32.

van den Assem, J. (1967). Territory in the three-spined stickleback, *Gasterosteus aculeatus* L. an experimental study in intra-specific competition. *Behaviour (Suppl.)*, **16**, 1–164.

van den Berghe, E.P. and Gross, M.R. (1989). Natural selection resulting from female breeding competition in a Pacific salmon. *Evolution*, **43**, 125–140.

van den Berghe, E.P., Wernerus, F. and Warner, R.R. (1989). Female choice and the mating cost of peripheral males. *Anim. Behav.*, **38**, 875–886.

Vincent, A., Ahnesjö, I. and Berglund, A. (1994). Operational sex ratios and behavioural sex differences in a pipefish population. *Behav. Ecol. Sociobiol.*, **34**, 435–442.

Wade, M.J. and Pruett-Jones, S. G. (1990). Female copying increases the variance in male mating success. *Proc. Natl. Acad. Sci. USA*, **87**, 5749–5753.

Ward, P.I. (1988). Sexual dichromatism and parasitism in British and Irish freshwater fish. *Anim. Behav.*, **36**, 1210–1215.

Warner, R.R. (1987). Female choice of sites versus mates in a coral reef fish, *Thalassoma bifasciatum*. *Anim. Behav.*, **35**, 1470–1477.

Warner, R.R. (1988). Sex change in fishes: hypotheses, evidence and objections. *Environ. Biol. Fish.*, **22**, 81–90.

Warner, R.R., Robertson, R.R. and Leigh, E.G. (1975). Sex change and sexual selection. *Science*, **190**, 633–638.

Wedekind, C. (1992). Detailed information about parasites revealed by sexual ornamentation. *Proc. R. Soc. Lond. B.*, **247**, 169–174.

Wootton, R.J. (1976). *The biology of the sticklebacks*. Academic Press, London.

Zahavi, A. (1975). Mate selection—a selection for a handicap. *J. Theor. Biol.*, **53**, 205–214.

11 *Parental care*

Robert Craig Sargent

11.1 Introduction

Parental care is generally defined as an association between parent and offspring
that enhances offspring development and survival (e.g. Sargent and Gross 1986).
Although fishes exhibit a considerable diversity of kinds of parental care, by far the
most common form of parental care is brooding the eggs on a substratum (Blumer
1982; Sargent and Gross 1986, 1993). Substrate brooding may be subdivided into
care performed directly at the eggs, such as aeration (e.g. fanning), and care that
can be provided at a distance from the eggs, such as protecting the eggs from
predators. After substrate brooding, the three next most common forms of parental care in descending frequency of occurrence are mouthbrooding, external egg
carrying and carrying the eggs in a brood pouch; however, these three forms of
parental care combined account for only a small fraction of the taxonomic families
that exhibit substrate brooding (Blumer 1982; Sargent and Gross 1986, 1993).

Fishes also exhibit a considerable diversity of parental care states. Approximately 78% of all fish families have no species that exhibit parental care. Of the remaining families in which at least one species exhibits parental care, approximately
50% exhibit care by the male parent only, or *paternal care*; 32% exhibit care by both
parents, or *biparental care*; and 18% exhibit female only care, or *maternal care*.
For reviews on the phylogenetic distribution of parental care in fishes, see Breder
and Rosen (1966), Blumer (1982) and Gross and Sargent (1985). For reviews of hypotheses to explain this distribution, especially the preponderance of paternal care,
see also Williams (1975), Maynard Smith (1977), Perrone and Zaret (1979), Baylis
(1981), Sargent and Gross (1986, 1993) and Clutton-Brock and Vincent (1991).

In this chapter I will focus on understanding the behavioural ecology of parental
care in fishes within the context of life-history evolution. First, I review parental
investment theory in terms of optimal resource allocation that is constrained by
trade-offs among fitness components. Secondly I review empirical evidence for
these trade-offs. Thirdly, I examine specific predictions from parental investment
theory. Fourthly, I review what is known about the energetics of parental investment. Finally, I review what I consider to be specific limitations in our understanding of parental care, and make suggestions for future research.

11.2 Parental investment theory: trade-offs among fitness components

Lack (1947, 1954) was one of the first ecologists to appreciate the importance of trade-offs among components of fitness in determining optimal life-history strategies. Lack proposed that what limited clutch size in birds was the amount of food in the environment and the rate at which parents could feed their young. He predicted that the observed clutch size will be the one that maximises the number of young that fledge, and empirical evidence supported his prediction (Lack 1954). Thus, Lack identified an important trade-off in life-history evolution, namely, the trade-off between offspring number and offspring survival, or between *quantity* and *quality*. Wisenden and Keenleyside (1995) provide evidence for Lack's trade-off in the convict cichlid, *Cichlasoma nigrofasciatum*. This trade-off may be very important in understanding the behavioural ecology of parental care.

Williams (1966) extended Lack's hypothesis to include a second trade-off, namely, a trade-off between *present* and *future* reproduction. Williams reasoned that an animal who maximises total reproductive success within a breeding season may pay a cost in terms of reduced reproduction in future breeding seasons. Thus, a parent who maximises the number of offspring that fledge (or otherwise reach independence) may actually have less than maximum lifetime reproductive success.

Williams introduced the concept of residual reproductive value, which in a stationary population (i.e. a population that is not growing and in a stable age distribution, Pianka 1994) is the total expectation of future reproduction of an individual beyond the current breeding episode. We assume natural selection maximises, at any point in an animal's lifetime, total expected reproduction from now until the end of the its life (i.e. remaining lifetime reproductive success), which is equivalent to age-specific reproductive value for a stationary population . The reproductive value of an individual of age x, in a stationary population, can be written as

$$v_x = \frac{1}{l_x} \sum_{t=x}^{\infty} l_t m_t , \qquad (11.1)$$

where t is the index of age classes, x is the current age class, v_x is reproductive value at age x, l_x is survival from birth to age x, l_t is survival from birth to age t, and m_t is expected fecundity at age t. Following Williams (1966) and Pianka (1994), one can move reproduction at age x (i.e. present reproduction) outside the summation sign to give age-specific reproductive value in terms of present and future reproduction. Thus, age-specific reproductive value becomes

$$v_x = m_x + \sum_{t=x+1}^{\infty} \frac{l_t}{l_x} m_t , \qquad (11.2)$$

where the ratio, l_t / l_x gives the survival between ages x and t. Thus, the total expectation of reproduction of an individual of age x is its fecundity at age x, or its present reproduction, plus its future reproduction, which is its survival from age *x* to age *x* + 1 times its fecundity at age *x* + 1, plus its survival from age *x* to *x* + 2 times its fecundity at age *x* + 2, and so on.

The residual reproductive value of an animal of age *x*, assuming a stationary population, is simply the probability that the animal survives from age *x* to *x* + 1, times its reproductive value at age *x* + 1 (Pianka 1994). This can be expressed as

$$v_x = m_x + \frac{l_{x+1}}{l_x} v_{x+1} ,$$ (11.3)

where the ratio, l_{x+1}/l_x represents survival from age *x* to *x* + 1. The first term on the right-hand side of equation 11.3 represents present reproduction at age *x*, or m_x. The second term on the right-hand side of this equation represents residual reproductive value, which is the total expectation of reproduction from age *x* + 1 to the end of the animal's life, or v_{x+1}, devalued by the probability of survival from age *x* to *x* + 1, or l_{x+1}/l_x. Thus, equation 11.3 is a simple recursion equation that states that the total expectation of lifetime reproductive success of an animal of age *x* can be expressed in terms of two episodes in time, age *x* and age *x* + 1.

Whereas Williams (1966) emphasised a trade-off between present and future reproduction (between ages *x* and *x* + 1), Lack (1947, 1954) emphasised a trade-off within present reproduction (i.e. within age *x;* in fact, within m_x). In fact, both trade-offs may operate simultaneously, particularly in fishes with paternal care (Sargent 1990; Sargent *et al.*, in press). Equation 11.3 can easily be modified to incorporate both Lack's and Williams' trade-offs, by simply indexing time over shorter durations than age classes. For example, by assuming time to be in units of days within a brood cycle, the term $v_x + 1$, includes offspring that hatch during the present brood cycle, future brood cycles in the present breeding season, and future breeding seasons (e.g. Sargent 1990; Sargent *et al.*, in press). Thus, we can explore the trade-offs between parental care and several components of future reproduction simultaneously.

Now let us turn to the behavioural ecology of parental care in fishes. Consider a parental fish caring for eggs. If the parent is a female, then she has two options: caring for her eggs and feeding. Caring for the eggs is an investment in the fitness of her brood, whereas feeding is an investment in her own survival, which may benefit the brood indirectly. If her eggs and her food are spatially separated, or if she cannot otherwise feed and care for her eggs simultaneously, then she faces conflicting demands. The fitness benefits obtained through parental care must be traded off against the benefits obtained through feeding. If the female leaves her eggs to forage, then the eggs may experience higher rates of predation and slower rates of development. If the female foregoes feeding for parental care, then she may reduce her expected future reproductive success by depleting her energy reserves, reducing her growth rate, or both. Both costs of parental care may lead to reduced survival, reduced future fecundity, and increased time to next breeding.

If the parent is a male, then he has the additional option of trying to obtain more matings, and thus add more clutches to his current brood. If the nest, food and potential mates are all separated spatially, then the male trades off each option against the others. For example, if the male leaves his eggs to court additional females, then his eggs are unprotected and he may forego feeding, at least while he is courting. If the male leaves his nest to feed, then again his eggs are unprotected, and he may forego courtship. If the male remains with his eggs and provides parental care, then he may forego both feeding and courtship. Thus, brood care is an investment in the current offspring that may result in both reduced brood size during the present brood cycle and lower expectation of reproduction in future brood cycles (Sargent 1990; Sargent *et al.*, in press).

A recent empirical example of the conflicting demands faced by a parent is Rangeley and Godin's (1992) study of the biparental convict cichlid. Rangeley and Godin examined parental foraging both in the presence and absence of a brood predator. They observed that, in the presence of the predator, parents spent considerable time in brood defence, less time foraging and consumed less food than parents in the absence of the predator. In addition, they observed conflicting demands among types of parental care. In the presence of the predator, parents spent less time in direct care of the brood (e.g. fanning) than parents in the absence of the predator.

11.3 Benefits of parental care: evidence

Parental care presumably improves offspring survival, and may also increase the rate of development (and/or growth) of the offspring (Sibly and Calow 1986; Clutton-Brock 1991). An obvious manipulation would be to examine the effects of parental removal on offspring survival and development. For example in the laboratory, if parental male threespine sticklebacks, *Gasterosteus aculeatus*, are removed from their nests, the eggs become mouldy and die (van Iersel 1953). The same holds true for fathead minnows, *Pimephales promelas* (personal observation). A particularly interesting species in this context is the peacock wrasse, *Symphodus tinca*, in which paternal care is facultative (Warner *et al.* 1995). Some males in this species build nests and defend the eggs spawned therein, whereas other males follow females and spawn with them away from any nests (van den Berghe 1990). The incidence of paternal care changes over the course of the breeding season; early in the season nesting males are much less frequent than in the middle of the season. Interestingly, egg survival outside nests is similar to egg survival inside nests early in the season, but is much lower than egg survival inside nests at the middle of the season (Warner *et al.* 1995). Thus, as the relative benefit of parental care changes during the breeding season, so does the frequency of males that provide parental care.

In the biparental convict cichlid, Keenleyside and Mackereth (1992) investigated the effects of removing the male parent at different stages of offspring development. They examined three treatments: natural male desertion of the female, artificial male removal, and controls with both parents present. The general finding was that offspring survival was lower with the male absent than with the male present, but that offspring survival was higher the later in the nesting cycle the male was removed. Thus, it appears that two parents may be better than one.

It would also be interesting to investigate how offspring development and survival covary with continuous variation in the amount of care that a parent provides. Although it may be impossible to directly manipulate a parent's level of care, the level of parental care may vary naturally, it may covary naturally with some ecological variable (e.g. predation risk), or it may covary in response to experimental manipulation of such an ecological variable. For example, van Iersel (1953) examined natural variation in the amount of time parental male threespine sticklebacks fanned their eggs, and observed that as the amount of fanning increased, time to hatching decreased. A similar result was obtained with threespine sticklebacks by Sargent (1985). In his study, parental males were assigned to one of two treatments: competitive or solitary. Competitive males were separated by transparent partitions, whereas solitary males were separated by opaque partitions. Competitive males spent more time 'fighting' across the transparent partition and less time fanning their eggs than solitary males. The eggs of competitive males took longer to hatch than the eggs of solitary males. These two studies provide evidence that parental care, in the form of fanning, increases the rate of offspring development.

What about offspring survival? In a study with fathead minnows, Sargent (1988) examined the paternal behaviour and egg survival of solitary males in the laboratory. Males were assigned to one of two treatments: with or without crayfish, a potential predator on both the eggs and the parental male. The number of eggs that a male defended varied according to the size and number of females with whom he had spawned. Paternal care, in the form of egg rubbing, and egg survival both increased with egg number. In addition, in the absence of crayfish predators, males exhibited both higher egg rubbing rates and higher egg survival than males in the presence of crayfish (Sargent 1988). Thus, this study suggests a causal relationship between the level of parental care and egg survival.

11.4 Costs of parental care: evidence

The basic costs of parental care are assumed to be:

(1) survival of the parent,

(2) time to next breeding, and

(3) number of future offspring or fecundity.

These costs are generally thought to reduce reproduction during future brood cycles; however, male parents may also realise mating costs within the present brood cycle. The costs of parental care may be mediated through reduced energy

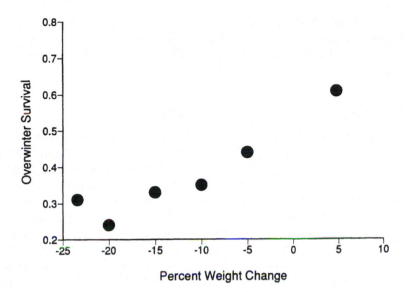

Fig. 11.1 Overwinter survival increases as percentage weight gain during the breeding season increases in the rock bass, *Ambloplites rupestris*. Each data point represents mean survival for six size classes of males; total sample size is 191. From Sabat (1994).

reserves and reduced growth rate. I will consider the evidence for each of these costs below.

11.4.1 Costs to parent survival

It has been well documented that parental fishes expend energy reserves during the breeding season (e.g. Unger 1983; Sargent 1985; Chellappa *et al.* 1989; Fitz-Gerald *et al.* 1989). I will focus on two studies that correlate these energy expenditures with reduced parental survival.

Chellappa *et al.* (1989) measured lipid, glycogen and protein reserves of male threespine sticklebacks throughout the year. They observed that all three measures peaked at the beginning of the breeding season. At the end of the breeding season, males that bred had lower energy reserves and higher mortality rates than males that failed to breed. The non-breeders, however, tended to lose their energy reserves and die during their second winter (Chellappa *et al.* 1989).

Sabat (1994) conducted a field study on the rock bass, *Ambloplites rupestris*, a species with paternal care. He augmented the broods of nesting males by assigning them to one of three treatments: small, medium, or large increases in brood size. Among other things, Sabat measured paternal weight loss over the brood cycle and whether or not a male was recaptured on the spawning grounds in the following breeding season. He observed that as brood size increased, the amount of weight that males lost tended to increase. In addition, the more weight that a male lost during the breeding season, the less likely he was to be recaptured in the following breeding season (Fig. 11.1). Thus, it appears that parental energy expenditures increased with increasing brood size, which resulted in higher overwinter mortality.

11.4.2 Costs to time elapsed until next breeding

For females, increases in the level of parental care provided may result in an increase in her interspawning interval. For example, Mrowka (1987) observed in his study of the maternal mouthbrooder, *Pseudocrenilabrus multicolor,* that females who cannibalised their eggs had shorter interspawning intervals than females who brooded their eggs. Similarly, Smith and Wootton (1994), in their study of the maternal mouthbrooding cichlid, *Haplochromis 'argens',* observed that females who brooded their young had longer interspawning intervals than females whose eggs were removed from their mouths after spawning. Part of the mechanism underlying the relationship between increased parental care and increased interspawning interval in female fishes may be an energy reallocation from the ovaries. Lavery and Keenleyside (1990a, b) reported a tendency for female ovary weight and brood size to be negatively correlated in the biparental convict cichlid. In addition, they observed that as brood size increased, female nest defence increased, but their foraging decreased. Thus, it appears that females with small broods invest less energy into parental care and more energy into their ovaries (and thus available for the next brood cycle) than females with large broods.

The costs of care to male convict cichlids are more immediate. The testes weights of males are not correlated with brood size (Lavery and Keenleyside 1990a, b), which suggests that males are capable of spawning more frequently than females. Indeed, many males in this species desert their mates and offspring, and exhibit bigamy (Keenleyside 1985; Keenleyside *et al.* 1990; Wisenden 1994). Thus, it appears that in species with biparental care, males trade off care against opportunities for immediate remating.

Parental care may also impose costs to male mating frequency in species with paternal care. For example, in the threespine stickleback, male fanning increases and the tendency to court females decreases as the number of female clutches in his nest increases and as the age of those clutches increases (van Iersel 1953). After some combination of clutch age and number, male sticklebacks cease courting females and switch into a 'parental phase' These males typically do not court females again until the eggs have hatched and the young are free swimming (van Iersel 1953). The more clutches a male has, the earlier he enters the parental phase (van Iersel 1953). Although nesting male sticklebacks typically average several female clutches per brood cycle (e.g. Pressley 1981), males will enter the parental phase with as few as one clutch in their nests (van Iersel 1953). Thus, males appear to trade off parental care against courtship.

11.4.3 Costs to future fecundity

To my knowledge, Balshine-Earn (1995) has conducted the only study that directly correlates parental care with a cost to future fecundity. She studied the Galilee St. Peter's fish, *Sarotherodon galilaeus*, a mouthbrooder with biparental care. Mouthbrooding severely reduces the parents' ability to feed. Balshine-Earn examined two treatments: parents whose eggs were experimentally removed from their mouths one day after spawning, and control parents who fully brooded their clutches. She found in both sexes that care for the first brood resulted in greater weight losses and longer interspawning intervals. In addtion, females that cared for their first brood had smaller second clutches, whereas females that did not care for their first brood had larger second clutches. Thus, females payed a future fecundity cost for parental care. In a similar study with the maternal mouthbrooding cichlid *Haplochromis 'argens'*, Smith and Wootton (1994) found that females who had their eggs removed from their mouths shortly after spawning had higher overall growth rates between their first and second clutches than did control females. However, they were unable to show a significant reduction in future fecundity in the control group.

11.5 Parental investment theory: predictions

Several parental investment models have been developed to increase our understanding of the behavioural ecology of parental care in fishes. Rather than review the 'nuts and bolts' of how these models work, I will describe their logic, their essential parameters and their predictions. In general, there are two conceptual frameworks for modelling in behavioural ecology: optimisation theory and game theory. Both approaches are similar in assuming that natural selection favours behaviours that maximise a fitness-related currency, such as lifetime reproductive success. Under both approaches the modeller must define the optimisation criterion (the currency being maximised), the strategy set (the range of behavioural options available to the animal), the state space (variables that depend on past behavioural decisions that help to determine the present optimal policy), and the constraints (limitations on the state space). Both approaches assume trade-offs among fitness components in terms of benefits and costs of parental care. The key difference between the two approaches is that under optimisation theory one assumes that the optimal behaviour of an individual animal is independent of the behaviour of other animals, whereas under game theory one assumes that an individual's best choice depends on the behaviour of others. Optimisation models have been used mainly to study the dynamics of parental care in species in which only one parent cares for the offspring (e.g. Pressley 1976; Sargent and Gross 1985, 1986, 1993; Sargent 1990, 1992), whereas game theory has been used to study the dynamics of biparental care (e.g. Coleman 1993; Wisenden 1994). Both approaches have yielded valuable insights into the behavioural ecology of parental care in fishes.

Two predictions from optimisation models of parental investment have been reviewed elsewhere (Sargent and Gross 1986, 1993) and are only summarised briefly here. First, parental care should increase as brood size increases, even if larger broods do not require more care (e.g. Sargent and Gross 1985, 1986, 1993; Sargent 1990). Second, early in the brood cycle, parental care should increase as brood age increases; however, as the larvae become free swimming and approach independence, parental care should decrease as offspring age increases (e.g. Sargent and Gross 1986, 1993; Sargent 1990). Next, I review the logic of these predictions.

Many optimisation models of parental care (e.g. Pressley 1976; Carlisle 1982; Sargent 1985; Sargent and Gross 1985, 1986, 1993) focus on parental resource allocation among fitness components. If these fitness components are additive and if there exists an optimal resource allocation where the parent invests energy into each fitness component, then at the optimal allocation of resources the parent enjoys *equal rates of return* on investment into each fitness component. The greater the rate of return on investment into one fitness component relative to the others, the greater is the optimal investment in that component. For example, as the rate of return on investment in parental care increases and the rates of return on investment into other fitness components remain constant, the optimal level of parental investment increases and the optimal investment into each other fitness component decreases (e.g. Sargent 1985; Sargent and Gross 1986, 1993). Thus, parental investment decisions should be based on the value of the brood at stake, relative to the value of the parent's expected future reproduction. These models predict that parental care will increase with increasing brood size, even if larger broods do not require more care. In addition, these models predict that parental care will increase with increasing brood age up to the point where parental care no longer appreciably improves offspring survival. As offspring mature and their survival without care increases, parental care should decrease with increasing offspring age (e.g. Sargent and Gross 1986, 1993; Sargent 1990). There is ample empirical evidence supporting both predictions (brood size effect: e.g. van Iersel 1953; Pressley 1981; Coleman *et al.* 1985; Sargent 1988; Ridgway 1989; Lavery and Keenleyside 1990b; brood age effect: e.g. van Iersel 1953; van den Assem 1967; Colgan and Gross 1977; Ridgway 1988; Lavery and Colgan 1991).

Next, I illustrate how parental investment theory provides insights into the following phenomena: brood cycling, filial cannibalism, and sexual conflict over parental care.

11.5.1 Brood cycling in species with paternal care

van Iersel (1953) suggested that the reduction and eventual cessation of courtship during the parental phase may reflect the increased demands for care (i.e. aeration) by increased egg number and age. I agree that this is part of the explanation for this pattern; however, another trade-off may also be a factor. The parental phase may also result from an ecological trade-off between courtship and offspring survival (Sargent 1990; Sargent *et al.*, in press). The nests of males are particularly vulnerable to predation by conspecifics during courtship in both threespine sticklebacks (Sargent and Gebler 1980) and fathead minnows (personal observation). If the eggs in a male's nest are under higher predation risk during courtship than during parental care, and if that risk is constant in time, then the benefits of additional courtship will decrease as clutch number and clutch age both increase. Early in a brood cycle, the benefits of male courtship may outweigh the costs. Expected brood gain from courting a prospective female equals the probability that a courted female spawns in a male's nest times one, that is, her single clutch to be gained. Expected brood loss during courtship equals the probability of nest predation times the number of clutches present in the nest. If the male has few or no clutches in his nest, then the expected gains from courtship probably outweigh the expected costs. However, as the number of clutches in a male's nest increases, then, after some number of clutches already in the nest, the expected losses through brood predation will equal and then exceed the expected gain by courting an additional female. At that point, the male should cease courting females and switch into the parental phase. Similarly, as the eggs in a male's nest grow older, their value to the male increases; loss of older eggs will be more costly to the male than loss of new eggs. Thus, as egg age increases and egg predation risk during courtship stays roughly constant, then one would expect courtship to decrease and parental care to increase. These are precisely the patterns that van Iersel (1953) observed for male threepsine sticklebacks.

A prediction of this line of reasoning is that, as the risk of nest predation during courtship increases, among populations or among experimental treatments, one should observe on average fewer clutches per brood cycle (Sargent 1990; Sargent *et al.*, in press). To my knowledge, this prediction has not been tested.

11.5.2 Filial cannibalism

Filial cannibalism, or parental consumption of offspring as food, is a particularly intriguing phenomenon in fishes with parental care. The basic ideas were first proposed in a verbal model by Rohwer (1978) and later elaborated by Sargent (1992) with a dynamic-programming model. Filial cannibalism may actually enhance a parent's lifetime reproductive success if feeding is traded off against offspring survival, and if offspring can be used as an alternative source of food. There are two kinds of filial cannibalism, *total* and *partial*. In total filial cannibalism, the parent consumes all of its clutch, whereas the parent consumes only a fraction of its clutch in partial filial cannibalism.

In terms of parental investment theory, total filial cannibalism can only represent an investment in the parent's future brood cycles. If the parent's residual reproductive value is greater than the value of its current clutch, then total filial cannibalism may be favoured. It may pay the parent to eat the current clutch and invest the resulting energy gain into future reproduction. One would expect the incidence of total filial cannibalism to increase as clutch size decreases (Rohwer 1978; Sargent 1992), if all else is equal.

Rohwer (1978) proposed that partial filial cannibalism may represent investment in the present brood cycle, by enabling the parent to survive through a long brood cycle and thus ensuring the survival of the remainder of its current clutch. This counter-intuitive hypothesis is realistic if clutch mortality due to filial cannibalism is less than the mortality that the clutch would experience if the parent leaves it undefended while feeding (Sargent 1992). If the parent's food requirements are independent of clutch size, then *per capita* clutch mortality due to filial cannibalism would decrease as clutch size increases. If the *per capita* mortality of an undefended clutch is constant and independent of clutch size, then as clutch size increases the cost of partial filial cannibalism decreases. Thus, one would expect the incidence of partial filial cannibalism to increase as clutch size increases (Sargent 1992), if all else is equal. Lastly, one would expect both forms of filial cannibalism to decrease as the age of the clutch increases and as parental energy reserve increases (Rohwer 1978; Sargent 1992).

Several studies corroborate the predictions on how filial cannibalism is expected to covary with clutch size and clutch age. In three paternal egg guarding species (*Oxylebius pictus*, DeMartini 1987; *Stegastes rectifraenum*, Petersen and Marchetti 1989; *Pimephales promelas*, Sargent 1988, 1989) and in one maternal mouthbrooder (*Pseudocrenilabrus multicolor*, Mrowka 1987), parents totally cannibalise small clutches, but partially cannibalise large clutches (Fig. 11.2), and younger clutches are more likely to be cannibalised than older clutches. Thus, the correlations between filial cannibalism and clutch size and clutch age support the predictions of parental investment theory.

Although filial cannibalism has been widely documented in fishes with parental care, there have been relatively few studies that support the prediction that the incidence of filial cannibalism is inversely related to parental energy reserve. Moreover, those studies that do exist are equivocal. For example, Whoriskey and FitzGerald (1985) reported that female and parental male threespine sticklebacks have similar food items in their guts. Thus, it appears that parental males have the same access to food as free-ranging females in salt marsh tide pools. In addition, Belles-Isles and FitzGerald (1991) examined parental male sticklebacks on three different rations in the laboratory, and observed that males in all three treatments consumed some of their own eggs and that the number of eggs consumed was independent of ration. These results suggest that filial cannibalism in this species may not be based on energy per se; however, there is the possibility that eggs contain nutrients not normally available to parental males.

Hoelzer (1992) conducted a field study on the Cortez damselfish, *Stegastes rectifraenum*, a species with paternal care. Parental males of this species typically eat algae from mats on their territories and eggs from their nests. Hoelzer observed that males whose diets were supplemented with eggs (from the nests of

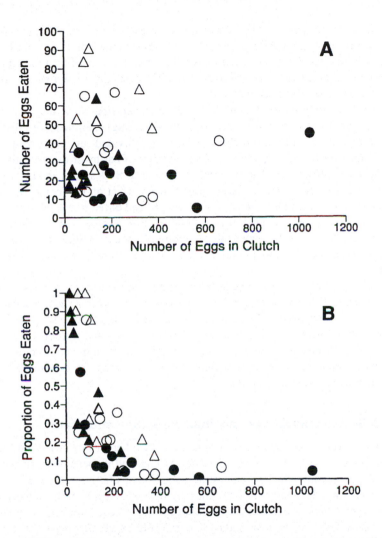

Fig. 11.2 Filial cannibalism versus clutch size in the fathead minnow, *Pimephales promelas*. Although the absolute number of eggs eaten was independent of clutch size (panel A), the proportion of the clutch eaten decreased as clutch size increased (panel B). Solid triangles represent males without crayfish ($n = 11$) and open triangles represent males with crayfish present ($n = 11$) in Sargent (1988); solid circles represent first clutches and open circles represent second clutches for 11 males who sired both clutches in Sargent (1989).

other males) were less likely to eat their own eggs than were control males. In another experiment, Hoelzer (1992) observed that the more eggs a male ate from his clutch, the more time he spent off his territory courting females; however, he was unable to demonstrate that filial cannibalism correlated with subsequent mating success. His study does suggest an energetic basis for filial cannibalism in the Cortez damselfish; however, the energetics may be complex. Although males that were fed eggs did exhibit lower rates of filial cannibalism, filial cannibalism still persisted. Moreover, males fed eggs did not exhibit a significant reduction in feeding on the algae on their territories. Thus, it appears that these fish are limited by more than one food-related resource.

In a recent study on the fantail darter, *Etheostoma flabellare*, a species with paternal care, Lindström and Sargent (unpublished data) examined the effects of feeding regime on filial cannibalism. Males were divided into two groups: experimental males that were fed an *ad libitum* diet of earthworms or small arthropods, and control males that were unfed. Similar to Belles-Isles and FitzGerald (1991), they observed that feeding regime had no effect on the incidence of filial cannibalism. Fed males tended to gain weight and unfed males tended to lose weight; however, males tended to gain more (or lose less) weight the more eggs that they ate. These results suggest that, although filial cannibalism seems to have energetic benefits, male fantail darters are limited by more than one food-related resource.

Clearly more work needs to be done to understand the energetic bases of filial cannibalism in fishes with parental care. It does seem apparent that the food that parents gain from filial cannibalism and that gained from other items in their diet may not be comparable on single energy axis. Future empirical studies will need to document the nutritional benefits of eating eggs versus eating other food items, and future parental investment models will need to account for the possibility of more than one food or energy axis.

11.5.3 Sexual conflict over parental care: biparental care

Conflict between the sexes exists whenever a member of one sex would have higher fitness if a member of the opposite sex changed its behaviour. In species with biparental care, either parent would have higher lifetime reproductive success if its partner did a larger share of the parental care. However, the partner who did the larger share of parental care would pay higher costs in other components of fitness. In fishes with biparental care, one would expect that males, with their immediate abilities to remate, would be more likely than females to shirk their parental duties and even desert their mates and offspring. The logic of mate desertion has been modelled extensively (e.g. Maynard Smith 1977; Lazarus 1990; Wisenden 1994). The general predictions of this theory are that males are more likely to desert their mates when the opportunities for immediate remating are high, when offspring survival rates are high, when offspring are advanced in age, and when offspring numbers are low. One would expect females to pay a cost for male desertion in terms of reduced offspring survival, reduced future reproduction, or both.

Keenleyside and colleagues have studied this phenomenon extensively, and have corroborated each prediction using cichlids with biparental care. Male convict

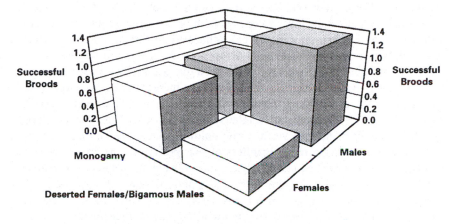

Fig. 11.3   Sexual conflict over male parental care in the convict cichlid, *Cichlasoma nigrofasciatum*. Bigamous males enjoy higher reproductive success than monogamous males ($n = 72$ spawnings; from Keenleyside (1985)); however, monogamous females enjoy higher reproductive success than deserted females ($n = 28$; from Keenleyside and Mackereth (1992)). The two data sets were standardised to give equal female and male reproductive success under monogamy.

cichlids who desert their mates benefit through remating and enjoy higher total reproductive success than males who do not desert (Keenleyside 1985; Fig. 11.3). The likelihood of male desertion increases as the operational sex ratio increases (i.e. the ratio of reproductive females to males) in the rainbow cichlid, *Herotilapia multispinosa* (Keenleyside 1983). Wisenden (1994) observed in a field study of the convict cichlid that male desertion was more likely at sites with higher offspring survival (lower predation pressure), when the offspring were close to independence, and when brood sizes were very small. Female convict cichlids that have been deserted have lower offspring survival than females that have not been deserted (Keenleyside and Mackereth 1992); thus, females do appear to pay a cost for male desertion (Fig. 11.3). These studies all document sexual conflict over the level of male parental care in species with biparental care. It would now be interesting to explore whether or not females exhibit any counter-strategies to induce higher levels of parental care from their mates or to prevent desertion.

Females usually exhibit greater parental investment than males in species with biparental care (e.g. Townshend and Wootton 1984; Rangeley and Godin 1992). However, this pattern may be due in part to the fact that adult males are generally larger than adult females, and that there is positive assortative mating by size (e.g. Keenleyside 1985; McKaye 1986; Wisenden 1995; personal observation). Preference for large mates would appear to be reproductively advantageous for both sexes. Males would gain larger and more fecund partners, who also may be better parents (e.g. Coleman 1993), whereas females may gain partners who are better brood defenders (Keenleyside *et al.* 1985).

Coleman (1993), however, observed in convict cichlids that the amount of brood defence that a male invests depends both on his size and that of his mate. Holding clutch size constant and well below the average female clutch size, Coleman (1993) observed that male brood defence, in the absence of the female, decreased as male body size increased. He argued that a constant clutch size should be of decreasing value to males as male body size increases, because large males can expect to mate with large females with large clutches. In a second experiment, Coleman (1993) compared the relative brood defence of small males paired with large females, and vice versa, while holding clutch size constant. He observed that the smaller parent, regardless of sex, performed the greater amount of brood defence. Coleman (1993) argued that this pattern results from a given clutch size being of greater value to a small parent than a large one. Males in both treatments, however, performed less defence than males without females. Thus, it appears that body size is an important variable in understanding the dynamics of parental investment, and that parents adjust their brood defence according to their own body size relative to the size (or level of brood defence) of their mate. It would be interesting to examine brood defence when both parents are small and when they are large; unfortunately, this was not done. Nevertheless, these kinds of parental investment dynamics have profound implications for sexual conflict; future research on this system should reveal interesting findings.

11.5.4 Sexual conflict over parental care: paternal care

In fishes with paternal care, sexual conflict over parental care appears to be primarily mediated through female choice and sexual selection. Kirkpatrick (1985) demonstrated analytically that female choice for male parental care can lead to higher levels of paternal care than would have been favoured by natural selection alone acting on the male. Thus, there exists a potential evolutionary mechanism for sexual conflict over parental care in species with paternal care.

An example of this kind of sexual conflict may be illustrated by the phenomenon of egg adoption in the fathead minnow (Unger 1983; Unger and Sargent 1988). In this species, males spawn with many females, and alone provide the parental care. Newly reproductive males will often contest males with eggs in their nests, evict these nesting males if possible, and then care for the adopted eggs (Unger 1983; Unger and Sargent 1988). This seemingly maladaptive behaviour exists even though there are many unused nest sites in the environment. Unger and Sargent (1988) hypothesised that male adoption of unrelated eggs may make evolutionary sense if females prefer to mate with males that already have eggs in their nests. They tested this hypothesis, and observed a striking preference by females for males with eggs (Unger and Sargent 1988). Why would females exhibit such a preference? Parental investment theory predicts that parental care will increase with increasing clutch size (Sargent and Gross 1985, 1986, 1993), and that egg mortality due to filial cannibalism will decrease with increasing clutch size (Rohwer 1978; Sargent 1992). If overall egg survival were to increase with clutch size, then we would have an adaptive explanation for female preference for males with eggs (Jamieson 1995). Sargent (1988) observed that paternal care and egg survival both increased with clutch size, which may explain the evolution of a female preference for males with eggs. Thus, it appears that natural selection acting on males led to the correlation between clutch size and egg survival, and that female preference for males with eggs led to the evolution of male adoption of unrelated eggs as a mating strategy.

An alternative explanation for the evolution of adoption is dilution of predation through safety in numbers (McKaye and McKaye 1977). Although this may be operating in the fathead minnow system, we have been unable to find any evidence for it. On the other hand, there is considerable evidence supporting this hypothesis in cichlids with biparental care (e.g. McKaye and McKaye 1977; Wisenden and Keenleyside 1992, 1994). A distinction between convict cichlids and fathead minnows is that care ends at hatching in minnows, but extends well after the offspring are free swimming in cichlids. Adoption in cichlids occurs after hatching, and the effects of dilution of predation have only been demonstrated at the free-swimming stage (Wisenden and Keenleyside 1992, 1994). Parental cichlids adopt fry that are smaller than their own, and reject fry that are larger than their own; within a brood larger fry have higher survival, which may be the selective force behind adopting small rather than large fry (Wisenden and Keenleyside 1992, 1994). In fathead minnows, however, adoption precedes breeding; thus, adopted eggs are older than eggs sired by the adoptive fathers. Egg mortality is highest when eggs are less than one day old; eggs sired by an adoptive father therefore may be at higher risk of predation than the older adopted eggs (Sargent 1988, 1989). Thus, it appears that female preference for males with eggs was the more important selective force favouring the evolution of egg adoption in the fathead minnow.

Filial cannibalism is another phenomenon fraught with sexual conflict in species with paternal care. One would expect a female to prefer a male that is unlikely to eat her eggs. Given that males are more likely to eat eggs that are young and clutches that are small, one might expect females to have evolved strategies to counter male filial cannibalism. Examples of such female counter-strategies may include: female preference for males with young eggs (e.g. damselfish, *Stegastes dorsopunicans*, Petersen 1990; fathead minnow, Unger and Sargent 1988; Sargent, unpublished data); female preference for males in good condition (i.e. with high energy reserves), who may be less likely to eat eggs (e.g. bicolor damselfish, *Stegastes partitus*, Knapp and Kovach 1991; fathead minnow, Unger 1983, Sargent, unpublished data); and female use of test eggs to test male parental quality (e.g. Mediterranean blenny, *Aidablennius sphynx*, Kraak and van den Berghe 1992). As females evolve counter-strategies to minimise the effects of filial cannibalism by males, one may expect males to evolve counter-strategies themselves. Further research into this seemingly endless 'battle of the sexes' should prove most interesting.

11.6 Energetics of parental investment

Several studies have demonstrated energetic costs of reproduction in species with parental care (e.g. Unger 1983; Sargent 1985; Chellappa *et al.* 1989; FitzGerald *et al.* 1989; Sabat 1994), which all indicate that parental energy reserves likely influence the dynamics of parental investment. For example, in the bicolor damselfish, a species with paternal care (Knapp and Kovach 1991; Knapp and Warner 1991), male courtship display rate is the best predictor of male mating success and egg survival. Interestingly, male courtship display rate is significantly correlated with fat reserve (Knapp and Kovach 1991). Although supplemental feeding experiments produced higher rates of courtship and increased levels of fat reserve over unfed males, fed males did not have any higher egg survival than unfed males (Knapp 1992). Thus, it appears that fat reserve may play a role in the dynamics of parental care and that males may honestly signal their fat reserves through their courtship display rates. However, the relationship between paternal fat reserve and egg survival is a complicated one.

Empirical studies of the effects of supplemental feeding on filial cannibalism (Belles-Isles and FitzGerald 1991; Hoelzer 1992; Lindström and Sargent, unpublished data) indicate that the energetics of feeding and filial cannibalism also may be complex. Only Hoelzer's (1992) study revealed any reduction in filial cannibalism due to supplemental feeding, and the supplemental food in that study was conspecific eggs. Collectively, these studies suggest that filial cannibalism may supply parents with nutrients that are not available in other food sources.

Supplemental feeding experiments generally do produce interesting changes in the dynamics of parental care, however. For example, Townshend and Wootton (1984, 1985) examined the effects of ration on reproduction and parental investment in the biparental convict cichlid. In their first experiment (Townshend and Wootton 1984), pairs of fish were allowed to spawn and then assigned to one of three different levels of ration (a mixture of commercial flake food, freeze-dried *Tubifex* worms and frozen fish): low, medium, or high. Immediately after spawning, the eggs were removed (rather than brooded naturally), and each pair was allowed to achieve as many spawnings as possible for a period of four months. As ration increased, the number of spawnings per female increased, fecundity per spawning increased, and female growth rate increased. Toward the second half of the experiment, female interspawning interval decreased as ration increased. Finally, at the end of the experiment, female total weight, liver weight, ovary weight and the proportion of oocytes that were vitellogenic all increased as ration increased. For males, only growth rate was correlated with ration. Thus, it appears that ration had positive effects on both present and future reproduction in female convict cichlids. Under the high ration, the partial correlation between fecundity and growth, holding body size constant, was significantly positive, which argues against a trade-off between present and future reproduction for this treatment. However, as ration decreased, this partial correlation became increasingly negative, and approached statistical significance ($P < 0.09$) for the low ration treatment (Townshend and Wootton 1984, p. 98). These results suggest that a trade-off between fecundity and growth is most pronounced under conditions of low food, and that this trade-off is weakened or disappears under conditions of high food. How ration physiologically affects this trade-off is not known.

What about parental behaviour? In their second experiment, Townshend and Wootton (1985) investigated the effects of ration on parental investment. They observed that fanning increased and time spent foraging decreased as ration increased. Thus, supplemental feeding appears to free parents from searching for food, thereby allowing them to spend more time in direct parental care. Unfortunately, the effects of ration on offspring survival were not measured.

In a fascinating study, Ridgway and Shuter (1994) investigated the effects of food supplementation on reproduction in the paternal egg-guarding smallmouth bass, *Micropterus dolomieui*. In this field study, the nests of free-ranging individual males were identified and assigned to one of two treatments: fed and unfed. Fed males were provided with up to ten dace (*Phoxinus eos*) as food every two or three days. Ridgway and Shuter tallied the number of males whose broods reached the juvenile stage, the duration of parental care after the young were free swimming, and the overwinter survival of the parental males. In the first year of the study, fed males had higher overwinter survival than unfed males; the two treatments did not differ in offspring survival or the duration of parental care (Fig. 11.4). In the second year of the study, quite a different pattern emerged. Fed males had lower overwinter survival, higher offspring survival, and longer duration of parental care than unfed males (Fig. 11.4). In both years, food supplementation significantly increased one fitness component; however, which fitness component was affected differed significantly between years. In the first year of the study, males apparently channelled their supplemental food into future reproduction through higher overwinter survival. In the second year of the study, fed males apparently channelled their supplemental food into present reproduction (longer parental care duration and higher offspring survival), at a cost to future reproduction (lower overwinter survival). Why were the responses so drastically different between years? The answer is not clear; however, Ridgway and Shuter (1994) hypothesised that it may be due to the differences between years in the density of breeding adults and in conditions that favour the growth of larvae and juveniles that were observed by Ridgway and Friesen (1992). In the first year of the study, nesting male densities were lower, which may indicate a more favourable environment for adult growth and survival (Ridgway and Friesen 1992; Ridgway and Shuter 1994). In the second year of the study, larval and juvenile growth rates were higher, which may indicate a more favourable environment for

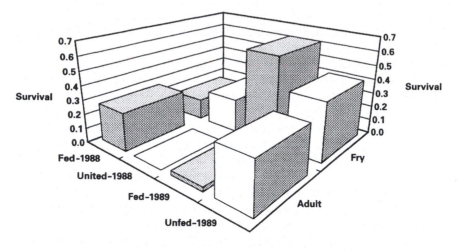

Fig. 11.4   Food supplementation versus adult and fry survival in the smallmouth bass, *Micropterus dolomieui*. In 1988, fed males ($n = 34$) had non-significantly higher fry survival, but significantly lower overwinter adult survival than unfed males ($n = 34$). However, in 1989, fed males ($n = 37$) had significantly lower fry survival and significantly higher overwinter adult survival than unfed males ($n = 37$). Thus, benefits and costs of food supplementation were different between the two years of the study. From Ridgway and Shuter (1994).

parental investment into offspring survival (Friesen 1990; Ridgway and Shuter 1994).

Although these studies constitute an excellent beginning to our understanding of the effects of food on reproductive fitness and parental care, we still do not fully understand exactly how feeding translates into lifetime reproductive success. Whereas increased ration increased both present and future reproduction in the study of Townshend and Wootton (1984), food supplementation increased either present or future reproduction, with potential costs to the other fitness component, in the study of Ridgway and Shuter (1994). Furthermore, we do not know to what extent ration may affect the trade-off between quantity and quality of offspring, for example in fishes with paternal care. Given the vast literature on optimal foraging (e.g. Stephens and Krebs 1986, Hart, Chapter 5; Hughes, Chapter 6), and on parental investment (e.g. Clutton-Brock 1991), it may well be time to integrate these two fields.

11.7 Conclusions

Life-history theory is a valuable tool for understanding the dynamics of parental care. The basic assumption of life-history theory is that natural selection favours behaviours that maximise lifetime reproductive success, subject to two trade-offs: (1) present versus future reproduction and (2), within present reproduction, quantity versus quality of offspring. Considerable empirical evidence exists that supports the assumption of trade-offs among fitness components. Moreover, life-history theory has been able to explain diverse parental-investment phenomena, such as brood cycling, filial cannibalism and adoption. There is, however, ample room for further research, and I suggest that two areas need particular attention.

First, more research on the energetics of parental investment is required. Although many empirical studies indicate that life-history trade-offs are mediated through energetics or energy limitation, it has become increasingly obvious that we still do not understand precisely how this takes place on a physiological level. We need more information on how nutrition translates into growth, survival and reproduction between juveniles and adults, between adult brood cycles, and between adult breeding seasons. We also need more information on multiple-resource limitation, complemented by new parental-investment models that are based on multiple resources.

A second phenomenon that requires further research is sexual conflict over parental care. Although there have been several game-theoretical models of parental investment, most modelling efforts have been based on optimisation theory and focused on species in which only one parent cares for the offspring. However, there exists considerable evidence that there is sexual conflict over parental care, not only in species with biparental care, but in species with uniparental care as well. Future models need to explicitly address the interactions between mate choice, sexual selection and parental care, where both sexes are choosy. In addition, we require empirical studies that explicitly investigate life-history strategies in general, and parental-investment strategies in particular, from a sexual conflict point of view.

Acknowledgements

I thank Michael Alfieri, Lee Dugatkin, Jean-Guy Godin, Kasi Jackson, Miles Keenleyside, Roland Knapp, Sarah Kraak, Kai Lindström, Mark Ridgway, Lock Rogers, Jennifer Sadowski, Susan Smith-Sargent, Robert Warner, David Westneat and Brian Wisenden for numerous discussions during the preparation of this manuscript. My research was supported in part by National Science Foundation grant BSR-8918871 and by National Science Foundation/Kentucky EPSCoR grant EHR-9108764.

References

Balshine-Earn, S. (1995). The costs of parental care in Galilee St. Peter's fish, *Sarotherodon galilaeus. Anim. Behav.*, **50**, 1–7.

Baylis, J.R. (1981). The evolution of parental care in fishes, with reference to Darwin's rule of male sexual selection. *Environ. Biol. Fish.*, **6**, 223–251.

Belles-Isles, J.-C. and FitzGerald, G.J. (1991). Filial cannibalism in sticklebacks: a reproductive management strategy? *Ethol. Ecol. Evol.*, **3**, 49–62.

Blumer, L.S. (1982). A bibliography and categorization of bony fishes exhibiting parental care. *Zool. J. Linn. Soc.*, **76**, 1–22.

Breder, C.M. Jr. and Rosen, D.E. (1966). *Modes of reproduction in fishes*. T.F.H. Publications, Neptune City, NJ .

Carlisle, T.R. (1982). Brood success in variable environments: implications for parental care allocation. *Anim. Behav.*, **30**, 824–836.

Chellappa, S., Huntingford, F.A., Strang, R.H.C. and Thomson, R.Y. (1989). Annual variation in energy reserves in male three-spined stickleback, *Gasterosteus aculeatus* L. (Pisces, Gasterosteidae). *J. Fish Biol.*, **35**, 275–286.

Coleman, R.M. (1993). *The evolution of parental investment in fishes*. Ph.D. thesis, University of Toronto, Toronto.

Coleman, R.M., Gross, M.R. and Sargent R.C. (1985). Parental investment decision rules, a test with bluegill sunfish. *Behav. Ecol. Sociobiol.*, **18**, 59–66.

Colgan, P.W. and Gross, M.R. (1977) Dynamics of aggression in male pumpkinseed sunfish (*Lepomis gibbosus*) over the reproductive phase. *Z. Tierpsychol.*, **43**, 139–51

Clutton-Brock, T.H. (1991). *The evolution of parental care*. Princeton University Press, Princeton.

Clutton-Brock, T.H. and Vincent, A. (1991). Sexual selection and the potential reproductive rates of males and females. *Nature, Lond.*, **351**, 58–60.

DeMartini, E.E. (1987). Paternal defence, cannibalism and polygamy: factors influencing the reproductive success of painted greenling (Pisces, Hexigrammidae). *Anim. Behav.*, **35**, 1145–1158.

FitzGerald, G.J., Guderley, H. and Picard, P. (1989). Hidden reproductive costs in the threespine stickleback (*Gasterosteus aculeatus*) *Exp. Biol.*, **48**, 295–300.

Friesen, T.G. (1990). *Growth and early life history of smallmouth bass (Micropterus dolomieui) under parental care*. M.Sc. thesis, University of Waterloo, Waterloo.

Gross, M.R. and Sargent R.C. (1985). The evolution of male and female parental care in fishes. *Amer. Zool.*, **25**, 807–822.

Hoelzer, G.A. (1992). The ecology and evolution of partial-clutch cannibalism by paternal Cortez damselfish. *Oikos*, **65**, 113–120.

Jamieson, I. (1995). Do female fish prefer to spawn in nests with eggs for reasons of mate choice copying or egg survival? *Am. Nat.*, **145**, 824–832.

Keenleyside, M.H.A. (1983). Mate desertion in relation to adult sex ratio in the biparental cichlid fish *Herotilapia multispinosa*. *Anim. Behav.*, **31**, 683–688.

Keenleyside, M.H.A. (1985). Bigamy and mate choice in the biparental cichlid fish *Cichlasoma nigrofasciatum*. *Behav. Ecol. Sociobiol.*, **17**, 285–290.

Keenleyside, M.H.A. and Mackereth, R.W. (1992). Effects of loss of male parent on brood survival in a biparental cichlid fish. *Environ. Biol. Fish.*, **34**, 207–212.

Keenleyside, M.H.A., Rangeley, R.W. and Kuppers, B.U. (1985). Female mate choice and male parental defence behaviour in the cichlid fish, *Cichlasoma nigrofasciatum*. *Can. J. Zool.*, **63**, 2489–2493.

Keenleyside, M.H.A., Bailey, R.C. and Young, V.H. (1990). Variation in the mating system and associated parental behaviour of captive and free-living *Cichlasoma nigrofasciatum* (Pisces, Cichlidae). *Behaviour*, **112**, 202–221.

Kirkpatrick, M. (1985). Evolution of female choice and male parental investment in polygynous species: the demise of the 'sexy son'. *Am. Nat.*, **125**, 788–810.

Knapp, R.A. (1992). *Female mate choice in the bicolor damselfish, Stegastes partitus: direct and indirect assessment of male and nest quality.* Ph.D. thesis, University of California, Santa Barbara.

Knapp, R.A. and Kovach, J.T. (1991). Courtship as an honest indicator of male parental quality in the bicolor damselfish, *Stegastes partitus. Behav. Ecol.,* **2**, 295–300.

Knapp, R.A. and Warner, R.R. (1991). Male parental care and female choice in the bicolor damselfish, *Stegastes partitus:* bigger is not always better. *Anim. Behav.,* **41**, 747–756.

Kraak, S.B.M. and van den Berghe, E.P. (1992). Do females assess paternal quality by means of test eggs? *Anim. Behav.,* **43**, 865–867.

Lack, D. (1947). The significance of clutch size. *Ibis,* **89**, 302–352.

Lack, D. (1954). *The natural regulation of animal numbers.* Oxford University Press, Oxford.

Lavery, R.J. and Colgan, P.W. (1991). Brood age and parental defence in the convict cichlid, *Cichlasoma nigrofasciatum* (Pisces: Cichlidae). *Anim. Behav.,* **41**, 945–951.

Lavery, R.J. and Keenleyside, M.H.A. (1990a). Filial cannibalism in the biparental fish *Cichlasoma nigrofasciatum* (Pisces: Cichlidae) in response to early brood reductions. *Ethology,* **86**, 326–338.

Lavery, R.J. and Keenleyside, M.H.A. (1990b). Parental investment of a biparental cichlid fish, *Cichlasoma nigrofasciatum,* in relation to brood size and past investment. *Anim. Behav.,* **40**, 1128–1137.

Lazarus, J. (1990). The logic of mate desertion. *Anim. Behav.,* **39**, 672–684.

McKaye, K.R. (1986). Mate choice and assortative pairing in the cichlid fishes of Lake Jiloá, Nicaragua. *J. Fish Biol.,* **29**, 135–150.

McKaye, K.R. and McKaye, N.M. (1977). Communal care and the kidnapping of young by parental cichlids. *Evolution,* **31**, 674–681.

Maynard Smith, J. (1977). Parental investment: a prospective analysis. *Anim. Behav.,* **25**, 1–9.

Mrowka, W. (1987). Filial cannibalism and reproductive success in the maternal mouthbrooding cichlid fish *Pseudocrenilabrus multicolor. Behav. Ecol. Sociobiol.,* **21**, 257–265.

Perrone, M. Jr. and Zaret, T.M. (1979). Parental care patterns of fishes. *Am. Nat.,* **85**, 493–506.

Petersen, C.W. (1990). The occurrence and dynamics of clutch loss and filial cannibalism in two Caribbean damselfishes. *J. Exp. Mar. Biol. Ecol.,* **135**, 117–133.

Petersen, C.W. and Marchetti, K. (1989). Filial cannibalism in the Cortez damselfish, *Stegastes rectifraenum. Evolution,* **43**, 158–168.

Pianka, E.R. (1994). *Evolutionary ecology,* 5th edn. Harper Collins College Publishers, New York.

Pressley, P.H. (1976). *Parental investment in the threespine stickleback, Gasterosteus aculeatus.* M.Sc. thesis, University of British Columbia, Vancouver.

Pressley, P.H. (1981). Parental effort and the evolution of nest-guarding tactics in the threespine stickleback, *Gasterosteus aculeatus* L. *Evolution,* **35**, 282–295.

Rangeley, R.W. and Godin, J.-G.J. (1992). The effects of a trade-off between foraging and brood defence on parental behaviour in the convict cichlid fish, *Cichlasoma nigrofasciatum. Behaviour,* **120**, 123–138.

Ridgway, M.S. (1988). Developmental stage of offspring and brood defence in smallmouth bass (*Micropterus dolomieui*). *Can. J. Zool.,* **66**, 1722–1728.

Ridgway, M.S. (1989). The parental response to brood size manipulation in smallmouth bass (*Micropterus dolomieui*), *Ethology*, **80**, 47–54.

Ridgway, M.S. and Friesen, T.G. (1992). Annual variation in parental care in smallmouth bass, *Micropterus dolomieui*. *Environ. Biol. Fish.*, **35**, 243–255.

Ridgway, M.S. and Shuter, B.J. (1994). The effects of supplemental food on reproduction in parental male smallmouth bass. *Environ. Biol. Fish.*, **39**, 201–207.

Rohwer, S. (1978). Parental cannibalism of offspring and egg raiding as a courtship strategy, *Am. Nat.*, **112**, 429–440.

Sabat, A.M. (1994). Costs and benefits of parental effort in a brood-guarding fish (*Ambloplites rupestris* Centrarchidae). *Behav. Ecol.*, **5**, 195–201.

Sargent, R.C. (1985). Territoriality and reproductive tradeoffs in the threespine stickleback, *Gasterosteus aculeatus*. *Behaviour*, **93**, 217–226.

Sargent, R.C. (1988). Paternal care and egg survival both increase with clutch size in the fathead minnow, *Pimephales promelas*. *Behav. Ecol. Sociobiol.*, **23**, 33–38.

Sargent, R.C. (1989). Allopaternal care in the fathead minnow, *Pimephales promelas*, step-fathers discriminate against their adopted eggs. *Behav. Ecol. Sociobiol.*, **25**, 379–385.

Sargent, R.C. (1990). Behavioural and evolutionary ecology of fishes: conflicting demands during the breeding season. *Ann. Zool. Fennici*, **27**, 101–118.

Sargent, R.C. (1992). Ecology of filial cannibalism in fish: theoretical perspectives. In *Cannibalism: ecology and evolution among diverse taxa* (eds. M.A. Elgar and B.J. Crespi), pp. 38–62, Oxford University Press, Oxford.

Sargent, R.C. and Gebler, J.B. (1980). Effects of nest site concealment on hatching success, reproductive success, and paternal behavior of the threespine stickleback, *Gasterosteus aculeatus*. *Behav. Ecol. Sociobiol.*, **7**, 137–142.

Sargent, R.C. and Gross, M.R. (1985). Parental investment decision rules and the Concorde fallacy. *Behav. Ecol. Sociobiol.*, **17**, 43–45.

Sargent, R.C. and Gross, M.R. (1986). Williams' principle, an explanation of parental care in teleost fishes. In *Behaviour of teleost fishes* (ed. T.J. Pitcher), pp. 275–293, Croom Helm, London.

Sargent, R.C. and Gross, M.R. (1993). Williams' principle, an explanation of parental care in teleost fishes. In *Behaviour of teleost fishes*, 2nd edn. (ed. T.J. Pitcher), pp. 333–361, Chapman and Hall, London.

Sargent, R.C., Crowley, P.H., Huang, C., Lauer, M., Neergaard, D. and Schmoetzer, L.A. In press. Dynamic program for male parental care in fishes: brood cycling and filial cannibalism. *Behaviour*

Sibly, R.M. and Calow, P. (1986). *Physiological ecology of animals*. Blackwell Scientific Publ., Oxford.

Smith, C. and Wootton, R.J. (1994). The cost of parental care in *Haplochromis 'argens'* *Environ. Biol. Fish.*, **40**, 99–104.

Stephens, D.W. and Krebs, J.R. (1986). *Foraging theory*. Princeton University Press, Princeton.

Townshend, T.J. and Wootton, R.J. (1984). Effects of food supply on the reproduction of the convict cichlid, *Cichlasoma nigrofasciatum*. *J. Fish Biol.*, **23**, 91–104.

Townshend, T.J. and Wootton, R.J. (1985). Adjusting parental investment to changing environmental conditions: the effect of food ration on parental behaviour of the convict cichlid, *Cichlasoma nigrofasciatum*. *Anim. Behav.*, **33**, 494–501.

Unger, L.M. (1983). Nest defence by deceit in the fathead minnow, *Pimephales promelas*. *Behav. Ecol. Sociobiol.*, **13**, 125–130.

Unger, L.M. and Sargent, R.C. (1988). Allopaternal care in the fathead minnow, *Pimephales promelas*: females prefer males with eggs. *Behav. Ecol. Sociobiol.*, **23**, 27–32.

van den Assem, J. (1967). Territory in the three-spined stickleback, *Gasterosteus aculeatus* L.: an experimental study in intra-specific competition. *Behaviour (Suppl.)*, **16**, 1–164.

van den Berghe, E.P. (1990). Variable parental care in a labrid fish: how care might evolve. *Ethology*, **84**, 319–333.

van Iersel, J.J.A. (1953). An analysis of parental behaviour of the male three-spined stickleback (*Gasterosteus aculeatus* L.). *Behaviour (Suppl.)*, **3**, 1–159

Warner, R.R., Wernerus, F., Lejeune, P. and van den Berghe, E.P. (1995) Dynamics of female choice for parental care in a species where care is facultative. *Behav. Ecol.*, **6**, 73–81.

Whoriskey, F.G. and FitzGerald G.J. (1985). Sex, cannibalism and sticklebacks. *Behav. Ecol. Sociobiol.*, **18**, 15–18.

Williams, G.C. (1966). Natural selection, costs of reproduction, and a refinement of Lack's principle. *Am. Nat.*, **100**, 687–690.

Williams, G.C. (1975). *Sex and evolution*. Princeton University Press, Princeton.

Wisenden, B.D. (1994). Factors affecting mate desertion by males in free-ranging convict cichlids (*Cichlasoma nigrofasciatum*). *Behav. Ecol.*, **5**, 439–447.

Wisenden, B.D. (1995). Reproductive behaviour of free-ranging convict cichlids, *Cichlasoma nigrofasciatum*. *Environ. Biol. Fish.*, **43**, 121–134..

Wisenden, B.D. and Keenleyside, M.H.A. (1992). Intraspecific brood adoption in convict cichlids: a mutual benefit. *Behav. Ecol. Sociobiol.*, **31**, 263–269.

Wisenden, B.D. and Keenleyside, M.H.A. (1994). The dilution effect and differential predation following brood adoption in free-ranging convict cichlids (*Cichlasoma nigrofasciatum*). *Ethology*, **96**, 203–212.

Wisenden, B.D. and Keenleyside, M.H.A. (1995). Brood size and the economy of brood defence: examining Lack's hypothesis in a biparental cichlid fish. *Environ. Biol. Fish.*, **43**, 145–151.

12 *Flexibility in fish behaviour: consequences at the population and community levels*

Lennart Persson, Sebastian Diehl, Peter Eklöv and Bent Christensen

12.1 Introduction

Flexible behaviour of organisms in response to food resources and predators has been a major topic in ecology during the last decade (Lima and Dill 1990; Persson and Diehl 1990; Werner 1992; see also Kramer *et al.*, Chapter 3; Grant, Chapter 4; Hart, Chapter 5; Hughes, Chapter 6; Smith, Chapter 7; Godin, Chapter 8; Sargent, Chapter 11). The purpose of this chapter is to review current theoretical and empirical knowledge of the consequences of flexible behaviour for population and community processes in fishes. Understanding these consequences is import-ant for both population and community ecologists and behavioural ecologists, because the behaviour of individuals influences population- and community-level processes, and vice versa.

   One basic component of optimisation models in general and foraging models in particular is the assumptions about constraints on behaviour (Stephens and Krebs 1986; see also Hart, Chapter 5 and Hughes, Chapter 6). It is therefore natural to first briefly consider size-specific and other morphological constraints that limit the performance of individual fish, and then consider the implications of flexible behaviour for population and community dynamics. The question of how flexible behaviour affects population and community dynamics has been most thoroughly addressed in theoretical models. We compare different types of population models to evaluate the potential importance of flexible behaviour (especially habitat selection) in individuals for population dynamics. Although the predictions of these models have not yet been rigorously tested, a num-ber of empirical examples exist in which flexible habitat use appears to have had major impacts on population and community dynamics. Our review of these examples also suggest that size-structured processes are overwhelmingly import-ant in fishes. Size structure in fish populations often implies the simultaneous presence of competitive and predatory interactions between pairs of species, comprising individuals of different body size, which may produce complex and highly asymmetric population interactions. We will discuss such asymmetries in interactions between fish species from a behavioural perspective. To be able to formally analyse size-structured interactions, physiologically-structured models have recently been developed. We will examine the different approaches that have been used and discuss how the flexible behaviour of individuals may be incorporated into these models. Due to the complex dynamical behaviours of these models, there is a need for several approaches emphasising different aspects of the relationship between size-structured processes and individual behaviour.

12.2 Mechanistic approaches to population/community ecology: definitions

The consideration of the flexible behaviour of individuals in population/community ecology is closely linked to the use of mechanistic approaches for understanding ecological processes (Schoener 1986; Koehl 1989; Persson and Diehl 1990). Mechanistic approaches have been contrasted to phenomenological approaches. Koehl (1989) differentiated between models that seek to understand the essential processes governing the dynamics of ecological systems and models that are phenomenological descriptions of the system, such as empirical regression models. The latter models may be useful to predict the fish biomass in a lake based on its phosphorus levels, for example (Mills and Schiavone 1982). Phenomenological models may be preferable to mechanistic models in that they may yield predictions over shorter time frames, but the predictions will be valid only if conditions do not change (Koehl 1989).

Phenomenological and mechanistic approaches are not incompatible. The former may help to organise observations from which mechanistic models can be formulated, and the latter can in turn be used to explain phenomenological patterns, such as productivity-biomass relationships. In advocating a more mechanistic approach to population and community ecology, Schoener (1986) viewed mechanistic models as being characterised by a high degree of lower-level derivability; that is, population and community properties are derived from the properties of individual organisms. Individual properties can in turn be related to physiology, functional morphology and behavioural ecology. In this chapter, we will adhere to Schoener's definition. It is important to note that mechanistic models also include non-mechanistic elements, and mechanistic and non-mechanistic models should not be viewed as exclusive alternatives (Schoener 1986). Mechanistic approaches to the population and community ecology of fishes have generally focused on the behaviour of individuals and its physiological determinants (e.g. individual energetics) (Mittelbach 1981; Werner *et al.* 1983a; Persson and Greenberg 1990b) or morphological constraints (O'Brien 1987; Wainwright 1988; Wainwright *et al.* 1991). Thus, mechanistic approaches provide means to explicitly link individual behaviour to population/community dynamics (Abrams 1984, 1992; Ives and

Dobson 1987; Sih 1987; Kotler and Holt 1989) and to investigate the evolution of the behaviour of individual organisms within the appropriate ecological context (Abrams 1986).

12.3 Mechanistic approaches and morphological constraints

The use of mechanistic approaches in population and community ecology necessitates the explicit consideration of morphological, energetic and behavioural constraints on the behaviour of individual organisms. An important constraint on the performance of an individual is its size. Body size affects many aspects of an organism's biology, such as its foraging rate, vulnerability to predators and fecundity (Werner 1988; Hart, Chapter 5; Hughes, Chapter 6; Godin, Chapter 8). Foraging capacity generally increases with body size as a result of increases in perceptive, digestive, locomotor and capture capacities (Persson and Diehl 1990, and references therein). Since metabolic demands also increase with body size (Brett and Groves 1979), the competitive interactions between organisms of different sizes will depend on their relative energetic gains and costs. Size-specific gains and costs will in turn depend on the density and size distribution of prey organisms.

The importance of body size in competitive interactions has been demonstrated in several studies on habitat shifts (Mittelbach 1981; Werner *et al.* 1983b; Werner and Hall 1988; Osenberg *et al.* 1992) and growth stunting (Persson 1987) in fishes. These studies focused mainly on short-term behavioural shifts, with the purpose of understanding static community patterns within and between systems. In comparison, longer-term population dynamics were explicitly considered in a study of a pelagic planktivore (vendace, *Coregonus albula*), where competition between fish of different size (age) classes was suggested to be the causal mechanism underlying a two-year population oscillation (Hamrin and Persson 1986).

The foraging capacity of an organism is not only affected by its body size, but also by other aspects of its morphology such as body shape, positioning of locomotor appendages and feeding apparatus (O'Brien 1987; Persson and Diehl 1990; Galis 1991; Wainwright *et al.* 1991; Hart, Chapter 5; Hughes, Chapter 6). For example, Webb (1984) found that esocids had a 60% higher prey capture efficiency than non-esocid piscivores. The differences in capture efficiency could be related to the body form and positioning of the caudal fin in esocids, which enables them to reach high maximum accelerations and strike efficiencies. Such findings suggest a tight relationship between morphology and foraging behaviour, which has been studied in detail in some cichlids and sunfishes. For example, size-specific feeding efficiency and prey choice in the cichlid *Haplochromis piceatus* have been related to ontogenetic changes in its pharyngeal jaw apparatus (Galis 1991). Abundance relationships among different species of sunfish may depend on both the relative availability of different resources in different lakes and on species-specific constraints in body form and feeding apparatus (Mittelbach 1984; Osenberg and Mittelbach 1989). In cichlids and sunfishes, polymorphism in the structure of the trophic apparatus has been described and related to environmental conditions (Ehlinger 1990; Wainwright *et al.* 1991). For example, the pharyngeal jaw muscles and jaw bones of pumpkinseed sunfish, *Lepomis gibbosus*, were larger in fish from a lake rich in snails compared with fish from a lake with low snail density (Wainwright *et al.* 1991). Whether the observed polymorphisms represent genetic divergence or ontogenetic phenotypic plasticity could not be definitely determined, but much evidence suggests that phenotypic plasticity plays a major role (Wainwright *et al.* 1991, and references therein). Further supporting such phenotypic plasticity in fishes is the report of a predator-induced change in body shape in the crucian carp, *Carassius carassius* (Brönmark and Miner 1992).

The above examples illustrate both the plasticity in the relationship between foraging behaviour and morphology and the constraints on behaviour set by morphology. These constraints substantially influence the behavioural repertoire of the individual. How the behaviour of the individual in turn influences population and community dynamics will be considered in the following.

12.4 Mechanistic approaches and flexible behaviour

There is compelling empirical evidence that flexibility in behaviour is widespread in many taxa (including fishes) and that individuals make adaptive behavioural decisions within species- and size-specific constraints (Lima and Dill 1990; see also the other chapters in this volume). The choice by individuals of habitats in which to forage, rest or produce offspring potentially affects the dynamics of their population and the structure of their community. Models of population interactions, which exclude flexible behaviour, have shown that indirect effects between populations arise primarily through the numerical outcomes of direct interactions (e.g. Abrams 1992). For example, a top consumer indirectly benefits a resource two trophic levels distant by reducing the abundance of the direct consumer (at the intermediate trophic level) of this resource. However, if individuals are flexible in their behaviour, the mere presence (and density) of a third species can alter the per capita consumer effects in a pairwise, direct interaction. For example, a top consumer may affect the activity level and/or habitat choice of an intermediate species causing changes in consumption rate of the resource. Hugie and Dill (1994) have also recently demonstrated interesting behavioural effects of consumers on three trophic-level dynamics using game theory. To take into account such behaviourally-mediated indirect effects requires both substantial changes in models of population dynamics (Abrams 1984; Kotler and Holt 1989) and in the protocol of experimental studies (Werner 1992).

In this section, we first discuss results obtained from population models incorporating flexible behaviour. The focus will be on the effects of diet and habitat choices, activity level and resource-searching mode of individuals. A number of empirical studies are then reviewed with the purpose of illustrating the implications of flexible behaviour on population and community dynamics in natural communities.

12.4.1 Population consequences of diet choice

The population and community consequences of flexible behaviour in individuals have been investigated both theoretically and empirically with respect to three main decision variables: diet choice, habitat choice and choice of activity level. The effect of adaptive diet selection by predators on population stability was treated theoretically by Fryxell and Lundberg (1994). Adaptive diet selection (see Hughes, Chapter 6) was shown to enhance stability properties of a predator–prey system only under a small range of parameter values; otherwise, adaptive diet choice was suggested not to be an important stabilising factor (Fryxell and Lundberg 1994). Gleeson and Wilson (1986) showed theoretically that a predator foraging on two competing prey species according to the optimal (energy maximising) diet model (Stephens and Krebs 1986; Hughes, Chapter 6) could prevent the weaker competitor from going extinct if the dominant competitor was the more profitable prey. However, this result does not differ substantially from predictions of models of frequency-dependent prey selection (switching behaviour) (see reviews in Persson and Diehl 1990; Hughes, Chapter 6). Empirical studies of fish in nature also suggest that the effects of morphological constraints (e.g. predator and prey sizes) on encounter rates and handling times are generally very important in determining a consumer's diet (see reviews in Persson and Diehl 1990; Mittelbach and Osenberg 1994; Hart, Chapter 5; Hughes, Chapter 6). In contrast, the addition of active choice in the consumer does not markedly improve the ability to predict diet, energy gain and individual growth rate. Still, investigators commonly note substantial differences between observed diets and those predicted by both optimal diet and encounter rate models. Such non-concordance may be attributed to the usual assumption that encounter rates are determined only by the size and type of predator and prey and by prey density. However, encounter rates are often strongly influenced by flexible predator and prey behaviours (Persson and Diehl 1990; Hart, Chapter 5; Smith, Chapter 7; Godin, Chapter 8). Behavioural decisions that affect encounter rates might therefore have a greater impact on population dynamics than the decision to attack an encountered prey or not (Persson and Diehl 1990; Sih and Moore 1990).

12.4.2 Population consequences of flexibility in habitat use and activity level—theoretical aspects

The behavioural decisions that most crucially influence encounter rates between predator and prey are probably their respective choices of habitat and activity level. Whereas empirical studies of fish have mainly dealt with the population consequences of habitat choice, theoretical population studies have concentrated on individual activity level (in the sense of the proportion of time spent foraging) as the major determining decision variable. However, the proportion of time spent foraging can be interpreted in terms of habitat choice when the time spent inactive is considered equivalent to resting in a refuge, with associated reduced mortality costs and foraging benefits. This view is also empirically justified, as changes in the foraging activity of fish are often combined with changes in habitat (Gilliam and Fraser 1987; Kramer *et al.*, Chapter 3). In the following, we will therefore treat collectively studies on habitat choice and on choice of the proportion of time spent foraging.

The population consequences of the habitat choice of individuals have been investigated in models in which spatially distinct habitats ('patches') differ in the densities of interacting species. In such models, random dispersal of individuals among patches promotes community persistence, even of intrinsically unstable communities of predators and prey (Caswell 1978). Recently, McCauley *et al.* (1993) showed, using an individual-based model incorporating spatial structure, that the stability of a predator–prey system is determined by the relative mobility of predators and prey, and that prey mobility in particular had strong effects on stability. Aggregative behaviours in consumers (including optimal patch choice according to the model of Charnov *et al.* 1976) and high variance in prey abundance always contribute to stability, as does some degree of habitat segregation between prey species if alternative prey species are present (Holt 1984; Comins and Hassell 1987). Using a more realistic model, McNair (1986) showed that several kinds of refuge can be either stabilising or destabilising.

The above predator–prey models focus on the effects of habitat choice in just one species, and assume that the species in question is either predator or prey. Most animals (including fishes) are simultaneously predators and prey during at least some life-history stages (Werner and Gilliam 1984). Consequently, population models need to make assumptions about how individuals trade off conflicts between their foraging activities and their avoidance of predation. For example, predator–prey models treating a particular number of prey as always being in refuge implicitly assume that individual prey are risk minimisers. However, if the growth rates of prey are substantially lower in a refuge than outside it, then they might actually achieve a higher net benefit outside the refuge. Therefore, if prey are benefit maximisers and can affect their own resources, they should make flexible use of the two habitats (Gilliam and Fraser 1988; Kramer *et al.*, Chapter 3). Such flexible habitat selection by prey tends to stabilise the predator–prey interaction by reducing predation rates at high predator densities, reducing population growth rates and dampening oscillatory tendencies (Ives and Dobson 1987; Sih 1987).

The above conclusions are based on models which lack an equation for the dynamics of the resource and which ignore the possibility that both species in a predator–prey interaction are able to mutually adjust their behaviours. The explicit inclusion of both the dynamics of all populations that interact with a behaviourally flexible species and the assumption of the presence of behavioural flexibility in more than one species may give rise to complex and sometimes counter-intuitive population dynamics. For example, in three- and four-trophic level models with foraging-effort maximisers at the intermediate trophic levels, rather large indirect effects may occur between the species at the most distant trophic levels. Furthermore, the overall effects between adjacent trophic levels may not necessarily have the same signs as the direct interactions (Abrams 1984, 1992). Flexible behavioural responses in both a predator and two alternative prey species to each other may also produce negative switching, that is, a decrease in the relative predation rate on a prey species with a subsequent increase in its relative abundance (Abrams and Matsuda 1993).

12.4.3 Population and community consequences of habitat choice—empirical evidence

Manipulations of piscivore populations have often resulted in effects across several trophic levels (e.g. through planktivores and zooplankton to phytoplankton; Carpenter and Kitchell 1988; Persson *et al.* 1993). Traditionally, these effects have been interpreted as a direct result of predator consumption. In contrast, behavioural studies of predator–prey interactions have emphasised non-lethal indirect effects (e.g. through prey activity and habitat use) of predators on community dynamics (Power 1987; Gilliam and Fraser 1988; Turner and Mittelbach 1990; Persson 1993). In the following, we review a number of case studies on the community consequences of flexible behaviour in fishes. In these studies, data on community-wide patterns were combined with experimental and observational data on mechanisms. Since most fish species are simultaneously both predators and prey, we focus on cases where at least three trophic levels have been included in the investigation of population/community consequences of risk-mediated foraging behaviour in fishes.

Armoured catfish (Loricariidae) distributed themselves approximately according to ideal free distribution theory (*sensu* Fretwell 1972) among pools in a Panamanian stream that differed in resource productivity (attached algae), thus demonstrating their ability to track resource availabilities (Power 1984a). At the same time, it was observed that striking differences in the standing stock of algae between shallow margins and deeper central areas within pools were not equalised by grazing catfish (Power 1984b). Observational and experimental data suggest that large (> 3 cm) catfish, although severely resource limited in the deeper parts of pools, avoided shallow areas because of the greater risk of predation by birds there (Power 1987). Small (< 3 cm) catfish were much more abundant in shallow water, where they were postulated to avoid predatory fishes but did not deplete algal resources (Power 1987). Overall, the evidence suggests that risk-sensitive foraging by large catfish produced a strong indirect interaction between avian predators and benthic algae. In contrast, risk-sensitive foraging by small catfish had no significant effect on algae, as resource limitation in adult catfish kept the densities of juveniles below those necessary to deplete algae (see also Kramer *et al.*, Chapter 3).

In small Scandinavian lakes, allopatric populations of crucian carp are generally dense (ca. 30 000 fish/ha) and dominated by many small individuals (4–10 cm). In contrast, in lakes with piscivores, crucian carp populations are small (25–250 fish/ha) and dominated by large individuals (15–35 cm) (Piironen and Holopainen 1988; Paszkowski *et al.* 1989). Through manipulations of sections of a lake previously only occupied by crucian carp, Tonn *et al.* (1992) investigated the direct and indirect (behaviourally-mediated) effects of a piscivore on the whole lake community. Stocked piscivorous perch reduced the numbers of young-of-the-year (YOY) crucian carp by 90%, decreased their activity levels and confined them to vegetated inshore areas. The exclusive use of vegetation as a refuge in the presence of piscivores also decreased the growth rates of YOY crucian carp compared with YOY in piscivore-free sections. On the other hand, competitive release from YOY allowed the invulnerable, larger size classes of carp to achieve higher growth rates in the piscivore sections. The effects of piscivores on crucian carp densities and behaviour may also have an impact on the zooplankton and phytoplankton communities. After a direct manipulation of carp densities in the same experimental system, mean biomass of zooplankton (mainly Cladocera) was lower and phytoplankton biomass was higher at high carp densities than at low densities (Holopainen *et al.* 1992).

Mittelbach (1981) found that the classical energy-maximising diet model (Stephens and Krebs 1986; Hughes, Chapter 6) was a good predictor of the diet of large size classes (adults) of the bluegill sunfish, *Lepomis macrochirus*, in a Michigan (USA) lake, but a poor predictor of the diet of small size classes (juveniles). Gape-limited piscivores in the open water force the vulnerable juveniles to stay in the vegetated littoral zone, where they experience reduced encounter rates with their preferred prey and consequently reduced individual growth rates (Werner *et al.* 1983b). The size at which bluegills move into the more profitable pelagic habitat varies between lakes, depending on ambient predation risk (Werner and Hall 1988). This flexible and adaptive habitat choice by individual bluegills results in complex indirect interactions between populations at several trophic levels that do not even share habitat. Since bluegills in the presence of piscivores use two different habitats during ontogeny and have the potential to reduce resource levels in both habitats (Mittelbach 1988; Turner and Mittelbach 1990), negative indirect interactions (i.e. apparent competition) occur between the resources of bluegills in these habitats. An increase in food resources for juveniles translates, via higher juvenile growth rates and survival, into increased numbers of adults and increased consumption rates on the adult's food resources, which ultimately leads to a decline in the latter. Similarly, an increase in food resources for adults increases juvenile abundance via increased adult fecundity, which then results in a decrease in the abundance of juvenile food resources (Fig. 12.1; Mittelbach and Chesson 1987).

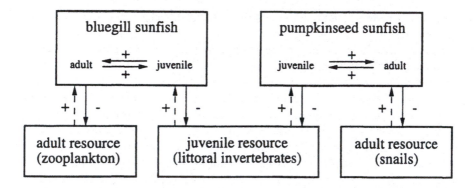

Fig. 12.1   Direct interactions between the juvenile and adult stages of bluegill and pumpkinseed sunfishes and their resources. Solid arrows with negative signs indicate negative effects of consumers on the abundances of their resources. Hatched arrows indicate positive effects of resource abundances on growth and/or fecundity of consumers, which translate into increased abundances of the life stages not consuming the resources themselves (solid arrows with positive signs). Juvenile bluegills are assumed to occupy the inshore vegetation zone due to the presence of gape-limited predators (largemouth bass). After Osenberg *et al.* (1992).

The extent to which piscivore-mediated habitat choice in bluegills affects lower trophic levels depends on the size structure of the bluegill population. In experiments using only vulnerable size classes, piscivores confined bluegills to vegetated habitats and thereby indirectly increased zooplankton abundances in the open water without causing any planktivore mortality (Turner and Mittelbach 1990). In contrast, in experiments with a broader range of bluegill sizes, the larger invulnerable individuals in the population reduced the zooplankton standing stock in the presence of piscivores to levels as similarly found in the treatments with no piscivores present and where zooplankton was consumed by all bluegill size classes (Werner *et al.* 1983b). Through their behavioural effects on planktivorous sunfish, piscivores may also affect the species composition and habitat distribution of zooplankton (Leibold 1991; Leibold and Tessier 1991).

In many lakes, further indirect community interactions are caused by the presence of pumpkinseed sunfish, *Lepomis gibbosus*. Adult pumpkinseed use different resources (littoral snails) than do adult bluegills, but juvenile pumpkinseeds prefer the littoral habitat even in the absence of piscivores (Osenberg *et al.* 1992). When piscivores force juvenile bluegills into the littoral vegetation, the resulting increase in pelagic zooplankton indirectly translates, via increased adult bluegill fecundity and juvenile bluegill density, into decreased juvenile pumpkinseed growth rate and survival. As a result, adult pumpkinseed abundance will decrease and snail abundance increase (Fig. 12.1; Mittelbach and Chesson 1987; Osenberg *et al.* 1992). The crucial point here is that the life-history stage that is not limited by its own resources, but rather by recruitment from the other life stage, is unable to fully respond numerically to increases in its own resource. Therefore, changes in the abundance of a resource population may have opposite effects on the abundance and per capita growth rate of the size class of fish that does not use the resource itself (Fig. 12.1). The abundances and growth rates of juvenile and adult fish in a population are ultimately determined by the productivities of their respective resources and also by the relative sizes of the littoral and pelagic habitats. Littoral macroinvertebrates are most often the resource that limits bluegill populations as a whole, resulting concurrently in negatively density-dependent growth rates in juveniles but positively density-dependent growth rates in adults (Mittelbach and Osenberg 1993).

In Scandinavian lakes, the composition of fish communities varies over a large gradient of lake primary productivities (Persson *et al.* 1991). While the biomasses of percids (mainly perch, *Perca fluviatilis*) and coregonids peak in lakes of inter-mediate productivity, cyprinids (mainly roach, *Rutilus rutilus*) increase over the whole range of productivities and become dominant in highly productive lakes (Fig. 12.2a). From moderately to highly productive lakes, a shift in the size struc-ture of the fish community towards smaller individuals occurs, which implies a drastic reduction in the relative and absolute abundances of piscivorous size classes of perch (Fig. 12.2b). Changes in lake productivity are generally correl-ated with major changes in physical habitat structure (abundance and growth form of macrophyte vegetation in the littoral zone) and in the relative productivities of littoral and pelagic invertebrates (Fig. 12.2c, d; Wetzel 1983). The observed shifts in size structure and relative abundance of species within Scandinavian fish communities can be related to these habitat changes, since the major fish species differ in their habitat-specific abilities to acquire resources and avoid predators. Roach perform best in open habitats where they are able to search and capture prey at a high cruising speed (Fig. 12.3; Persson 1988). Therefore, in simple, un-structured environments, roach are superior foragers on planktonic prey (Winfield 1986; Persson 1988). Consequently, in highly productive lakes, which have a high relative and absolute productivity of pelagic invertebrates, intense competition for zooplankton from roach induces an early ontogenetic habitat shift in YOY perch from zooplankton feeding in the pelagic habitat to macroinvertebrate feeding in the benthic habitat. YOY perch then compete with larger size classes of perch for benthic invertebrates, resulting in much reduced growth rates in benthic-feeding perch (Persson 1986; Persson and Greenberg 1990a, b). Juvenile perch, which use a more saltatory search mode (see Hart, Chapter 5), are superior to juvenile roach when foraging in structurally complex, vegetated habitats (Fig. 12.3; Win-field 1986; Diehl 1988; Persson 1991, 1993). Because juvenile perch and roach forage in vegetation refuges when the risk of predation is high in the open water, an abundance of both pelagic piscivores and submerged littoral vegetation in lakes of intermediate productivity may actually reverse the outcome of competi-tive interactions between juveniles in favour of perch (Fig. 12.3; Persson 1991, 1993; Persson and Eklöv 1995). The picture is even more complex because roach

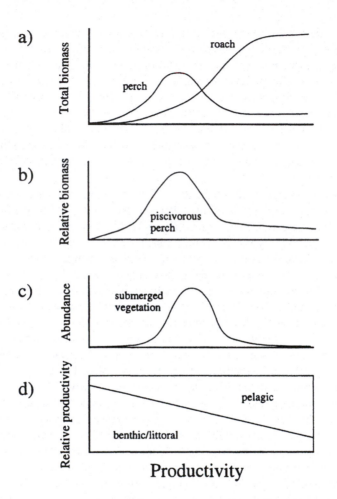

Fig. 12.2   Schematic representation of the changes along the productivity gradient of temperate European lakes in (a) the biomasses of Eurasian perch and roach in gill net catches, (b) the proportion of piscivorous perch of the total fish biomass in gill net catches, (c) the abundance of submerged vegetation, and (d) the relative productivities of the pelagic and the littoral/benthic habitats, respectively. Very shallow, highly productive lakes may be covered by macrophytes and then show similar characteristics to lakes of intermediate productivity. After Wetzel (1983) and Persson *et al.* (1991).

are relatively effective at avoiding predators in vegetated areas (Christensen and Persson 1993).

12.4.4 Population consequences of search mode and intraspecific interactions

Predators can trade off the costs and benefits of foraging not only by adjusting the

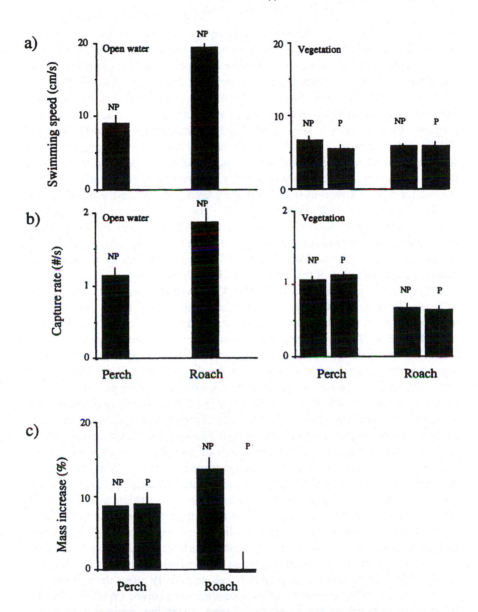

Fig. 12.3 Mean (+ SE) swimming speed (a), feeding rate on planktonic prey (b), and percentage mass increase (c) of perch and roach in the open-water and vegetation habitats of experimental tanks. P = predators (two piscivorous perch) were present in the open water habitat, NP = predators were absent. Both species almost totally avoided the open-water habitat in the presence of predators. Data from Persson (1991).

proportions of time spent foraging or occupying risky habitats, but also by adjusting the absolute activity level while foraging (Werner and Anholt 1993; Hart, Chapter 5). Models of optimal search mode have been developed for predators facing a trade-off between encounter rate with prey and energy expenditure (Norberg 1977; Krebs and McCleery 1984; Pyke 1984; Hart, Chapter 5). These models ignore the potential need to avoid one's own predators and thus may be most appropriate for top predators such as piscivores. In models developed by Norberg (1977) and Speakman (1986), the optimal activity levels for foraging predators were categorised as either sit-and-wait or active search, depending on resource levels. An intermediate category of activity pattern has more recently been described for zooplanktivorous fish as 'saltatory search' or 'pause-travel' (Ehlinger 1989; O'Brien *et al.* 1990; Hart, Chapter 5). The above models relate predator search modes only to encounter probabilities between predators and prey. However, search mode will also be related to the predator's morphology and behavioural flexibility, to the nature of intraspecific interactions during foraging, and to the physical structure of the habitat (see Hart, Chapter 5).

Eklöv (1992) and Eklöv and Diehl (1994) showed that two piscivores, perch and pike, *Esox lucius*, with different morphologies differed in the flexibility of their search mode and foraging efficiency under different environmental conditions (Fig. 12.4). The body form of piscivorous perch is better adapted to active search than ambushing, and their prey capture rate and per capita growth in open water is greater when they are foraging in groups than solitarily (Webb 1984, 1986; Eklöv 1992). The presence of a prey refuge caused a decrease in prey capture rate for both piscivorous species, but the decrease in efficiency was much greater for perch than for pike (Fig. 12.4). In the presence of prey refuges, perch changed foraging tactic and adopted a sit-and-wait foraging mode (cf. Hart, Chapter 5), which yielded a lower foraging efficiency than that of pike. In comparison, pike used a sit-and-wait foraging mode in both the presence and absence of a prey refuge, and suffered a decreasing per capita efficiency with increasing predator density. The latter was a result of size-dependent interference competition, by which small pike were restricted to small areas farthest away from the food resource (Eklöv 1992). These experiments illustrate how differences in morphology, behaviour and spatial distribution of piscivores can affect variables, such as prey mortality and predator growth rates, that are important for population dynamics. In addition, intraspecific interactions among predators can potentially affect the equilibrium densities and stability properties of the populations involved (Wollkind 1976). Interference interactions among predators tend to stabilise predator–prey interactions and increase prey equilibrium densities but decrease predator densities. For example, Grant and Kramer (1990) found that the upper population density limit of juvenile salmonids in streams could be predicted from the size of individual feeding territories, which suggests that interactions between individuals are important in determining the population density of predators. Conversely, mutualistic interactions between predators can destabilise predator–prey interactions

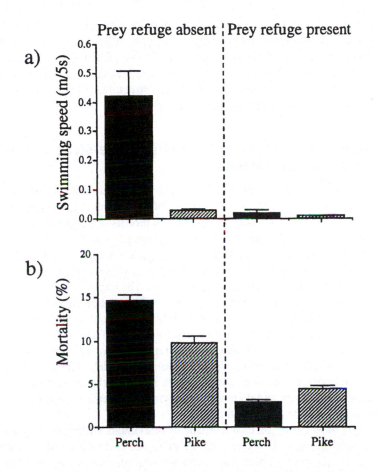

Fig. 12.4 Mean (+ SE) swimming speed (a) of piscivorous perch and pike while hunting on age 0+ prey fish (perch), and (b) prey mortality, expressed as percentage of prey fish released into experimental enclosures that were eaten, in the absence (*n* = 3) and presence (*n* = 4) of a vegetation refuge for the prey. Data from Eklöv (1992) and Eklöv and Diehl (1994).

and reduce prey densities, whilst the predator population densities may increase or decrease (Wollkind 1976).

Several conclusions can be drawn from this section. First, empirical studies have shown that flexible behaviour in fishes can produce important indirect effects between populations that are several trophic levels apart and even occupy different habitats. Second, the consideration of flexible behaviour necessitates substantial modifications in theoretical, experimental and descriptive studies of population interactions. Third, empirical studies suggest that behaviours affecting encounter rates between consumers and resources (i.e. choice of habitat and/or activity level) are likely to have strong effects on population interactions. Finally, theoretical studies suggest that flexible behaviour in individuals often enhances the possibilities for species coexistence. Although there is a paucity of data to test predictions of models of population dynamics, empirical studies of fishes have emphasised the importance of individual body size for behavioural decisions. This suggests that theoreticians should include population size structure into models of populations comprised of behaviourally flexible individuals.

12.5 Individual variation, population size structure and mechanistic models

Modelling population and community dynamics requires simplifying assumptions that are often not compatible with the reality of natural ecological systems. One of the most important assumptions in traditional models of populations is that all individuals within populations are identical (DeAngelis and Rose 1992). This assumption contradicts some basic tenets in ecology, especially evolutionary ecology, which state that individuals differ in behaviour, physiology, morphology, etc., due to genetic and environmental influences. The neglect of individual variation is, however, not only present in descriptive population models but also in the majority of mechanistic models, including those incorporating flexible individual behaviour. Lomnicki's (1988) studies on the effects of intrapopulation phenotypic variation on population stability represent one attempt which explicitly introduces individual variation into population models. His studies largely focused on individual variation in competitive ability in the context of scramble and contest competition, and do not treat individual properties as explicitly as do the models that will be considered below.

The single most important characteristic of individuals within fish populations is their body size. Population size structure gives rise to a mixture of competitive and predatory interactions, both within and between species. In the following, we discuss some examples which illustrate this mixture of interactions, and discuss how flexible behaviour in individuals can affect the outcome of these interactions. The empirical examples suggest that incorporating individual size into population models is essential if the complex population interactions are to be adequately described. We review possible approaches to incorporate such detail in population models.

12.5.1 Population size structure, flexible behaviour and mixed predation–competition interactions

Because traditional models in population and community ecology treat individuals within populations as identical, pairwise species interactions can be defined by a single combination of signs (e.g. +/–, –/– or 0/0) and communities represented by static food webs or interaction matrices (Werner 1992). However, size-structured populations, in which individuals vary in body size due to ontogenetic processes, are extremely widespread among animal taxa, including fishes. Therefore, the major source of individual variation within populations is likely due to size differences between individuals (Werner and Gilliam 1984; Ebenman and Persson 1988; Werner 1988). When populations are size-structured, the signs and intensities of species interactions may not be identical for all size classes, and the overall effects of interacting populations will likely be more complex than in non-structured systems. For example, a species may, when small, be preyed upon by another species, compete with the other species during another ontogenetic period, and finally outgrow the other species and prey on it (Wilbur 1988). Size structure also often incorporates a new, important variable—individual growth rate—into population dynamics, thus linking the dynamics of size classes that occupy different ontogenetic niches. This has been recognised in the notion that predation can reduce competition among prey through numerically-mediated effects. For example, if a resource-limited prey population is subjected to predation, then the surviving prey may experience competitive release and accrue benefits in terms of increased individual growth and fecundity. This increase may more than outweigh the direct numerical effects of the predator on the population as a whole (Wilbur 1988).

Perhaps more importantly, both population size structure and individual growth rates dramatically increase the number of possible effects of flexible behaviour on population dynamics. These effects include an increased potential for alternative community states. In the case of predation on a size-structured population, flexible prey behaviour may either enhance or reduce the direct negative effects of the predator on prey abundance and its indirect positive effects on prey growth rates within a particular size class of prey (Fraser and Gilliam 1992). In the previous example of piscivorous perch and crucian carp (Tonn *et al*. 1992), piscivores not only had a direct, consumptive effect on YOY carp but also caused them to shift towards a safer, but less productive, habitat. Thus, YOY carp experienced both reduced survival and reduced individual growth rates. The latter should affect survival further by prolonging the time prey fish are vulnerable to gape-limited predators. In comparison, the larger size classes of carp were invulnerable to piscivores, did not shift habitat, and increased individual growth rates owing to the numerically- and behaviourally-mediated release from conspecific (YOY) competition. Such differences in the sign of the impact of one species on different size classes of another species are commonplace in fish communities, and may fundamentally affect the outcome of population interactions, as exemplified by the contrasting patterns of total abundance and size distribution of crucian carp in lakes with and without piscivores (Paszkowski *et al*. 1989).

The above discussion illustrates that the effects of flexible behaviour in individuals on the dynamics of fish populations can often only be understood in the context of population size structure. Furthermore, individual growth introduces an important feedback mechanism to the interaction between individual behaviour and population dynamics. It also illustrates that individuals in size-structured populations experience highly variable, size-specific selective regimes during on-togeny. Traits that are expressed over several ontogenetic stages in an individual are often highly correlated because of pleiotropic mutations (Ebenman 1992). Such genetic covariances between size (age)-specific traits constrain local evolutionary adaptations of different stages/sizes to specific niches during ontogeny (Ebenman and Persson 1988; Werner 1988; Ebenman 1992). In taxa such as fishes, changes in body morphology with age are largely due to increases in body size (except for metamorphosing larvae), which often necessitate a series of ontogenetic niche shifts toward larger prey sizes and shifts in habitat to allow for an optimal net energy intake (Werner and Gilliam 1984; Persson 1987; Persson and Greenberg 1990b; Kramer *et al.*, Chapter 3). For example, piscivorous fish can undergo three to four fairly discrete niche shifts during ontogeny. In such cases, genetic covariances will set profound constraints on how efficient an individual can be at each of these ontogenetic stages, which typically results in ontogenetic trade-offs (Persson 1988; Werner 1988).

Given the costs of ontogenetic trade-offs, it has been suggested that a fish species undergoing substantial niche shifts during its life will have a lower maximum efficiency in any of the ontogenetic niches it utilises than another fish species undergoing less substantial niche shifts (Werner 1986; Persson 1988). In pisciv-orous species, for example, planktivorous juveniles may be constrained by mor-phological structures and behaviours that are adapted for piscivory in the adult stage. If the larger size classes of a given species are potential predators of another smaller species, then the interaction between such a species pair will often be greatly asymmetric. Increased abundances of the competitively superior, smaller species will have a negative effect on the growth and survival of the juvenile size classes of the potentially larger species (= competitive asymmetry). In contrast, there will be a positive effect on the growth and fecundity of the piscivorous adult size classes, resulting in a negative effect of the piscivores on the smaller species (= predatory asymmetry, Persson 1988). The outcome of such complex interactions will be determined by the relative strengths of intra- and interspecific predation and competition. Several cases have been described in the literature where competi-tive asymmetry among juveniles is intense, leading to competitive juvenile bot-tlenecks in the recruitment of piscivorous size classes (Johannes and Larkin 1961; Persson 1988, and references therein). For example, an invasion of redside shiner, *Richardsonius balteatus*, into a lake inhabited by rainbow trout, *Oncorhynchus mykiss*, foraging on a mixture of plankton, benthos and terrestrial insects led to a shift to piscivory in adult trout and increased their growth rates. However, compe-tition from shiners decreased the growth rate of juvenile trout, which negatively affected the recruitment of piscivorous size classes (Johannes and Larkin 1961).

For some species interactions, it has been suggested that alternative states exist depending on environmental conditions. For example, in the perch–roach system, predominance of either the competitive or predatory asymmetry has been sug-gested to depend on the productivity and/or spatial heterogeneity of the system

(Persson 1988; Persson *et al.* 1991, 1992). A predominance of the competitive asymmetry between juvenile perch and roach in structurally simple, moderately to highly productive systems is indicated by a negative correlation between the biomasses of roach and piscivorous perch (Fig. 12.2a, b). In less productive and often structurally more complex systems, positive correlations between piscivorous perch biomass and both individual size and total biomass of roach indicate a predominance of the predatory asymmetry. Increased habitat complexity in the form of submerged, littoral vegetation may contribute to the predominance of the predatory asymmetry both directly, by differentially affecting the vulnerability of juvenile perch and roach to predation from piscivorous perch (Christensen and Persson 1993; Persson and Eklöv 1995), and indirectly, by positively affecting the recruitment of piscivorous size classes of perch (Diehl 1993; Persson 1993; Persson and Eklöv 1995). Thus, habitat-specific asymmetries in both the foraging abilities and predation vulnerabilities of juvenile perch and roach may interact with the relative sizes and productivities of different habitats to determine the abundances and size distributions of the two species. The dynamics of several other species interactions have been proposed to be driven by opposing forces of competition and predation, which suggests that such asymmetric interactions are common in fish communities (Werner 1986; Persson 1988; Osenberg *et al.* 1994).

12.5.2 Modelling size-structured populations and the potential to include flexible behaviour

The simultaneous occurrence of competitive and predator–prey interactions illustrates the complexity of interactions in size-structured populations. Actually, it is only during the past 5–10 years that systematic attempts have been made to develop theoretical frameworks dealing with the dynamics of size(stage)-structured populations, which are currently recognised as the domain of physiologically-structured models (Metz and Diekmann 1986; Ebenman and Persson 1988; DeAngelis and Rose 1992). Several different approaches have been used to build an explicit mechanistic link between individual performance and population dynamics, all of which take the individual as the basis of the models (see DeAngelis and Rose 1992). In this respect, there is a direct link between individual behaviour and population dynamics. However, due to the complexity of the dynamics caused by size structure per se and limits to mathematical tractability in some approaches, the extent to which behavioural flexibility can be introduced in models may be limited.

In physiologically-structured models, two states are distinguished: an *i*-state representing the state of the individual and a *p*-state describing the frequency distribution over all possible *i*-states (Metz and Diekmann 1986; Metz *et al.* 1988; Caswell and John 1992; DeAngelis and Rose 1992). There are a number of possible *i*-state variables, such as size, age and energy reserves. The number of *i*-states that need to be included in a model depends on how well the *i*-states reflect the important characteristics of the individual. However, including more *i*-state variables render analyses more difficult. In most cases, the *i*-state is in the form of a set of equations describing the individual's energetics. The model output usually consists of quantities at the population level, such as population number and population size structure. The best studied organism with respect to physiologically-structured population models is *Daphnia* sp. (Metz *et al.* 1988), but models have also been developed for fishes (DeAngelis *et al.* 1991; Breck 1993; Rice *et al.* 1993; Tyler and Rose 1994, and others).

Two major approaches, distribution models and configuration models (Caswell and John 1992, DeAngelis and Rose 1992; Gross *et al.* 1992; Tyler and Rose 1994), have attempted to link individual performance to population phenomena. In configuration models, each individual in the population (or a subset of the population) is followed over its life cycle, whereas distributions are followed in distribution models (DeAngelis *et al.* 1991; DeAngelis and Rose 1992). In a distribution model, a size cohort may, for example, be represented by one *i*-state. Distribution models may be more appropriate for qualitative and long-term predictions, and configuration models for quantitative and short-term predictions (DeAngelis and Rose 1992). Configuration models have been suggested to be the only realistically useful under three conditions:

(1) small populations where stochastic events may be important for the dynamics,

(2) situations where complex *i*-states are needed to describe individual performance, and

(3) situations where local (neighbourhood) interactions are important (Caswell and John 1992).

For example, for sessile and territorial species in which local interactions are important, it may be necessary to follow individuals explicitly.

DeAngelis *et al.* (1991) considered certain important factors underlying the rationale of configuration models (several of which are also relevant to distribution models):

(1) each individual in a biological population is unique with regards to genotype, age, sex, size and experience,

(2) in populations with plastic individual growth, fast-growing individuals may be much more important for the dynamics than the 'average individual',

(3) behavioural decision-making based on day-to-day circumstances may be included, and

(4) the effects of short-term variability in the environment, as well as small-scale heterogeneity, are incorporated.

In a simulation model of a cohort of smallmouth bass, DeAngelis *et al.* (1991)

showed that a small proportion (the largest individuals) of the initial cohort had a disproportionally large influence on its fate over a season. This model included flexible behaviour in that selection of pelagic prey was dependent on whether the individual's daily consumption was at or below its maximum daily ration. By following individuals, configuration models can readily account for flexible and adaptive behaviour. Optimal foraging was, for example, included in a model for first-year growth of bluegill sunfish (Breck 1993). However, configuration models of fish populations are presently restricted to the time scale of within-season events. Modelling longer-term population dynamics will necessarily complicate the simulation programmes, because individuals of several age classes of both predator and prey need to be tracked in time and space.

Distribution models are exemplified by population models that are based on partial differential equations (Metz and Diekmann 1986; Metz *et al.* 1988; DeRoos *et al.* 1992). This mathematical approach has been elaborated with respect to long-term population dynamics (Metz *et al.* 1988; DeRoos *et al.* 1992) and is suitable for continuously growing organisms like fish. Analyses of *Daphnia* population dynamics show that size structure may produce cohort-driven suppressions of new cohorts, causing time delays which are not present in non-structured predator–prey models. To date, approaches based on distribution models have ignored the influence of flexible behaviour. For analytical solutions, it is intractable to include flexible behaviour, but for numerical solutions it should, for example, be possible to allow energy gain to affect search rate.

To conclude, the inclusion of size structure into theoretical models seems to be essential to understand the complex population dynamics of size-structured populations such as fish populations. Physiologically-structured models are appropriate to accomplish this, because they link individual performance and population dynamics. However, analyses of the long-term dynamics of populations resulting from size-structured interactions are still at a very early stage, and such analyses may need to be explored further before the implications of flexible behaviour on population dynamics can be fully appreciated.

12.6 Conclusions

In this chapter, we have emphasised the effects of the behaviour of individuals on the dynamics and stability of populations and communities, and illustrated the potential for community level effects of flexible behaviour with case studies of piscivores affecting habitat choice and activity of prey fish. To take into account flexible behaviour, traditional predator–prey population models of the Lotka–Volterra type have been modified, and many of them have demonstrated the potential effects of the behaviour of individuals on population processes.

In most cases, predators will have both a direct and an indirect behavioural effect on their prey (see example of the perch-crucian carp interaction reviewed above). Although certain experiments have demonstrated short-term effects of predators on prey behaviour (e.g. Werner *et al.* 1983b; Turner and Mittelbach 1990), these were carried out on limited temporal and spatial scales and thus cannot address long-term and large-scale numerical effects of flexible behaviour. Moreover, these particular experiments involved only one predator species. In situations with multiple predator species using different foraging strategies, prey often face conflicting antipredator demands (Lima 1992), which considerably increases the potential for a variety of direct and indirect behavioural effects (see Matsuda *et al.* 1993 for a theoretical treatment). Therefore, direct and indirect behavioural effects of predators, and especially interactions between these effects, deserve more attention in future research (see also Kramer *et al.*, Chapter 3; Smith, Chapter 7; Godin, Chapter 8).

Recently, non-structured population models have been modified to take into account the flexible behaviour of individuals. However, the implications of size-structured interactions on the dynamics of fish populations have largely been ignored until recently, as noted above. Size-structured competitive and predator–prey interactions, in combination with resource-dependent individual growth, have the potential to enhance the diversity of behaviours, both with respect to individual and population processes. Flexibility in behaviour, such as habitat use (see also Kramer *et al.*, Chapter 3; Grant, Chapter 4; Smith, Chapter 7; Godin, Chapter 8), may affect both individual growth rate and predation risk. Hence, the potential for flexible behaviour to affect the dynamics of fish populations is substantial. However, since our knowledge of the influence of size-structured interactions on population dynamics is very scanty, it is difficult to predict to what extent and in what ways flexible behaviour will affect such dynamics.

Given the complexity of configuration models and continuous size distribution models, it will be difficult in many situations to distinguish the effects of flexible behaviour from those caused by size structure per se. To ascertain more explicitly the effects of flexible behaviour on the dynamics of fish populations, it would be useful if continuous size distributions could be collapsed into a small number (2–3) of discrete stages such as juveniles and adults, for example. Such models might then serve as references for the interpretation of more complex models, and also yield important insights into the behaviour of systems that approximately conform to a few discrete stages. For example, in the bluegill–pumpkinseed interaction, the dynamics may be represented by a two life-stage model (Fig. 12.1; Mittelbach and Chesson 1987) that has been successful in predicting static population patterns (Osenberg *et al.* 1992; Mittelbach and Osenberg 1993). However, this model neither includes behavioural flexibility nor identifies the two life stages by size. In fact, the model implicitly assumes that juvenile bluegills and pumpkinseeds inflexibly occupy the littoral vegetation habitat as a result of the presence of piscivorous largemouth bass, and that the switch to the adult open-water habitat occurs at a fixed age. On the other hand, empirical data indicate that the behaviour of individual bluegill may be variable in the presence of predators, resulting in considerable intra-cohort variation in size (Turner and Mittelbach 1990), and that the habitat shifts of bluegills are dependent on both the their body size and the abundance of piscivores in the open-water habitat (Werner and Hall 1988).

Recently, Osenberg *et al.* (1994) have suggested that a few life-history stages might characterise fish populations sufficiently well if the rates of vital parameters (growth, mortality, etc.) are similar within stages but different between stages. The commonness of this situation needs to be investigated empirically before general conclusions about such an approach can be made. Furthermore, even though discrete ontogenetic niche shifts may turn out to be relatively common in fish populations, the rates at which shifts between ontogenetic stages occur are dependent on individual growth rates. For example, in a two-stage situation, a juvenile may not only die or become adult over a given time unit, but also remain as a juvenile if it grows slowly. Thus, recruitment rates from one stage to another will necessarily be functions of individual growth rates. By explicitly introducing individual growth rate into stage-based models, a mechanistic link would be made with adaptive individual behaviour with respect to foraging and predation risk (which both affect growth and mortality rates). However, simplifying size-structured processes into a small number of stage-structured processes must be done with caution, because so many processes (foraging rate, metabolism, fecundity, predation risk) are continuous functions of body size in fishes.

It is likely that a productive research agenda, focused on the interactions between individual behaviour and population dynamics in fishes, will have to make use of both life-stage and 'continuous' approaches (configuration and distribution models), in which generality and the potential to investigate population dynamics will often have to be traded off against the potential to introduce flexible behaviour. Pluralistic approaches will be especially needed when processes are studied in which both predator and prey populations are size-structured, such as in the asymmetric interactions discussed above. What can be firmly stated at present is that the consequences of flexible behaviour for the dynamics of fish populations can rarely be understood outside the context of their size (stage) structures.

Acknowledgements

We thank P.A. Abrams, W.M. Tonn, R.E. Vandenbos, an anonymous reviewer, and especially J.-G.J. Godin for comments on previous versions of this chapter. The research on which this review is based has been sponsored by the Swedish Natural Science Research Council and the Swedish Council for Agricultural and Forestry research to L. Persson.

References

Abrams, P.A. (1984). Foraging time optimization and interactions in food webs. *Am. Nat.*, **124**, 80–96.

Abrams, P.A. (1986). Adaptive responses of predators to prey and prey to predators. The failure of the arms-race analogy. *Evolution*, **40**, 1229–1247.

Abrams, P.A. (1992). Predators that benefit prey and prey that harm predators: unusual effects of interacting foraging adaptations. *Am. Nat.*, **140**, 573–600.

Abrams, P.A. and Matsuda, H. (1993). Effects of adaptive predatory and anti-predator behavior in a two-prey-one-predator system. *Evol. Ecol.*, **7**, 312–326.

Breck, J.E. (1993). Hurry up and wait: growth of young bluegills in ponds and in simulations with an individual-based model. *Trans. Am. Fish. Soc.*, **122**, 467–480.

Brett, J.R. and Groves, T.D.D. (1979). Physiological energetics. In *Fish physiology*, Vol. 8 (eds. W.S. Hoar, D.J. Randall and J.R. Brett), pp. 280–352. Academic Press, New York.

Brönmark, C. and Miner, J.G. (1992). Predator-induced phenotypical change in body morphology in crucian carp. *Science*, **258**, 1348–1350.

Carpenter, S.R. and Kitchell, J.F. (1988). Consumer control of lake productivity. *BioScience*, **38**, 764–769.

Caswell, H. (1978). Predator mediated coexistence: a non-equilibrium model. *Am. Nat.*, **112**, 127–154.

Caswell, H. and John, A.M. (1992). From the individual to the population in demographic models. In *Individual-based models and approaches in ecology—populations communities and ecosystems* (eds. D.L. DeAngelis and L.J. Gross), pp. 36–61. Chapman and Hall, New York.

Charnov, E.L., Orians, G.H. and Hyatt, K. (1976). Ecological implications of resource depression. *Am. Nat.*, **110**, 247–259.

Christensen, B. and Persson, L. (1993). Species specific antipredatory behaviours: effects on prey choice in different habitats. *Behav. Ecol. Sociobiol.*, **32**, 1–9.

Comins, H.N. and Hassell, M.P. (1987). The dynamics of predation and competition in patchy environments. *Theor. Popul. Biol.*, **31**, 393–421.

DeAngelis, D.L and Rose, K.A. (1992). Which individual-based approach is most appropriate for a given problem? In *Individual-based models and approaches in ecology—populations communities and ecosystems* (eds. D.L. DeAngelis and L.J. Gross), pp. 67–87. Chapman and Hall, New York.

DeAngelis, D.L., Godbout, L. and Shuter, B.J. (1991). An individual-based approach to predicting density-dependent dynamics in smallmouth bass populations. *Ecol. Modell.*, **57**, 91–115.

DeRoos, A.M., Metz, J.A.J. and Diekmann, O. (1992). Studying the dynamics of structured population models: a versatile technique and its application to *Daphnia*. *Am. Nat.*, **139**, 123–147.

Diehl, S. (1988). Foraging efficiency of three freshwater fish: Effects of structural complexity and light. *Oikos*, **53**, 207–214.

Diehl, S. (1993). Effects of habitat structure on resource availability, diet and growth of benthivorous perch, *Perca fluviatilis*. *Oikos*, **67**, 403–414.

Ebenman, B. (1992). Evolution in organisms that change their niches during the life cycle. *Am. Nat.*, **139**, 990–1021.

Ebenman, B. and Persson, L. (1988). Dynamics of size-structured populations—an overview. In *Size-structured populations: ecology and evolution* (eds. B. Ebenman and L. Persson), pp. 3–9. Springer-Verlag, Berlin.

Ehlinger, T.J. (1989). Learning and individual variation in bluegill foraging: habitat-specific techniques. *Anim. Behav.*, **38**, 643–658.

Ehlinger, T.J. (1990). Habitat choice and phenotype-limited feeding efficiency in blue-gill: individual differences and trophic polymorphism. *Ecology*, **71**, 886–896.

Eklöv, P. (1992). Group foraging versus solitary foraging efficiency in piscivorous predators: the perch, *Perca fluviatilis*, and pike, *Esox lucius*, patterns. *Anim. Behav.*, **44**, 313–326.

Eklöv, P. and Diehl, S. (1994). Piscivore efficiency and refuging prey: the importance of predator search mode. *Oecologia*, **98**, 344–353.

Fraser, D.F. and Gilliam, J.F. (1992). Nonlethal impacts of predator invasion: facultative suppression of growth and reproduction. *Ecology*, **73**, 959–970.

Fretwell, S.D. (1972). *Populations in a seasonal environment*. Princeton University Press, Princeton.

Fryxell, J.M. and Lundberg, P. (1994). Diet choice and predator–prey dynamics. *Evol. Ecol.*, **8**, 407–421.

Galis, F. (1991). *Interactions between the pharyngeal jaw apparatus, feeding behaviour and ontogeny in the cichlid fish Haplochromis piceatus—a study of constraints in evolutionary ecology*. Ph. D. thesis, University of Leiden, Leiden.

Gilliam, J.F. and Fraser, D.F. (1987). Habitat selection under predation hazard: test of a model with foraging minnows. *Ecology*, **68**, 1856–1862.

Gilliam, J.F. and Fraser, D.F. (1988). Resource depletion and habitat segregation by competitors under predation hazard. In *Size-structured populations: ecology and evolution* (eds. B. Ebenman and L. Persson), pp. 173–184. Springer-Verlag, Berlin.

Gleeson, S.K. and Wilson, D.S. (1986). Equilibrium diet: optimal foraging and prey coexistence. *Oikos*, **46**, 139–144.

Grant, J.W.A. and Kramer, D.L. (1990). Territory size as a predictor of the upper limit to population density of juvenile salmonids in streams. *Can. J. Fish. Aquat. Sci.*, **47**, 1724–1737.

Gross, L.J., Rose, K.A., Rykiel, E.J. Jr., van Winkle, W. and Werner, E.E. (1992). Individual-based modeling: summary of a workshop. In *Individual-based models and approaches in ecology—populations communities and ecosystems* (eds. D.L. DeAngelis and L.J. Gross), pp. 511–522. Chapman and Hall, New York.

Hamrin, S.F. and Persson, L. (1986). Asymmetrical competition between age classes as a factor causing population oscillations in an obligate planktivorous fish species. *Oikos*, **47**, 223–232.

Holopainen, I.J., Tonn, W.M. and Paszkowski, C.A. (1992). Effects of fish density on planktonic communities and water quality in a manipulated forest pond. *Hydrobiologia*, **243/244**, 311–321.

Holt, R.D. (1984). Spatial heterogeneity, indirect interactions and the coexistence of prey species. *Am. Nat.*, **124**, 377–406.

Hugie, D.M. and Dill, L.M. (1994). Fish and game: a game theoretic approach to habitat selection by predators and prey. *J. Fish Biol.*, **45** (*Suppl.* A), 151–169.

Ives, A.R. and Dobson, A.P. (1987). Antipredator behaviour and the population dynamics of simple predator–prey systems. *Am. Nat.*, **130**, 431–447.

Johannes, R.D. and Larkin, P.A. (1961). Competition for food between redside shiners and rainbow trout in two British Columbia lakes. *J. Fish. Res. Board Can.*, **18**, 203–220.

Koehl, M.A.R. (1989). Discussion: From individuals to populations. In *Perspectives in ecological theory* (eds. J. Roughgarden, R.M. May and S.A. Levin), pp. 39–53. Princeton University Press, Princeton.

Kotler, B.P. and Holt, R.D. (1989). Predation and competition: the interaction of two types of species interactions. *Oikos*, **54**, 256–260.

Krebs, J.R. and McCleery, R.H. (1984). Optimization in behavioural ecology. In *Behavioural ecology: an evolutionary approach*, 2nd edn. (eds. J.R. Krebs and N.B. Davies), pp. 91–121. Sinauer Associates, Sunderland, MA.

Leibold, M.A. (1991). Trophic interactions and habitat segregation between competing *Daphnia* species. *Oecologia*, **86**, 510–520.

Leibold, M.A. and Tessier, A.J. (1991). Contrasting patterns of body size for *Daphnia* species that segregate by habitat. *Oecologia*, **86**, 342–348.

Lima, S.L. (1992). Life in a multi-predator environment: some considerations for antipredatory vigilance. *Ann. Zool. Fennici*, **29**, 217–226.

Lima, S.L. and Dill, L.M. (1990). Behavioural decisions made under the risk of predation: a review and prospectus. *Can. J. Zool.*, **68**, 619–640.

Lomnicki, A. (1988). *Population ecology of individuals*. Princeton University Press, Princeton.

McCauley, E., Wilson, W.G. and de Roos, A.M. (1993). Dynamics of age-structured and spatially structured predator–prey interactions: individual-based models and population-level formulations. *Am. Nat.*, **142**, 412–442.

McNair, J.N. (1986). The effects of refuges on predator–prey interactions: a reconsideration. *Theor. Popul. Biol.*, **29**, 38–63.

Matsuda, H, Abrams, P.A. and Hori, M. (1993). The effect of adaptive anti-predator behavior on exploitative competition and mutualism between predators. *Oikos*, **68**, 549–559.

Metz, J.A.J. and Diekmann, O. (1986). *The dynamics of physiologically structured populations*. Springer-Verlag, Berlin.

Metz, J.A.J., de Roos, A.M. and van den Bosch, F. (1988). Population models incorporating physiological structure: A quick survey of the basic concepts and an application to size-structured population dynamics in waterfleas. In *Size-structured populations: ecology and evolution* (eds. B. Ebenman and L. Persson), pp. 106–126. Springer-Verlag, Berlin.

Mills, E.L. and Schiavone, A. (1982). Evaluation of fish communities through assessment of zooplankton populations and measures of lake productivity. *N. Amer. J. Fish. Manage.*, **2**, 14–27.

Mittelbach, G.G. (1981). Foraging efficiency and body size: a study of optimal diet and habitat used in bluegill sunfish. *Ecology*, **62**, 1370–1386.

Mittelbach, G.G. (1984): Predation and resource partitioning in two sunfishes (Centrarchidae). *Ecology*, **65**, 499–513.

Mittelbach, G.G. (1988). Competition among refuging sunfishes and effects of fish density on littoral zone invertebrates. *Ecology*, **69**, 614–623.

Mittelbach, G.G. and Chesson, P.L. (1987). Predation risk: indirect effects on fish populations. In *Predation: direct and indirect impacts on aquatic communities* (eds. W.C. Kerfoot and A. Sih), pp. 315–332. University Press of New England, Hanover, NH.

Mittelbach, G.G. and Osenberg, C.W. (1993). Stage-structured interactions in bluegill: consequences of adult resource variation. *Ecology*, **74**, 2381–2394.

Mittelbach, G.G. and Osenberg, C.W. (1994). Using foraging theory to study trophic interactions. In *Theory and application in fish feeding ecology* (eds. D.J. Stouder, K.L. Fresh and R.J. Feller), pp. 45–59. University of South Carolina Press, Columbia.

Norberg, R.Å. (1977). An ecological theory on foraging time and energetics and choice of optimal food searching method. *J. Anim. Ecol.*, **46**, 511–529.

O'Brien, W.J. (1987). Planktivory by freshwater fish: thrust and parry in the pelagia. In

Predation: direct and indirect impacts on aquatic communities (eds. W.C. Kerfoot and A. Sih), pp. 3–16. University Press of New England, Hanover, NH.

O'Brien, W.J., Browman, H.I. and Evans, B.I. (1990). Search strategies in foraging animals. *Amer. Sci.*, **78**, 152–160.

Osenberg, C.W. and Mittelbach, G.G. (1989). The effects of body size on predator–prey interaction between pumpkinseed sunfish and gastropods. *Ecol. Monogr.*, **59**, 405–432.

Osenberg, C.W., Mittelbach, G.G. and Wainwright, P.C. (1992). Two-stage life histories in fish: the interaction between juvenile competition and adult performance. *Ecology*, **73**, 255–267.

Osenberg, C.W., Olson, M.H. and Mittelbach, G.G. (1994). Stage structure in fishes: resource productivity and competition gradients. In *Theory and application in fish feeding ecology* (eds. D.J. Stouder, K.L. Fresh and R.J. Feller), pp. 151–170. University of South Carolina Press, Columbia.

Paszkowski, C.A., Tonn, W.M., Piironen, J. and Holopainen, I.J. (1989). An experimental study of body size and food size relations in crucian carp, *Carassius carassius*. *Environ. Biol. Fish.*, **24**, 275–286.

Persson, L. (1986). Effects of reduced interspecific competition on resource utilization of perch (*Perca fluviatilis*). *Ecology*, **67**, 355–364.

Persson, L. (1987). The effects of resource availability and distribution on size class interactions in perch (*Perca fluviatilis*). *Oikos*, **48**, 148–160.

Persson, L. (1988). Asymmetries in predatory and competitive interactions in fish populations. In *Size-structured populations: ecology and evolution* (eds. B. Ebenman and L. Persson), pp. 203–218. Springer-Verlag, Berlin.

Persson, L. (1991). Behavioral response to predators reverses the outcome of competition between prey species. *Behav. Ecol. Sociobiol.*, **28**, 101–105.

Persson, L. (1993). Predator-mediated competition in prey refuges: the importance of habitat dependent prey resources. *Oikos*, **68**, 12–22.

Persson, L. and Diehl, S. (1990). Mechanistic, individual-based approaches in the population/community ecology of fish. *Ann. Zool. Fennici*, **27**, 165–182.

Persson, L. and Eklöv, P. (1995). Prey refuges affecting interactions between piscivorous perch and juvenile perch and roach. *Ecology*, **76**, 70–81.

Persson, L. and Greenberg, L.A. (1990a). Juvenile competitive bottlenecks: the perch (*Perca fluviatilis*)-roach (*Rutilus rutilus*) interaction. *Ecology*, **71**, 44–56.

Persson, L. and Greenberg, L.A. (1990b). Interspecific and intraspecific size class competition affecting resource use and growth of perch, *Perca fluviatilis*. *Oikos*, **59**, 97–106.

Persson, L., Diehl, S., Johansson, L., Andersson, G. and Hamrin, S.F. (1991). Shifts in fish communities along the productivity gradient of temperate lakes—patterns and the importance of size-structured interactions. *J. Fish Biol.*, **38**, 281–293.

Persson, L., Diehl, S., Johansson, L., Andersson, G. and Hamrin, S.F. (1992). Trophic interactions in temperate lake ecosystems: a test of food chain theory. *Am. Nat.*, **140**, 59–84.

Persson, L., Johansson, L., Andersson, G., Diehl, S. and Hamrin, S.F. (1993). Density dependent interactions in lake ecosystems—whole lake perturbation experiments. *Oikos*, **66**, 193–208.

Piironen, J. and Holopainen, I.J. (1988). Length structure and reproductive potential of crucian carp (*Carassius carassius* (L.)) populations in some small forest ponds. *Ann. Zool. Fennici*, **25**, 203–208.

Power, M.E. (1984a). Depth distributions of armored catfish: predator-induced resource avoidance? *Ecology*, **65**, 523–528.

Power, M.E. (1984b). Habitat quality and the distribution of algae-grazing catfish in a Panamanian stream. *J. Anim. Ecol.*, **53**, 357–374.

Power, M.E. (1987). Predator avoidance by grazing fishes in temperate and tropical streams: importance of stream depth and prey size. In *Predation: direct and indirect impacts on aquatic communities* (eds. W.C. Kerfoot and A. Sih), pp. 333–351. University Press of New England, Hanover, NH.

Pyke, G.H. (1984). Optimal foraging theory: a critical review. *A. Rev. Ecol. Syst.*, **15**, 523–575.

Rice, J.A., Miller, T.J., Rose, K.A., Crowder, L.B., Marschall, E.A., Trebitz, A.S. and DeAngelis, D.A. (1993). Growth rate variation and larval survival: inferences from an individual-based size-dependent predation model. *Can. J. Fish. Aquat. Sci.*, **50**, 133–142.

Schoener, T.W. (1986). Mechanistic approaches to community ecology: a new reductionism? *Amer. Zool.*, **26**, 81–106.

Sih, A. (1987). Prey refuges and predator–prey stability. *Theor. Popul. Biol.*, **31**, 1–12.

Sih, A. and Moore, R.D. (1990). Interacting effects of predator and prey behaviour in determining diets. In *Behavioural mechanisms of food selection* (ed. R.N. Hughes), pp. 771–796. NATO ASI Series, Vol. G20, Springer-Verlag, Berlin.

Speakman, J.R. (1986). The optimum search speed of terrestrial predators when feeding on sedentary prey: a predictive model. *J. theor. Biol.*, **122**, 401–407.

Stephens, D.W. and Krebs, J.R. (1986). *Foraging theory*. Princeton University Press, Princeton.

Tonn, W.M., Paszkowski, C.A. and Holopainen, I.J. (1992). Piscivory and recruitment: mechanisms structuring prey populations in small lakes. *Ecology*, **73**, 951–958.

Turner, A.M. and Mittelbach, G.G. (1990). Predator-avoidance and community structure: interactions among piscivores, planktivores, and plankton. *Ecology*, **71**, 2241–2254.

Tyler, J.A. and Rose, K.A. (1994). Individual variability and spatial heterogeneity in fish population models. *Rev. Fish Biol. Fish.*, **4**, 91–123.

Wainwright, P.C. (1988). Morphology and ecology: functional basis of feeding constraints in Caribbean labrid fishes. *Ecology*, **69**, 635–645.

Wainwright, P.C., Osenberg, C.W. and Mittelbach, G.M. (1991). Trophic polymorphism in the pumpkinseed sunfish (*Lepomis gibbosus* Linnaeus): effects of environment on ontogeny. *Funct. Ecol.*, **5**, 40–55.

Webb, P.W. (1984). Body and fin form and strike tactics of four teleost predators attacking fathead minnow *Pimephales promelas* prey. *Can J. Fish. Aquat. Sci.*, **41**, 157–165.

Webb, P.W. (1986). Locomotion and predator–prey relationships. In *Predator-prey relationships: perspectives and approaches from the study of lower vertebrates* (eds. M.E. Feder and G.V. Lauder), pp. 24–41. University Press of Chicago, Chicago.

Werner, E.E. (1986). Species interactions in freshwater fish communities. In *Community ecology* (eds. J. Diamond and T.J. Case), pp. 344–358. Harper and Row, New York.

Werner, E.E. (1988). Size, scaling and the evolution of life cycles. In *Size-structured populations: ecology and evolution* (eds. B. Ebenman and L. Persson), pp. 60–81. Springer-Verlag, Berlin.

Werner, E.E. (1992). Individual behavior and higher-order species interactions. *Am. Nat.*, **140**, S5–S32.

Werner, E.E. and Anholt, B.R. (1993). Ecological consequences of the trade-off between growth and mortality rates mediated by foraging activity. *Am. Nat.*, **142**, 242–272.

Werner, E.E. and Gilliam, J.F. (1984). The ontogenetic niche and species interactions in size-structured populations. *A. Rev. Ecol. Syst.*, **15**, 393–425.

Werner, E.E. and Hall, D.J. (1988). Ontogenetic habitat shifts in bluegill: the foraging rate-predation risk trade-off. *Ecology*, **69**, 1352–1366.

Werner, E.E., Mittelbach, G.G., Hall, D.J.and Gilliam, J.F. (1983a). Experimental tests of optimal habitat use in fish: the role of relative habitat profitability. *Ecology*, **64**, 1525–1539.

Werner, E.E. , Gilliam, J.F., Hall, D.J. and Mittelbach, G.G. (1983b). An experimental test of the effects of predation risk on habitat use in fish. *Ecology*, **64**, 1540–1548.

Wetzel, R.G. (1983). *Limnology*. Saunders, Philadelphia.

Winfield, I.J. (1986). The influence of simulated aquatic macrophytes on the zooplankton consumption rate of juvenile roach, *Rutilus rutilus*, rudd, *Scardinius erythophthalmus*, and perch, *Perca fluviatilis*. *J. Fish Biol.*, **29**, 37–48.

Wilbur, H.M. (1988). Interactions between growing predators and growing prey. In *Size-structured populations: ecology and evolution* (eds. B. Ebenman and L. Persson), pp. 157–172. Springer-Verlag, Berlin.

Wollkind, D.J. (1976). Exploitation in three trophic levels: an extension allowing intraspecific carnivore interactions. *Am. Nat.*, **110**, 431–447.

Author index

Sikkel P.C. 238, 239
Sinclair A.F. 61
Sinclair M.S. 11, 14, 15
Sise T.E. 60
Skadsen J.M. 109, 113, 124, 125,
    193, 194, 207, 208, 221
Skúlason S. 149, 150
Slaney P.A. 91
Slatkin M. 51
Smith C. 298, 299
Smith M.J. 175, 179, 180, 208
Smith R.J.F. 10, 166, 167, 171,
    173, 174, 175, 179, 180,
    182, 183, 207, 208, 210,
    215, 220, 222, 225
Smith S.A. 158
Smith S.J. 42, 44
Smith T.B. 149, 150
Smythe N. 179
Snyder R.J. 26
Sokal R.R. 83
Southwood T.R.E. 38
Speakman J.R. 328
Sprague L.M. 67
Sproul C.D. 207, 208, 210
Stearley R.F. 18, 30
Stearns S.C. 246, 257
Stein R.A. 164, 166, 171, 207,
    208, 212, 221
Steiner W.W. 69
Stenseth N.Chr. 121
Stephens D.W. 4, 5, 38, 45, 50,
    59, 104, 121, 126, 135,
    136, 310, 316, 320, 323
Stepien C.A. 68
Stewart A.J. 212
Stillman R. 240
Stoner G. 169, 247, 248
Stott B. 50
Suarez S.D. 207
Suboski M.D. 174, 179, 180, 182, 183
Sutherland W.J. 57, 59, 60
Suthers I.M. 43, 65
Swain D.P. 51, 59, 62
Swearer S.E. 254
Syarifuddin S. 85

Symons P.E.K. 96

Taborsky M. 247, 248
Talbot A.J. 60, 95
Tallmark B. 212
Taylor E.B. 87
Taylor L.M. 206
Taylor M.H. 12
Tegeder R.W. 214, 217, 218, 219
Templeton J.J. 183
Terborgh J. 217
Tessier A.J. 324
Thomas G. 210
Thomson D.A. 176
Thorpe J.E. 95
Thorpe L.M. 51
Thresher R.E. 82, 91, 246
Tierney J.F. 210
Tinbergen L. 147
Tinbergen N. 269
Tinker S.W. 167
Tonn W.M. 323, 331
Torio A.J. 278
Torricelli P. 2
Toth L.A. 63
Townsend C.R. 50
Townshend T.J. 242, 305, 309
Tregenza T. 57
Treherne J.E. 215, 220
Tremblay D. 217
Tricas T.C. 91
Trivers R.L. 180
Tulley J.J. 208
Turner A.K. 97, 269
Turner A.M. 322, 323, 324, 335,
    336
Turner G.F. 158, 215, 272
Turner J.R. 171
Tyler J.A. 52, 59, 334
Tyler P.A. 122

Unger L.M. 96, 281, 297, 306,
    308
Urban T.P. 42

Species index

Note: page numbers in *italics* refer to figures and tables

Subject index

Note: page numbers in *italics* refer to figures and tables

CPSIA information can be obtained
at www.ICGtesting.com
Printed in the USA
LVOW01s2111291015

460352LV00002B/6/P